The Age of Disruption

To the memory of Gérard Granel and Nicolas Auray.
For Thomas Berns.
For Michel Deguy.
For the animals of the forest.

Bernard Stiegler

The Age of Disruption

Technology and Madness in
Computational Capitalism

followed by

A Conversation about Christianity

with Alain Jugnon, Jean-Luc Nancy and Bernard Stiegler

Translated by Daniel Ross

polity

First published in French as *Dans la disruption. Comment ne pas devenir fou?* © Les Liens qui Libèrent, 2016

This English edition © Polity Press, 2019

This book is supported by the Institut français (Royaume-Uni) as part of the Burgess Programme

Polity Press
65 Bridge Street
Cambridge CB2 1UR, UK

Polity Press
101 Station Landing
Suite 300
Medford, MA 02155, USA

ISBN-13: 978-1-5095-2926-1
ISBN-13: 978-1-5095-2927-8 (pb)

A catalogue record for this book is available from the British Library.

Library of Congress Cataloging-in-Publication Data
Names: Stiegler, Bernard, author.
Title: The age of disruption : technology and madness in computational
    capitalism / Bernard Stiegler.
Description: English edition. | Medford, MA, USA : Polity Press, [2019] |
    Includes bibliographical references and index.
Identifiers: LCCN 2018046809 (print) | LCCN 2018050869 (ebook) | ISBN
    9781509529292 (Epub) | ISBN 9781509529261 | ISBN 9781509529278 (pbk)
Subjects: LCSH: Technological innovations--Social aspects. |
    Capitalism--Social aspects. | Computers and civilization.
Classification: LCC T14.5 (ebook) | LCC T14.5 .S747 2019 (print) | DDC
    303.48/3--dc23
LC record available at https://lccn.loc.gov/2018046809

Typeset in 10 on 11 Sabon
by Fakenham Prepress Solutions, Fakenham, Norfolk NR21 8NL
Printed and bound in Great Britain by TJ International Limited

For further information on Polity, visit our website:
politybooks.com

Contents

Inch'Allah is a French transcription of the Arabic phrase هللا ءاش نإ (In Shaa Allah), which means 'God willing'. In Arab countries this term is used by Christians and Muslims with the same meaning, as inspired by the epistle of James: 'If the Lord will, we shall live, and do this, or that' (James 4:15).

Muslims believe it is obligatory to pronounce this formula when they evoke an action to be realized in the future. This belief is founded on the reading of the Sūrat al-Kahf (The Cave):

23. Do not say of anything: 'I will do it tomorrow.'

24. Without adding: 'If Allah wills.' When you forget, remember your Lord and say: 'May Allah guide me and bring me nearer to the truth.'

The expression is also used more generally to mark the desire and the hope to see realized an event in the future, equivalent to current secular sayings such as 'touch wood' or 'fingers crossed'.

There is also the expression Maa Shaa Allah (هللا ءاش ام), which means approximately, 'as God has willed'.

Finally, another expression, Law Shaa Allah, meaning 'If God wills/wishes', is used to express a desire or wish that cannot be reached.

The Portuguese word *oxalá* and Castilian *ojalá*, meaning 'Hopefully', are both derived from the Arabic Inshallah.

French Wikipédia

We should [...] deny our We and draw from this, not foreseeable dissolution, but a burst of refounding lucidity. We need to confront its defeat
because it is formed in exaltation, and, each time, in forgetting that it is
highly perishable, which does not mean that it is illusory. We, here, fall
into silence and contemplate an abyss. We close our eyes and clench our
teeth in order to avoid having to uselessly pronounce: Who am I? We
know that it would destroy that for which it asks. We think that our We
should choose collective union in despair, but can we make despair a
combative bond?

Bernard Noël[1]

And, after Solon's speech denouncing what is taking place and criticizing
his fellow citizens, the Council replies that in fact Solon is going mad
(*mainesthai*). To which Solon retorts: 'You will soon know if I am mad ...
when the truth comes to light.'

Michel Foucault[2]

But we are probably speaking at cross purposes and it will take centuries
before our differences are settled.

Sigmund Freud to Ludwig Binswanger[3]

Μή, φίλα ψυχά, βίον ἀθάνατονσπεῦδε,
τὰν δ᾽ ἔμπρακτον ἄντλει μαχανάν.

Pindar[4]

Now you ask nothing more because you have nothing more...
apart from what you remember.

Derya
(at the slam workshop held in the Glacis housing project, Belfort, and led
by Dominique Bourgon)

Part One

The *Epokhē* of My Life
Philosophizing So as Not to Go Mad

1

Disruption: A 'New Form of Barbarism'

1. The loss of reason

At 4:30 p.m. on 11 September 2001, I began delivering a lecture at the Université de technologie de Compiègne in which I introduced the theme of the industry of cultural goods, formulated by Theodor Adorno and Max Horkheimer in 1944 in a text that, in 1947, became the chapter of *Dialectic of Enlightenment* entitled 'The Culture Industry'.[1] Their chapter described a profound and dangerous transformation of Western societies, and the key part played in it by this new industry. Its rise, according to Adorno and Horkheimer, would be accompanied by a 'new kind of barbarism',[2] caused by the *inversion* of the Enlightenment project that had laid the foundations of modernity.[3]

On 11 September 2001, between 4:30 p.m. and 5:00 p.m., I began explaining to my students that the world that took shape after the Second World War, a world that took the 'American way of life' as its model, a world globally 'rationalized' and 'Westernized', was, according to Adorno and Horkheimer, actually in the course of losing its reason. I emphasized the remarkable foresight of these two German philosophers: taking refuge from Nazism in the United States, they saw this 'new kind of barbarism' emerging even before the end of the Second World War, first in New York, and then in California.[4] I then drew their attention to the following three points:

- in 1997, fifty years after the publication of *Dialectic of Enlightenment*, it was estimated that the world contained one billion television sets;
- on 3 April 1997, the US Federal Communications Commission (FCC) announced that the federal government would in 2006 shut off the analogue frequencies that were currently being used by 3,800 American radio and television stations, all of whom were advised to switch to digital by 2003;[5]
- in the spring of 1997, Craig Mundie, then a senior vice-president

at Microsoft (a company represented on the board of the FCC), declared during a European visit that his company, which at that time dominated the information industry (now called the digital industry), would launch a bid to dominate the multimedia business, taking advantage of the opportunity presented by the convergence of information, media and telecommunications technologies.

2. From the slums of Temara to the presidency of the Université de technologie de Compiègne

On 11 September 2001, at around 5:30 p.m., I explained to my UTC students that the one billion television sets that existed in 1997 had grown to cover almost the entire population of the planet, and that programmes are often watched by millions of viewers simultaneously. I offered the example that in the late 1980s, in a slum lying between Temara and Skhirat, south of Rabat, I had seen a crowd of parents and children watching, on a big screen, programmes produced by a recently-privatized French network.

I then invited these engineering students to reflect on what might be going on in the minds of these thousands of people dwelling under scraps of cardboard, sheet metal and recycled materials, who had gathered together at primetime to listen to Patrick Sébastien pour forth his nonsense.[6] I asked them what could have been going through the minds of these children and their parents deprived of just about everything, confronted with the images of showbiz politics, with omnipresent advertising and with the rapid rise of 'trash TV'.

It was then that the frightened face of the UTC general secretary appeared at the entrance of the auditorium and shouted to me: 'Come quickly, something unbelievable is happening!' Astonished and annoyed, I broke off my lecture and followed Luc Ziegler into the office of the university president, François Peccoud, who, eyes riveted to the screen, was beholding Manhattan's Twin Towers ablaze.

On 11 September 2001, between 5:30 p.m. and 6:00 p.m., we watched these images in the president's office, as people undoubtedly did in Temara – which, since my visit in the late 1980s, had seen the arrival of satellite dishes.

In February 2014, according to the Moroccan newspaper *Le Matin*, this slum was still home to 34,091 people.[7]

3. From Richard Durn to Jean-Marie Le Pen: primordial narcissism of the *I* and reason for living

Six months and sixteen days later, on 27 March 2002, Richard Durn, 'an environmental activist, former member of the Socialist Party before joining the Greens [...], and also an activist in the League of Human Rights',[8] murdered eight members of the Nanterre city council and wounded nineteen others. The following day he committed suicide by leaping from a window at the police station where he was being questioned. Less than a month later, on 21 April, Jean-Marie Le Pen finished ahead of Lionel Jospin in the first round of the presidential election. On 5 May, Jacques Chirac was elected with 82.21 per cent of the vote.

After 11 September 2001 and 21 April 2002, I delivered two lectures at Cerisy-la-Salle, in the framework of two seminars organized by Édith Heurgon and Josée Landrieu.[9] In the first lecture, I tried to understand what was at stake in the 9/11 event, and in the second, to imagine what could have being going through Durn's mind on 27 March 2002. I argued that in our 'epoch', which should be understood as the fulfilment of the new barbarism anticipated by Adorno and Horkheimer, what is occurring amounts to a *murderous dis-articulation* of the *I* and the *we*.

We have now also passed through the crisis of 2008, and this epoch has shown itself for what it is: *the epoch of the absence of epoch*, the meaning of which will be clarified in what follows.

In pointing out, during my second lecture at Cerisy and after 21 April, that, three weeks before the massacre, Durn had written of having 'lost the feeling of existing', I tried to show that the processes of psychic and collective individuation[10] characteristic of the life of the mind and spirit have slowly but surely been wiped out by the culture industries, now exclusively operating in the service of the market and the organization of consumption, and that the export of this state of affairs around the world was clearly one of the key factors lying behind the growth of Al-Qaeda.

In France itself, this situation was firmly entrenched in 1986, when François Mitterrand allowed the privatization of television, giving Silvio Berlusconi and Jérôme Seydoux the licence to operate a network that would be named La Cinq. Jacques Chirac and François Léotard, who would later demand that the Hersant group acquire a stake in La Cinq, would soon after arrange the privatization of TF1.[11] In competition with M6, which also appeared in 1987, TF1 quickly began to enter the path of systematically drive-based television, while La Cinq, which failed, ceased broadcasting in 1992.[12]

In 2003, I turned these two lectures into a book.[13] I dedicated it to those who voted for the National Front, and I argued that Durn had been stripped of his 'primordial narcissism' by a process of the same kind as that implemented by the industry of cultural goods, which, according to Adorno and Horkheimer, destroys what, in *Critique of Pure Reason*, Kant called the transcendental imagination.[14]

The destruction of primordial narcissism leads to madness, that is, to the loss of reason, and, more precisely, to the loss of this *reason for living* that creates and gives the feeling of existing. This is why I stated in the conclusion of that book:

> If we do not enact an *ecological critique* of the technologies and industries of the spirit, if we do not show that the unlimited exploitation of spirits as markets leads to a ruin comparable to that which the Soviet Union and the great capitalist countries have been able to create by exploiting territories or natural resources without any care to preserve their habitability to come – the future – then we move ineluctably toward a global social explosion, that is, toward absolute war.[15]

Today, this explosion is imminent. All of us now know it and fear it, but also repress it and deny it, and we do so in order to continue living with dignity [*dignement*]. This is, however, something that can no longer be repressed: in the stage we have now entered, this becomes, precisely, unworthy [*indigne*], and literally cowardly.

4. A 'new kind of barbarism' and algorithmic governmentality

The FCC's announcement on 3 April 1997, followed by Craig Mundie's European tour, was the beginning of a federal policy that would completely reshape the American audiovisual industry, through a process of digitalization[16] giving a brand new twist to this 'new kind of barbarism'. This FCC policy – coming after the World Wide Web entered the public domain on 30 April 1993 through a decision (made by Europeans) that gave the internet a completely new and revolutionary dimension, and after the Clinton government had granted tax exemptions to a set of businesses that would go on to become the 'giants' of the web – created the conditions for the rise, in the United States, of an industry that would be *fully digital*.

In this way, a path was laid out for what would become a new American hegemony – embodied by the Big Four: Amazon (created in 1993), Google (1997), Facebook (2004) and Apple.[17] Between 2007 and 2015, Apple sold 700 million iPhones, around which 900,000 'apps' were developed for sale on the App Store. In 1996 I was appointed deputy director general of the Institut national de l'audiovisuel (INA) in charge of the innovation department, that is, of research, production, training and publishing. I closely followed the developments that led to the emergence of Google, and recommended that the government build a new audiovisual policy focused on the web.[18] My recommendation was ignored and I resigned from the INA in 1999.

By completely reconfiguring telecommunications, and thereby constituting reticular society, the integration of the analogue communication industries, journalism and the editorial function in general into the digital information industries – of which the 1997 FCC decision was the first step[19] – continued and radicalized the process that Adorno and Horkheimer had analysed in 1944. But, at the same, this reconfiguration introduced *absolutely new* factors.

This absolute novelty is what Thomas Berns and Antoinette Rouvroy are trying to think today with the concept of *algorithmic governmentality*.[20] What is new is the systematic exploitation and physical reticulation of interindividual and transindividual relations – serving what is referred to today as the 'data economy', itself based on data-intensive computing, or 'big data', which has been presented as the 'end of theory'.[21] This amounts to the full realization of barbarism in Adorno and Horkheimer's sense, but they could surely never have imagined how far this would extend onto the noetic plane.

Reticulated society is based on smartphones and other embedded mobile devices (chips, sensors, GPS tags, cars, televisions,[22] watches, clothing and other prostheses), but also on new fixed and mobile terminals (urban territory becoming the infrastructure and architecture of constant mobility and constant connectivity). As such, it contains unprecedented powers of automation and computation: it is literally *faster than lightning* – digital information circulates on fibre-optic cables at up to two thirds of light speed, quicker, then, than Zeus' lightning bolt, which travels at only 100 million metres per second (one third of the speed of light). Automatic and reticulated society thereby becomes *the global cause of a colossal social disintegration*.

The automatic power of reticulated disintegration extends across the face of the earth through a process that has recently become known as *disruption*. Digital reticulation penetrates, invades, parasitizes and ultimately destroys social relations at lightning speed, and, in so doing, neutralizes and annihilates them from within, by outstripping, overtaking and engulfing them. Systemically exploiting the network effect, this *automatic nihilism* sterilizes and destroys local culture and social life like a neutron bomb: what it dis-integrates,[23] it exploits, not only local equipment, infrastructure and heritage, abstracted from their socio-political regions and enlisted into the business models of the Big Four,[24] but also psychosocial energies – both of individuals and of groups – which, however, are thereby depleted.

These individuals and groups are thus transformed into data-providers, de-formed and re-formed by 'social' networks operating according to new protocols of association. In this way, they find themselves disindividuated: their own data [*données*], which also amounts to what we call (in the language of the Husserlian phenomenology of time) *retentions*,[25] enables them to be dispossessed of their own *protentions*[26] – that is, *their own desires, expectations, volitions, will* and so on.

5. Always too late

'Desires, expectations, volitions, will and so on': everything that for individuals forms the horizon of *their* future, constituted by *their* protentions, is outstripped, overtaken and progressively replaced by *automatic* protentions that are produced by intensive computing systems operating between one and four million times quicker than the nervous systems of psychic individuals.[27]

Disruption *moves quicker than any will, whether individual or collective*, from consumers to 'leaders', whether political or economic.[28] Just as it overtakes individuals via digital doubles or profiles on the basis of which it satisfies 'desires' they have most likely never expressed – but which are in reality herd-like substitutes depriving individuals of their own existence by always preceding their will, at the same time emptying them of meaning, while feeding the business models of the data economy – so too disruption outstrips and overtakes social organizations, but the latter recognize this only after the fact: *always too late*.

Disruption renders will, wherever its source, *obsolete in advance*: it always arrives too late. What is thereby attained is an extreme stage of rationalization, forming a threshold, that is, a limit. What lies beyond this limit remains unknown: it destroys reason not only in the sense that rational knowledge finds itself eliminated by proletarianization, but in the sense that individuals and groups, losing the very possibility of existing (for their existence depends on being able to express their will), losing therefore all reason for living, become literally mad, and tend to despise life – their own and that of others. The result is the risk of a global social explosion consigning humanity to a nameless barbarism.

In the epoch of reticulated and automated disruption, the 'new kind of barbarism' induced by the loss of the feeling of existing no longer involves only isolated and suicidal individuals, whether Richard Durn or Andreas Lubitz, who crashed his passenger-laden aircraft into a mountain, or the suicidal perpetrators of 9/11. On 22 December 2014, Sébastien Sarron drove his van into a crowd at the Christmas market in Nantes. When reason is lost, all those technological powers that we hold in our hands as 'civilizational progress' become weapons of destruction through which this 'civilization' reveals the barbarism it contains. This is the key pharmacological question to be addressed in the epoch of disruption.[29]

The loss of the feeling of existing, the loss of the possibility of expressing one's will, the correlative loss of all reason for living and the subsequent loss of *reason as such*, a loss that Chris Anderson glorifies as the 'end of theory', are what now strike entire groups and entire countries – and it is for this reason that the far right is on the rise around the world, and especially in Europe, which, since the tragedy of Greece and the massacres in France, is undergoing significant deterioration.

But these losses also and especially strike *an entire generation*: that of

Florian. Florian is the name of a young man of fifteen, whose statements were published in *L'Effondrement du temps*:

> You really take no account of what happens to us. When I talk to young people of my generation, those within two or three years of my own age, they all say the same thing: we no longer have the dream of starting a family, of having children, or a trade, or ideals, as you yourselves did when you were teenagers. All that is over and done with, because we're sure that we will be the last generation, or one of the last, before the end.[30]

2

The Absence of Epoch

6. Before the end

Florian believes his generation will be the last, 'or one of the last, before the end'. Such is the state of Florian's *morale* [*moral*] – and I will return to this in Chapter 8, on the question of what ties 'thought' to so-called 'morale', which we either 'have' or 'do not have', which equally ties thinking to melancholy, which is also to say, to madness. Hence we will ask what *morale* means (from *moralis*, 'related to mores, manners'), and, more generally, what there is of morality in the fact of 'having [good] morale', and about *demoralization*. The *last* generation, or one of the last, *before the end*: such is the *extreme demoralization* of Florian and his generation.

In the horizon of *becoming* [*devenir*], Florian sees no possible *future* [*avenir*] for his generation – which is also to say, for the human species. He formulates in clear, simple and terrifying terms what everyone thinks, but which everyone represses – except a few who hurtle into the Twin Towers by plane, or into mountains, or into Christmas markets, or through the window of a police station after having killed or injured twenty-seven people (we should also mention Columbine, Breivik and many others – and it will be necessary to discuss the Kouachi brothers).

I will return to this repression, and the denial to which it leads, in Chapter 13.

Expressing this in the language of phenomenology, and returning to questions emerging from Martin Heidegger's existential analytic, we could say that for Florian, no positive collective protention is possible: there is no protention other than *the end of all protention*, that is, *the end of all dreams* and any possibility of realizing them. Florian's vision of the world and of his future is entirely subject to *an absolutely negative protention*: the complete disappearance of humankind.

We can try to imagine what the complete disappearance of humankind means for Florian. It could be envisaged as the self-extermination of humanity through a total and final world war. It could occur through a series of apocalyptic accidents. It could also be the outcome of climate change and its adverse effects on life in general and human life in particular. The last of these possibilities was the subject of a United Nations conference held in Paris from 30 November to 12 December 2015, which everyone knew would achieve next to nothing.

No doubt all these possibilities get mixed together for members of the younger generations, in various ways and with many other factors and causes for despair, in particular on the economic level, and more so still when this level is found to be massively subject to the disruptive madness of full and generalized automation.[1]

In 2015, the accumulation of these disasters that have affected men and women since the beginning of the twenty-first century[2] became conjoined to the attenuation of every form of will, and the result has been the proliferation of barbaric behaviour – all this gives everyone, and not only Florian's generation, every reason to believe that the world is on a path to ruin, and in short order.

It is then a question of understanding how it is possible that, at the very moment it becomes apparent to everyone that humanity and life in general are threatened by the madness that currently governs the world in partnership with systemic stupidity (or 'functional stupidity'[3]), people find themselves seemingly unable to create the conditions for a radical bifurcation – not the disruptive 'radical innovation' of the kind claimed by those startup entrepreneurs who present themselves as *'new barbarians'*,[4] but, on the contrary, a bifurcation taking account of the radicality of this disruption from the perspective of a *new public power*, such that it could *once again create an epoch*.

7. Negative teleology and end without purpose

It is impossible to live in a society without positive collective protentions, but the latter are the outcome of intergenerational and transgenerational transmission. Such protentions – which belong to what the Greeks in the age of Hesiod called *elpis* (ἐλπίς), a word that means *expectation* [*attente*], both as hope and as fear,[5] and which is the condition of *attention* – are the boundaries and boundary markers of the care that must be taken of the world (κόσμος).

Inhabited by this 'unsettling' [*inquiétant*] being that is the human,[6] this κόσμος is always exposed to *hubris* (ὕβρις), collective protentions of which open up a 'general economy' – in Georges Bataille's sense of this notion,[7] conceived in a fundamental relationship to sacrifice – through being inscribed into calendarities and cardinalities, each time specific, of one civilization or another.

These cardinalities and calendarities have been not only upset, but literally *overturned* by the advent of the culture industry, and yet more by digitalization as the convergence of telecommunications, the audio-visual and computing, a convergence that leads to reticulated, automatic society.

Today, the Christian calendar has been imposed throughout the entire world by all those clocks that synchronize every digital device – billions of devices, a huge number of which can be found in the pockets of terrestrial inhabitants connected by the industry of 'cloud computing', data centres, geostationary satellites and the algorithms of intensive computing, together forming what Heidegger called *Gestell*.

In so doing, the Christian calendar short-circuits every other form of calendarity, while itself becoming completely secularized as the system becomes purely computational – totally secularized, as Max Weber understood, and which Jacques Derrida described as 'globalatinization' [*mondialatinisation*].[8]

In such a *purely* computational context, individual as well as collective protentions *fade away*. Such is our 'desolate time'.[9] And such is the incommensurable tragedy of Florian and his generation. In the time of this generation, which is also that of 'digital natives',[10] nobody seems capable of producing intergenerational and transgenerational collective protentions, except ones that are purely negative – such a *negative teleology thereby reaches its end without purpose* (and not that purposiveness without end that provides the motives of Kantian reason).[11]

As such, Florian and his generation, and *us* – who are surviving with them, and among them, rather than truly living with them, since to live, for a noetic soul, is to exist by sharing ends, that is, *collectively* projecting dreams, desires and wills – we *all*, as and with Florian, we all, insofar as we *are*, find ourselves *thrown* into and thrown *out by* the epoch of the absence of epoch.

In earlier works (and in my first book[12]), I have tried to understand the meaning of an epoch via what philosophers call the *epokhē*. This Greek word, ἐποχή, refers to both 'a period of time, an era, an epoch', and to an 'arrest', an 'interruption', a 'suspension of judgement', a 'state of doubt'.

It is as such a suspension of judgement that the *epokhē* has become an element of philosophical vocabulary – used in particular by the Stoics and the Sceptics. And it was in these terms that, at the beginning of the twentieth century, it was revived by Edmund Husserl and placed at the centre of phenomenology – as a noetic method, that is, a path of thinking.

In a singular situation and by a path that I will retrace in summary in Chapter 5, in particular in §§28 and following, I came to the point of myself positing that what the philosophers call the *epokhē* – such that it lies at the origin of a *conversion of the gaze*, of a *change in the way of thinking*, and, through that, of a transformation of what Heidegger called 'the understanding that there-being (*Dasein*) has of its being'[13] (which, as we will see, consists in the individual and collective production of 'circuits of transindividuation') – this philosophical and more generally noetic *epokhē* (produced by a new form of thinking in general) is *always the outcome of a techno-logical upheaval*, itself derived from what Bertrand Gille described as a change in the technical system.[14]

8. *Epokhē* and disruption

A change of technical system always initially entails a disadjustment between this technical system and what Bertrand Gille called the social systems,[15] which had hitherto been 'adjusted' to the preceding technical system, and which had therein formed, *along with it*, an 'epoch' – but where the technical system as such fades into the background, forgotten as it disappears into everydayness, just as, for a fish, what disappears from view, as its 'element', is water.

Heidegger describes this vanishing of the technical element into everydayness (its forgetting) in §§12–18 of *Being and Time*.[16] What he shows is that the facticity of the world and of the epoch in which it presents itself becomes obvious and inevitable when there is an interruption of the technical element. This occurs, for example, when a tool we are using becomes broken: what is thereby revealed is the *fragility of the technical element*.

Heidegger's analysis must be carried over to another plane: not that of the tool, but of the technical system, which Heidegger himself thought in terms of a 'system of reference' (§17) and as phenomena related to what he calls 'relevance' or 'involvement' (*Bewandtnis*), as a complex of tools or a 'technical ensemble', as Simondon described it, and which, becoming in the twentieth century entirely globalized (as what Jacques Ellul would describe as the 'technological system'[17]), develops into what Heidegger will in 1949 begin to call *Gestell*.[18]

When a change of technical system occurs – in Bertrand Gille's sense – the epoch from which it originated comes to an end: a new epoch emerges, generally at the cost of military, religious, social and political conflicts of all kinds.

But the *new epoch* emerges only when – on the occasion of these conflicts, and due to the loss of the salience of the preceding epoch's knowledge and powers of living, doing and conceiving – new ways of thinking, new ways of doing and new ways of living take shape, which are 'new forms of life' in Georges Canguilhem's sense, on the basis of precursors *reconfiguring the retentions inherited from the earlier epoch into so many new kinds of protention.*

These new kinds of protention are new expressions of will, which we must understand here in the sense of the Greek βουλή (which is both the will of the citizen and that of the city), and constitute new forms of expectation (ἐλπίς) – that is, of desire and of the *economy* from which it stems: the libidinal economy, from which emerges, then, a new epoch. An epoch is always a specific configuration of the libidinal economy, organized around the ensemble of tertiary retentions (that is, around the technical supports of collective retention) that form, through their arrangement, a new technical system, which is always also a retentional system.

A libidinal economy is an economy of desire insofar as it is always both individual and collective. Desire is structured by a field of protentions that one inherits and then projects in a *singular* way, on the basis of collective retentions transmitted by the intergenerational play that is regulated by models of education at the different stages of life.

When tertiary retentions have adjusted to social systems, they tend always to be forgotten, just as water is forgotten by the fish. Nevertheless, in intergenerational processes of transmission, tertiary retentions radically condition the relationships between psychic individuals, and, through them, between collective individuals – between the mother and the *infans*, between the child and his or her siblings as well as other children, between the adolescent and the social milieu, between adults, between adults and new generations, and hence between generations, and, through the generations, between social groups, and so on.

In the contemporary epoch of the absence of epoch, the role of digital tertiary retentions in the intergenerational (non)relationship, and in the (non)formation of collective retentions and protentions, is both *perfectly obvious and totally escapes* comprehension – because there is no longer any adjustment between the new technical system and the social systems. Far from adjusting the social systems by reshaping them to suit a 'new epoch', the technical system short-circuits them and, ultimately, *destroys them*.

When a technical system engenders a new epoch, the emergence of new forms of thinking is translated into religious, spiritual, artistic, scientific and political movements, manners and styles, new institutions and new social organizations, changes in education, in law, in forms of power, and, of course, changes in the very foundations of knowledge – whether this is conceptual knowledge or work-knowledge [*savoir-faire*] or life-knowledge [*savoir-vivre*]. But this happens only in a second stage, that is, *after* the techno-logical *epokhē* has taken place.

This is why an epoch always occurs through a doubly epokhal redoubling:

- *double* because it always occurs in *two stages* – on the one hand, the technological *epokhē*; on the other hand, the *epokhē* of knowledge as forms of life and thought, that is, the constitution of a new transindividuation (characteristic of a particular time and place);
- *redoubling* because, starting from the *already there* forms of technics and time that are constituted as this or that established epoch, a new technical reality and a new historical reality (or, more precisely, historial[19] – *geschichtlich*) redoubles and through that relegates to the

past that which has engendered it, which seems, therefore, precisely to be *the past*;
- *epokhal* because it is only as an *interruption* inaugurating a *recommencement* and a *new current present* that this double redoubling occurs, eventually by firmly establishing itself as what we call, precisely, an *epoch*.

The *disruption* that is the *digital* technical system is one such *epokhē*: disruption is one such suspension of *all* previous ways of thinking, which were elaborated by appropriating previous changes of technical systems (and of the mnemotechnical and hypomnesic systems[20] that must be understood as processes of grammatization, which I will not discuss here[21]). But this *epokhē* is disruptive precisely in that *it gives absolutely no place to the second moment*, nor therefore to *any thinking*: it gives rise only to an *absolute emptiness of thought*, to a kenosis so radical that Hegel himself would not have been able to anticipate it.[22] It is, however, what Nietzsche would later see coming 'on doves' feet' – as the ordeal of nihilism.

The grotesque dimension of so-called 'intellectual debate', in France especially, which the French media discussed in autumn 2015, is a pathetic symptom of this fact.

In the midst of disruption, the second stage of the doubly epokhal redoubling fails to occur: there is no transindividuation. And hence there arises no new form of thinking capable of being translated into new organizations, new institutions, new behaviours and so on – through which *an epoch properly speaking* could be constituted. Behaviours, as *ways of living*, are being replaced by *automatisms* and *addictions*. At the same time, intergenerational and transgenerational relations are unravelling: transmission of knowledge has been prevented, and there are no protentions of desires that would be capable of bringing about a growth of transgenerational experience – of which ritual, religious or civil calendarities were hitherto the frameworks.

The age of disruption[23] is the epoch of the absence of epoch, announced and foreshadowed not just by Adorno and Horkheimer as the 'new kind of barbarism', but by Heidegger as the 'end of philosophy', by Maurice Blanchot as the advent of 'impersonal forces', by Jacques Derrida as 'monstrosity', and, before all of these, by Nietzsche as nihilism.[24] From around 1990, Deleuze broached this question, along with Guattari, in terms of the question of control societies and the 'dividuation' of individuals. Simondon didn't see it at all.

9. Epochs and collective protention

An epoch is what enables *collective protentions to be established* through the constitution of new circuits of transindividuation. Forms of thinking and forms of life are thereby metastabilized,[25] transindividuated by the psychic individuals of the epoch, through which new processes of collective individuation form, and thus new social groups and social systems, new social organizations and so on.[26] Circuits emerge through *affective relations* of various kinds – transitional, filial, friendship, familial, cooperative, recreational, religious, relations of power or knowledge – forging dreams, goals, objectives and common horizons, for which *close friends and family* play an *indispensable* role.

There are collective protentions only to the extent that there are collective retentions. The latter constitute forms of knowledge. They are transmitted collectively through educational organizations, and acquired over the course of life in its various stages – as elementary motor and language retentions, then as sayings, representations, formulas, rules, skills, doctrines, dogmas, narratives, ideas and theories. All these are what provide those capabilities by which the past can be interpreted, and it is from such interpretations that psychic and collective projections of the future can arise.

Heidegger transformed Husserlian phenomenology into an existential analytic (presenting itself as a development of phenomenology, one that takes the fundamental axioms of the Husserlian *epokhē* into account, while at the same time reforming them) when he explicitly and absolutely articulated psychic retentions (the mnesic elements forming the *psychē* of this or that individual, the individual being here what Heidegger called Dasein) and collective retentions.

Heidegger thus showed the following:

- All of Dasein's retentional activity is inscribed in the retentional activity of an epoch, which this Dasein inherits as it's already there, and which constitutes what I myself call collective secondary retentions (I will return to this[27]).
- Such an inheritance can occur only in the futural mode of a future [*futurition d'un avenir*]: as Heidegger will later say, 'the human [is] the one who awaits [*der Wartende zu sein*]',[28] this expectant awaiting being that of a future that comes to inscribe a difference in becoming (this difference being a *différance* that, as process of individuation, produces a bifurcation[29]).
- The futurity of the future is primordially constituted in Dasein by an *archi-retention* – 'archi' in the sense that it is always already known and '*remembered*' by Dasein – that is also an archi-protention (which is always already known and *fore-seen* by Dasein), namely, *the death of Dasein: Dasein knows first and foremost that it will die*, it knows this singular piece of knowledge [*savoir insigne*]. But this singular and primordial knowledge always conceals itself *through processes of denial of all kinds* belonging to what Heidegger called *Besorgen* ('busyness', 'pre-occupation').[30] It knows its end, most of the time,

only in the mode of this constant denial [*dénégation*].[31] Its whole existence is a *way* of knowing, which is also to say, most of the time (in busyness and preoccupation, *Besorgen*), a way of *refusing* to know. All the knowledge possessed by Dasein amounts to versions of this singular and primordial knowledge – but always in the partial way of a *différance* (a postponement[32]) that can never *quite* be known.

This knowledge is, in other words, the knowledge of a default, and a default of knowledge. It is a knowledge *by default*.

On the basis of these considerations, which emerged from a reading of *Being and Time* and *The Basic Problems of Phenomenology*, I have tried to extend the Husserlian concepts of retention and protention, and at the same time the Heideggerian concepts of the *already there*, *epoch*, *historiality* and *spatiality*, by forming the concept of tertiary retention – and, more recently, and in discussion with the work of Yuk Hui, of tertiary protention.[33]

Tertiary retention is, as we shall see, what compensates for the *default* of retention – which is also to say, the loss of both memory and knowledge. But it is also what *accentuates* this loss (this default): it is a *pharmakon*.[34]

Tertiary retentions and protentions allow us to understand what Heidegger investigated under the names of 'datability' and 'utility'.[35] Fields of collective retentions and protentions are thus shaped by the retentional systems of calendarity and cardinality[36] that underpin the epochs and, usually, *traverse* epokhalities – hence many epochs can belong to a single *era*, such as, for example, the epochs of the Christian era.[37]

10. Disruption and sharing

Dasein can receive the retentions that it inherits from an already-there past as its own retentions (by adopting them[38]) only because the latter are inscribed in the factical and technical space of the world (including as language), thereby constituting what, at the end of *Being and Time*, Heidegger called Dasein's 'world-historiality' (*Weltgeschichtlichkeit*), that is, the fact that temporality (and its historiality, *Geschichtlichkeit*) is already there before it in the world, as relics, monuments, stories, as *its* past that it nevertheless did not *live*.

This is what Heidegger shows in §76 of *Being and Time* in order to account for the possibility of historiography. But this is, before anything else, what conditions what he describes in §6, namely, that 'the past of Dasein always already precedes it'. This is possible, however, only because:

1. this past is not *only* its own – which means, in my own terminology, that it is formed from collective secondary retentions;[39]

2. it is *inscribed* in this world (which we see, Heidegger tells us, with
   relics, monuments and stories[40]) – which means that these collective
   retentions are made possible by *tertiary* retentions.

Dasein's psychic retentions are made possible by tertiary retentions
that are collective thanks to the very fact that they are exteriorized and
spatialized. Dasein is thus able to share, with other psychic individuals,
collective tertiary retentions that it apprehends as its *own* retentions, and
which belong to *the same epoch* (and to the same 'culture') as those with
whom this Dasein *shares* these retentions. From this it follows, too, that
individuals of the same epoch and the same culture have, if not quite
the same expectations, at least a *common horizon of the convergence
of their expectations*, forming *at infinity* the common protention of a
common future – the undetermined unity of a horizon of expectation
– which is also ultimately the future of humankind, that is, of noesis as
worthy of being lived in a non-inhuman way.

   We have seen, then, that such sharing constitutes the background
or the *funds* [*fonds*] of an epoch (and more precisely what Simondon
called its preindividual funds). *Digital tertiary retention, however, which
constitutes the digital technical system, is disruptive because it takes
control of this sharing*. This is what I have called, in pursuing the reflections of Gilles Deleuze and Félix Guattari, societies of hyper-control.

   These societies, however, are no longer quite societies, if it is true that
a society is constituted only within an epoch: they are aggregations of
individuals who are increasingly disindividuated (disintegrated). More
and more, this is leading to the rise of that new kind of barbarism
glimpsed in 1944, the contemporary realization of which is what we are
here calling disruption.

   The reconstitution of a true automatic *society* can occur only by
establishing a true economy of sharing – whereas what the current
disruption produces is, on the contrary, a *diseconomy* of sharing, that
is, *a destruction of those who share by the means of what they share*.

   Along with Ars Industrialis, I call this true sharing economy the
economy of contribution, which is the subject of the two volumes of
Automatic Society, where what is absolutely shared is knowledge as
negentropic potentiality. And it is shared as work, in the sense that the
father Schaeffer said to his son, Pierre:

Work at your instrument.[41]

3

Radicalization and Submission

11. Ὕβρις and aboulia

The horizon of expectation common to psychic individuals who live in the same epoch presents itself to them positively as that which contains their future in potential, insofar as this is something constantly renewed, and as such always new, thereby constituting the future properly speaking inasmuch as it is always unlike the present or the past. As such, the future [*avenir*] is unpredictable, bearing the improbable and the unknown that Heraclitus called *anelpiston* – the unexpected, the unhoped-for. And it does not reduce merely to becoming [*devenir*], which today we understand to be the entropic fate of the universe: *anelpiston* is the *différance of a becoming* that is itself entropic,[1] that is, a foregone conclusion, where everything will return to dust, and where 'unto dust shalt thou returne'.[2]

This horizon of expectation common to an epoch and to a generation is that of which Florian's generation has been deprived – 'blank', as the punks already said, presenting themselves as the 'blank generation'[3] – if we believe Florian. For expectation as the projection of a possible common future is always the expectation of an unexpected. Florian expects nothing: he expects nothing but the 'end', that is, the fulfilment of a becoming for which there is, precisely, no longer any future – a negative protention that is the absence of protention within an absence of epoch.

This deprivation of protention comes about from a deprivation of the possibilities of identification and idealization that precede it, and it participates directly in the new kind of barbarism installed by the culture industries. I attempted to analyse this in *Taking Care of Youth and the Generations* by showing how Canal J, a television network aimed at children, tries to eject parents and grandparents from the adolescent process of becoming adult, by short-circuiting the id that conditions identification, just as the Baby First channel, and television aimed at very young children, destroys transitional space and the processes of primary identification.[4]

What allows the *interiorization of collective secondary retentions* are primary and secondary identifications. Although collective secondary retentions are not simply 'mine', they *are* mine in the sense that they are *those of my epoch*, because I receive them from *within my intergenerational ancestry or through the friendship of my peers*: friendship is a fundamental vector of secondary identification through which the *philia* characteristic of an epoch is formed.

The new kind of barbarism heralded by Adorno and Horkheimer is characterized by the liquidation of these possibilities of identification and related possibilities of idealization. The liquidation of primordial narcissism – the liquidation of the *I* as well as of the *we* – is possible only on this basis. This deprivation of the possibility of identification and idealization, however, is *radicalized* by disruption: it is carried to its *breaking point* [*point de rupture*].

The radical rupture induced by dis-ruption makes *evident* that the epoch is missing [*fait défaut*], that it is merely the *absence* of epoch: disruption is what, in the geological era of the Anthropocene, and as its very *impasse*, *structurally* prevents the formation of collective protentions bearing a future charged with new potential. And it does so at a moment when the imminent possibility of an *excessively and definitively fatal* ὕβρις is gripping hold of and strangling any projection into the immensity of the improbable, and, in so doing, is sending us mad – mad with *sadness*, mad with *grief*, mad with *rage*.

The liquidation of protentions occurs in a structural way insofar as, as we have already seen,[5] psychic and collective protentions are being replaced by purely computational automatic protentions – eliminating the unhoped-for, essentially destroying every expectation of the unexpected, and thereby attenuating every form of desire (if desire, which is not simply drive, is always desire for the singular, that is, for the unexpected but awaited improbable).

The liquidation of protentions equally attenuates every kind of *will* – that is, all *power to bifurcate on the basis of knowledge derived from previous bifurcations*, knowledge that becomes collective retention through the processes of transindividuation characteristic of epochs. The outcome of this liquidation is abject aboulia.

Inasmuch as it always calls for an inscription into a more broadly shared protention, protention is always *bound* to a structure which is that of a promise, and as such to a mutual engagement that infinitely exceeds the psychic individual. This is what *Being and Time* ultimately fails to take genuinely into account:[6] the brilliant analyses it contains never explain how it is that Dasein always projects itself *beyond* its end,[7] and lives its mortality only in the primordial projection of a continuation of the world after its own end: in its *beyond*.

12. Speed and vanity

Disruption – in an age of ultra-libertarian capitalism where it amounts to a completely original form of ideology, and all the more so in that it states a *reality* that everyone otherwise denies – substitutes a blind becoming for this future desired in common, a future that is as such *wanted, in however small a way*: wanted *by* and *as* this '*in common*'. This is what blinds our fellow men and women today – a blindness wrapped in the highly complex, tortuous and devious 'storytelling'[8] of transhumanism, within which the absence of epoch wallows.

Throughout the epochs of the 'historial' form of life – inasmuch as we can, more or less badly, more or less well, know or deduce it from the documents, relics, monuments and stories received since the Upper Palaeolithic and up until the most recent data from the historiography of the Anthropocene – positive protentional horizons have existed. These horizons were shared as collective protentions across the most varied ways of life – via ritualizations capturing and forming the attention in which retentions and protentions are woven according to the conditions of retentional and protentional systems of all kinds: from Magic to Progress, via messianisms, redemptions, salvations and emancipations to come. Although these have been received from all cultures, those of the tragic Greeks, like *kleos* (κλέος), deserve particular attention.

As attentional formations, these retentional and protentional systems amounted to epochs of care [*soin*], *souci* [*Sorge*], as solicitude for the world, always exposed to the ὕβρις that facticity contains – which is an ὕβρις that can only be *contained* by this facticity, which can itself be factical only by always containing ὕβρις within it, which is also expectation, that is, *elpis* (ἐλπίς), and as such *curiosity*: this is the meaning of the jar of Pandora, woman-becoming-woman *through her being adorned in jewels*.

With disruption, *such systems can no longer be elaborated*: on the contrary, the barbarism specific to the absence of epoch consists in always *outstripping and overtaking* such systems, so that they seem always already futile, *vain*, the ruined remnants of what would have been only *pure vanity*, where care and attention arrive always too late – in vain. (Here we should obviously linger on the vanities that accompany protentions starting from the Baroque age, especially in Flemish painting.)

It is this vanity that haunts nihilism, weaving a dangerous form of contemporary melancholy that particularly strikes the younger generation, who do not deny (but who are confronted with the denials of those belonging to other generations) the radicalization of their discredit (and their 'disbelief'[9]) compared to the previous generation – taking this discredit and disbelief to a breaking point, a *point of rupture* that is the explosive counterpart of 'disruption'.

Hence arises Florian's terribly quiet desperation, *which in truth affects and disaffects all of us*,[10] including and firstly *in the mode of denial*, which here becomes a *modality of cowardice*. It afflicts all of us *so long as we are still capable* – in the abject aboulia that is this disaffection and this withdrawal [*désaffectation*] – of *wanting a future* that could wear away and pierce through the iron wall of becoming and cross the threshold that leads beyond the Anthropocene, in thus becoming the Neganthropocene.[11]

Only the prospect of a Neganthropocene – where one finds no virgins, it being neither the paradise of the desperate nor the brothel of Dominique Strauss-Kahn – can give to life its reasons for living at a moment when, on all sides, scientific reports produced by the international scholarly community make clear the irreversible character of the destructive process that began two centuries ago and that has significantly accelerated with the spread of consumerist capitalism across the whole planet.[12]

This 'planetarization' – which is the concretization of the Anthropocene (of human activity having become a geological factor) heading towards its limit, of which the IPCC report and the 2050 deadline now accepted as a tipping point are aspects – began with the culture industries that bore within them this new kind of barbarism.

13. Retention and disruption

Primary and secondary retentions are psychic realities – the first belonging to the present time of perception and the second to the past time of memory. Tertiary retentions are artificial retentions, not psychic but technical, such as archives, recordings and technical reproductions in general.

Richard Durn lived in the 'epoch' of *industrial temporal objects* produced by the industry of cultural goods, which, *in spreading the quotidian interiorization of analogue tertiary retentions to the whole world*, effected a major transformation of the way retentions and protentions are organized in that 'epoch' – at the cost of the disappearance of the very notion of the epoch as sharing, as heritage, as belonging and so on.

One of the main aspects of this epokhal transformation lies in the way that analogue broadcasting makes it possible to synchronize consciousnesses. With 'broadcast analogue tertiary retention',[13] the *industrial-temporal-object-consciousness* adheres to its object, and is at the same time synchronized with other consciousnesses, who adhere to it from their side[14] – frequently in the millions, sometimes in the tens or hundreds of millions.

Analogue tertiary retentions possess this synchronizing power to such an extent that they end up profoundly modifying the secondary retentions constituting psychic individuals – who *are* nothing other than their own secondary retentions inasmuch as they *singularly* project secondary protentions. Viewers, who are synchronized with each other by repeatedly watching the same programmes as one another, tend thereby to find their secondary retentions homogenized. In this way, they tend to lose the singularity of the criteria by which they select the primary retentions that they see in the programmes that they interiorize,[15] their protentions being transformed little by little into behavioural stereotypes concretely expressed in the form of purchasing behaviour.

The more viewers see the same thing, the less the criteria with which they are selecting what they retain in what they see varies from those who, together with them, compose the 'audience', that is, the mass of viewers.[16] In 1997, there were one billion televisions in the world, and such industrial tertiary retentions were being interiorized by almost all inhabitants of planet earth, including by Richard Durn.

In this way, it became possible to massify behaviour and to short-circuit the collective protentions constitutive of an epoch – because this retentional interiorization leads to processes (triggered by marketing) of 'identification' with the behaviours, brands and labels that typify this absence of epoch, in so doing ruining processes of psychic and collective individuation.

By massively modifying the processes by which collective secondary retentions are interiorized, where the latter are themselves methodically and industrially produced according to the dictates of the behavioural models conceived by marketing, the industry of cultural goods itself became the prescriber of the circuits of transindividuation constituting the 'second epokhal moment' of the techno-logical *epokhē* produced by the technical system based on analogue tertiary retention.

This prescription of circuits of transindividuation was functionally subject to the media economy, itself subject to the consumerist economy of which it was only a secular function – and was so at the cost of a structural de-symbolization of the mediatized masses, who thereby found themselves subjected to true symbolic poverty. With the analogue 'second epokhal moment', therefore, collective protentions had already been largely ruined, because social systems had been short-circuited along with the relations of primary and secondary identification that condition processes of psychic and collective individuation. And so it is that primordial narcissism suffered and regressed.

And so it is that Richard Durn – deprived of the 'feeling of existing' by the industrial synchronization and standardization of the attentional modes of the psychic secondary retentions that were his own, as well as the collective secondary retentions that bore protentions typical of an epoch – went mad and became homicidal.

Since the publication of 'To Love, to Love Me, to Love Us: From September 11 to April 21', I have described[17] the countless regressive processes that have been brought about by the massive interiorization of industrial analogue retentions, which has amounted to the destruction of attention by capturing it in the form of audiences subject to the criteriology of ratings. This is what has since come to be known as the attention economy, now 'refined' and radicalized by the data economy, which, as Frédéric Kaplan has shown, is an economy of expression made possible by digital tertiary retention.[18]

I have also argued in the *Symbolic Misery* series and the *Disbelief and Discredit* series that:

1. like any tertiary retention, analogue retention is a *pharmakon* (as Frank Capra insisted with respect to cinema[19]), and that it therefore does not *inevitably* lead to the inversion of the *Aufklärung*; this is why I have tried to show in *Technics and Time, 3* that the analysis of Adorno and Horkheimer was insufficient in the way it took up, without taking a step back from, the Kantian thesis of the schematism;
2. the new *pharmakon* that arose with *digital* tertiary retention brought with it new opportunities, fundamentally transforming analogue tertiary retention itself by integrating it into the process of digitalization, making it possible to go beyond the industrial model founded on the functional opposition between producers and consumers;
3. such opportunities will develop only provided that they are assisted (a) by a *European industrial policy* explicitly oriented in this direction (and where this is something we cannot expect from the United States, which on the contrary saw in the digital the possibility of reviving its own consumerist model) and (b) by implementing a *new macroeconomic organization* serving an economy of contribution capable of overcoming the impasses of consumerism.

This last point was developed collectively and systematically when I, along with George Collins, Marc Crépon, Catherine Perret and Caroline Stiegler, founded Ars Industrialis, positing in principle, in a manifesto published in 2005,[20] that digital tertiary retention is, *like* analogue tertiary retention, a *pharmakon* that must be socialized in Europe (which lay at the origin of the web) through a transformation of those institutions that emerged from literate (lettered) tertiary retention, and that this must be done within a broad European policy of the industrial technologies of the spirit, and so as to constitute *a new form of public power*.

In this lies the future of Europe, we said. And we made clear that unless measures are taken, we should fear the worst. After the crisis of 2008, in 2010 we published a new manifesto that scrutinized this slide towards the worst.[21]

14. Despair and submission

On the basis of the analyses of, on the one hand, Jonathan Crary, and, on the other hand, Thomas Berns and Antoinette Rouvroy, I have endeavoured in the first volume of *Automatic Society* to describe the way in which this 'worst' currently underway produces not only, as with analogue tertiary retention, a standardization of psychic secondary retentions and a loss of the primordial narcissism of the *I*s and the *we*s that television aims to tele-vise,[22] but the *elimination of individual and collective protentions*. These are replaced by automatic protentions derived from the automatic analysis of the retentions self-produced by internet users, and decomposed through a process of the automated 'dividuation'[23] of the digital traces produced by everyone. Hence it is that the data economy comes to replace the industry of cultural goods.

This replacement, which is a disruption of what was already disruptive, but by something *much more rapid and violent*, is demanded by those who, through a programme eloquently entitled 'Les barbares attaquent',[24] intend to promote, in France, something that does indeed present itself as the *radicalization* of a 'new kind of barbarism'.

In so doing, the 'disappointment' described by André Comte-Sponville in 1984 has long since given way to despair – and to the extreme violence that is its inevitable accompaniment when it becomes a major social and historical agent.[25] It is in this desperate context that the absence of epoch seems condemned to rush headlong to its end, not as the beginning of a new epoch but as the 'last generation'.

Whereas the industrial production of analogue tertiary retentions 'massified' [*massifiait*] psychic secondary retentions by replacing them with standardized collective secondary retentions, thereby eliminating the dia-chronic play that primary retentions make possible (a play that amounts to primary selections and as such to an interpretation that is each time singular[26]), psychic individuals *themselves* are the producers of digital tertiary retentions.

Psychic individuals therefore find themselves in the position of producing and expressing what amounts to the preindividual funds shared on the web and platforms. Reticulated digital tertiary retention, then, gives the appearance of being essentially participatory, collaborative and contributory. This is why, with Ars Industrialis, we posit that reticulated digital tertiary retention is a techno-logical *epokhē* that amounts to a new organological and pharmacological state of fact on the basis of which it is crucial to form a new macroeconomic and epokhal framework constituting a general economy of contribution.

Europe has failed – politically, economically, scientifically, artistically and socially – to develop an alternative model to the disruption promoted by the Californian model. It thereby utterly *submits* to this disruptive doctrine, and finds itself *overrun* by the pharmacological toxicity of digital tertiary retention.

Digital retention may indeed bring with it new and unprecedented protentional opportunities because it de-massifies the production of traces. Nevertheless, the disruption systematically explored and exploited by the new reticulation industry has in fact created a new, subtler stage of massification – that is, of the absence of epoch, giving rise to a new kind of barbarism, and doing so by creating a point of rupture, a breaking point.

What is massified today is no longer the criteriology by which primary retentions are selected, which was achieved by standardizing secondary retentions: it is the formation of circuits between secondary retentions via intensive computing, capable of treating gigabytes of data simultaneously, so as to extract statistical and entropic patterns that short-circuit all genuine circuits of transindividuation – where the latter would always be negentropic, that is, singular, and as such incalculable: intractable.[27]

15. What we must not lose

It was in 2005 that Florian expressed the statement that in 2006 became the epigraph of *L'Effondrement du temps*.[28] At that stage of the absenting of the epoch, social networks did not yet exist. Since then, we have witnessed the unfolding of countless disasters, including the 2008 crisis, and everything that has led to what is now described as a state of barbarism whose origin, so we are led to believe, is Islam – a description that amounts to a typical *causal inversion*, as is always typical of any ideology.[29]

Islam does not lie at the *origin* of the state of barbarism within which we are ever more obviously living. Rather, it is the spread of a 'new kind of barbarism' that has occurred with the rationalization of the *Aufklärung* – inverting its sense throughout the entire world, in so doing discrediting all Western culture, and at the same time the project of modernity, as well as the affirmation of secular principles, the right to education, economic rights and the protection of fundamental political freedoms – it is *all this* that has generated reactions that are themselves, *indeed, ever more barbaric*, especially in the Near East and the Middle East, where for decades the West has perpetuated a policy that is completely irresponsible.

This new kind of barbarism as generalized consumerism and venality no longer takes *any care* of the world in which consumers and speculators must nevertheless live. It is this blind stupidity leading to the madness of those it strips of the feeling of existing – that is, of being themselves worthy of respect, and of understanding themselves as such – it is this that has provoked the explosion of barbarism amongst those who do not respect life, including those who present themselves as 'Islamists' and who now channel the movement that has proclaimed a caliphate in the Sham region, to which Yassin Salhi, the 'psychiatric case' who beheaded his employer in Saint-Quentin-Fallavier in June 2015, claimed allegiance.

Before beginning the next chapter, we should try to think, if it is possible to do so (and I posit *in principle* that it *is* possible[30]), what is happening everywhere as so many abominable confirmations of the words spoken by Florian. For this, we must pursue a deeper understanding of what occurs in a general way with the destruction of psychic and collective retentions and protentions – and, along with this destruction, the destruction of all diachronies, all singularities, all desires *inasmuch as they constitute the negentropic capabilities of non-inhuman being qua Neganthropos*. In losing these negentropic capabilities, *non-inhuman being loses* reason insofar as reason is, precisely, always and uniquely *that which must not be lost* in order to live, noetically, the consistence of existence.

16. Neganthropy

We experience the meaning of Schrödinger's negentropy[31] when in a sunbeam we suddenly see, for the first time since the previous year, the explosion of the colours of spring – the fertility of everything that is *renewed again* in the light and heat that we had forgotten. As the release from the colourlessness of winter, spring is the ordinary experience of resurrection.

When we travel, we re-energize ourselves through the diversity of ways of life and the singularity of those cultures – that is, epochs – that *constitute* what we call the world by *cultivating* it. In this way, travel can provide a clear and immediate perception of that in which negentropy consists, which charms us, becoming what it is now a matter of thinking (that is, of thinking care-fully, *panser*[32]) with the name 'neganthropy', and through a neganthropology both philosophical[33] and positive.[34]

When we feel uneasy in front of a wasteland, a room in disarray, a depressed economic zone, what grips hold of us is anthropy. But it is a neganthropic promise that we feel when, crossing the threshold to enter a home, we encounter traces of everydayness unlike any other – which Italo Calvino described as the 'things' of his Reader in *If On a Winter's Night a Traveller*.[35]

A library (including that of the Reader) is a collection of *neganthropic potentialities* awaiting their reading so as to be actualized, noetically singularizing life as the neganthropy constituted by the anamnesis of pre-ceding neganthropies.

When we pay attention to them, and when we experience them, negentropy in general and the neganthropy that bifurcates from it organologically provide us with access to the extra-ordinary, which means not only that we, as Gilles Clément said, *always invent life*[36] – life that is within what we call 'nature' just as it is within what we call 'culture', which stems from what Georges Canguilhem described as a technical form of life – but also that we *discover* a plane of consistences through which the future is projected by *noetically differing from and deferring* [*différant noétiquement*] becoming, whether entropic or negentropic (that is, vital, qua natural selection).

Neganthropological différance, in other words, cannot be reduced to the plane of subsistence that governs life in 'nature'. In technical life – which is Dasein's existence – another kind of bifurcation occurs that is not just vital but, as Simondon said, psychosocial, and such that *the différance in which the vital process of differentiation consists becomes not just anthropic but neganthropic.*

In this way, 'culture' is something *more* than negentropy (in Schrödinger's sense): through exosomatization and the organogenesis in which it fundamentally consists, this neganthropy bears within it the ὕβρις of facticity, that is, a colossal *acceleration* of both the negentropic and entropic possibilities of so-called nature, of which the degraded anthropized milieus that abound in the Anthropocene are as traces left on the landscape.

In their attempts to integrate Schrödinger's ideas, Shannon and Wiener, biologists, and complex systems theorists such as Henri Atlan and Edgar Morin, all end up running into paradoxes. Combined with problems posed by Prigogine's dissipative structures, these paradoxes have led to confusion concerning what opportunities there are for thinking the future by incorporating the notions of entropy and negentropy, a confusion that is broadly reflected in the theoretical models of bioeconomics.

Because it is primordially exosomatic, organological and pharmacological, *Neganthropos* bears within it the *possibility of the inhuman* (which Heraclitus called injustice, Ἀδικία) as the *condition* of its being *non-inhuman* – the condition, in this sense, of its surpassing. To conceive surpassing as transhumanist 'enhancement' or 'augmentation' has nothing to do with neganthropology. Like both negentropy and neganthropy, the extra-ordinary belongs to the consistence towards which noetic existences project themselves and through which they raise themselves above their subsistence.

For many centuries (at least three hundred, at least since the Upper Palaeolithic, and until almost the beginning of the twentieth century), the way to access this plane of consistence offering hospitality to the extra-ordinary, which occurs also in artistic experience (and which is its condition), was via experiences that were either magical, mystagogical, spiritual or religious. Art was able to become detached from these experiences only with the advent of modernity. But once that occurred, it was not long before art was appropriated as 'aesthetic experience' by the industry of cultural goods, as a function of the capture of attention, and by the speculative market of venal collectors and hyper-philistines:[37]

hence begins what in the eyes of Adorno and Horkheimer amounted to a 'new kind of barbarism' (of which 'postmodernity' is one name).

A psychic individual encounters the *necessity* of the extra-ordinary *in and through its very default* (in its radical *absence*, or what theologians call dereliction, which struck Christ himself on the cross), which creates the *negative* experience of the extra-ordinary, or what we also call despair. It is inevitable and therefore necessary (if not very reassuring) that this psychic individual, struck thus by the feeling of abandonment, would be tempted to turn firstly to what humanity has for centuries and millennia proclaimed to be the condition of a *conversion* of the gaze.

In this search for a plane of consistence by overturning a way of life that suddenly seems absolutely vain and futile,[38] the candidate for conversion affirms the necessity of the extra-ordinary by seeking to gain access to it – the extra-ordinary inasmuch as it escapes the ordinary, such as the supernatural, the religious and all forms of spirituality that amount to so many eras in the succession of epochs throughout which collective protentions are formed as the *condition* of psychic protentions.

After the 7 July 2005 attacks by four 'suicide bombers' in London that left fifty-six dead and 700 wounded, I tried in *Uncontrollable Societies of Disaffected Individuals* to show that adolescent youth has a highly specific relationship to the super-ego, of which the tragedy of Antigone is the first formulation, and as a kind of ideal type – where the psychic individual in the course of becoming adult, who is thus said to be adolescent, turns upon his or her ancestors in order to reproach them for their infidelity with respect to the prescriptions that they claim to be transmitting to the next generation.[39]

In such periods, adolescence, which is often a time in which one experiences despair, can also be one of acting out [*passer à l'acte*] in myriad ways – and in particular by practising what I have called negative sublimation. In the epoch of the absence of epoch that is Florian's – it was in 2005 that he declared what Foucault might have called his *parrhēsia* (παρρησία)[40] – such possibilities are literally *exasperated*, and they are bound to proliferate, unless there is a genuine address to the new generations, and through them to us, responding to the *parrhēsia* of Florian with a discourse itself elaborated *on the basis* of this *parrhēsia* as such, that is, recognizing it *as such*.

Social groups struck by collective disindividuation are, and will increasingly be, prone to losing every reason for living, hence to losing the very notion of reason qua *convergence of protentions* – and to losing the notion of the value of life itself, especially when this noetic life, which is thoroughly organological and pharmacological, reveals itself to be such.

This loss of the reason for living, of the 'meaning of life', and therefore of its value, this form of madness, because it is the loss of the reason to *live*, is expressed above all by suicide. Hence we should not be surprised if the number of deaths by suicide in France soon exceeds the number of car accident fatalities, striking especially young people (10,000 'successful' suicides per year, and 200,000 'attempts', survivors of despair).

These social groups and forms of solidarity are woven, in highly variable ways, through the affective relations in which they consist. Processes that transmit the knowledge accumulated by the generations consolidate the reasons individuals have for living by inscribing these reasons within the horizon of the collective protentions that they engender and that they maintain by cultivating them – it is precisely this that we call 'culture'. Social 'cults' *maintain* neganthropy by cultivating its extra-ordinary variety.

With 'social engineering', or 'social networking', social groups are, as never before, struck by collective disindividuation – the social is being dis-integrated at its very root, that is, starting from *psychic* secondary retentions, which themselves lie at the origin of collective secondary retentions. They do so by depriving them in advance of any opportunity to form psychic and singular protentions, which is also to say of any *projective* capacity within identification processes that would in turn open onto idealization processes.

Just as the *I* is founded on a primordial narcissism that must be maintained and protected, and firstly against the pathological forms of 'secondary narcissism', so too there is a narcissism of the *we* that is formed through processes of collective individuation, stemming from collective protentions without which no psychic protentions could be cultivated. Weakening processes of collective individuation to the point of exhaustion can only have tragic consequences.

The fear of such consequences is what, in 2003, I expressed at the end of 'To Love, to Love Me, to Love Us: From September 11 to April 21'.

17. Identification, idealization and sublimation in the mutual admiration of the *we*

When the narcissism of the *we* is brutally harmed, one can expect only the worst. Like the narcissism of the *I*, it is always possible for the primordial narcissism of the *we* to become pathological, and to generate collective 'neurotic' or 'psychotic' forms of regression or disintegration – of which the *ressentiment of the average man* is often a harbinger. Yet this narcissism of the *we* remains indispensable. *Dangerously* indispensable, given that, in its collective forms (these too being extraordinarily varied), narcissism is *eminently pharmacological*.

In *The Ego and the Id*, Freud showed that to produce a process of sublimation – itself founded on a process of identification and capable of spreading to all objects of the world the process of idealization that first and foremost characterizes the constitution of the sexuated love object (this idealization being already a form of desexualization of its object) – it is necessary for the ego to become its own object of love, and for this to establish what I call primordial narcissism:

> [The question arises] whether all sublimation does not take place through the mediation of the ego, which begins by changing sexual object-libido into narcissistic libido and then, perhaps, goes on to give it another aim.[41]

This 'other aim' is one that *sublimates* its objects – and there are such sublimation and idealization processes operating *between* cultures, that is, between social narcissisms, of which the Western anthropology that arose in the nineteenth century is one case.

Even if it is a factor producing the detestable 'narcissism of minor differences'[42] that feeds parochialism and chauvinism, nevertheless only the narcissism of the *we* is capable of providing the feeling of the grandeur of a culture *by projecting itself into other cultures*, conferring their capacity for *mutual admiration* that is also the case for a healthy psychic narcissism constituted in the service of recognition.

Mutual admiration, which is indispensable to civilization, is always founded on this ability to recognize other cultures – which forms the conditions of 'peaceful co-existence' much more profoundly than does the balance of power.

Because it always threatens to turn pathologically into its opposite, the primordial narcissism of the *we* – which never stops transforming – can and even must be disturbing, if not frightening. Yet if we take Simondon seriously when he states that the psychic individual (the *I*) can individuate itself effectively only by participating in collective individuation (the *we*), then it is indeed necessary for there to be collective individuation, which can constitute itself only by distinguishing itself from other collective individuations, that is, through this collective identification that a *we* forms. Such collective individuations, however, never achieve completion: they are always metastable, and therefore amount not to an identity but *above all to this alterity to itself that constitutes its future as that which remains to come*, and that is promised to it in the collective protentions it cultivates.

18. Individuation, admiration and insubordination

It becomes a question, therefore, of understanding why it is that social networks have not given rise to other forms of the *we*, or other epochs of the *we*. It is precisely this possibility that we at Ars Industrialis *posit* as a *first* principle, when we say that an industrial politics of technologies of the spirit must constitute a new form of public power, which we further relate to the question of what Marcel Mauss called the 'internation'.[43]

But to reason in this way is precisely *not to submit* to the disruption promoted by Californian 'digital business'. For there is such a *submission*, not yet to come, as the poisonous fantasies distilled by a culture and a *business* of fear would have it, but rather right now, as resignation in the face of the diktat of 'radical innovation' in the service of the 'ecosystem' centred on Californian 'digital business'.

Clearly, a collective individuation constitutes itself (and can *only* do so) by exceeding the *directly closest* collective individuation of the psychic individual who is individuating: I can individuate myself psychically *only* by participating in a collective individuation that is dreamed *beyond the immediate collectivity closest to me* – that into which I was born, my 'family', my 'fraternity', my 'community', my 'country' (in the sense where the country [*pays*] is, for the peasant [*paysan*], the place he knows and where he lives to the extent he is capable of moving), and so on.

The arrangement of tribal relationships in Baruya society, studied by Maurice Godelier, is an example of the way such horizons of psychosocial individuation are embedded.[44]

It is, however, by *starting from* this original proximity of my *maternal (or paternal) facticity*, of which language is the most remarkable mark, it is starting from this original proximity of my local culture, giving rise to the idiom that I embody not only by speaking but in everything, by ek-sisting, and that I try to be as an improbable singularity, it is only by *starting* from this proximity that I can *begin* to encounter this strange and therefore 'foreign' alterity, where I find the other in myself – what I have in the past referred to as myself-an-other [*moi-l'autre*] (beyond myself [*moi-même*]).[45]

To start this way is to traverse the idiom that I embody, and to be traversed by it, which accommodates and enables the encounter with other, just as improbably singular idioms, provided that care has been taken in neganthropology – this care, that is, this culture, never being merely a conservation but always an individuation, in particular under the effect of the epokhal redoubling, irrespective of the pace of its being effected.[46]

It is from this tension between what is closest to me and what is contained there *already* as the most distant (in the experience of what Walter Benjamin called the 'aura'), it is through that which is most *idios* (ἴδιος) in my idiom, that is, most neganthropic, and that I encounter in the idiomaticity of other idioms or other idiosyncrasies, responding in their idiolect[47] to what is *already* contained in my idiom, and as the closest (and which is its *default*[48]), it is only thus that psychic individuation is possible.

Google, inscribing all its projects into the context of transhumanism, which is ὕβρις par excellence, takes the statistical and probabilistic calculation of averages as its standard, and thereby in fact eliminates idiomatic linguistic difference, that is, diachronic and idiosyncratic variability. The digital reticulation of all noetic life hence becomes the new programme of artificial intelligence whose goal is to eliminate all (de)faults – starting with the (de)faults of language, that is, of speakers.

This project is mad precisely in that it claims to eliminate the (de)fault that is *necessary* for desire to occur – not only as sexual differentiation and libidinal attachment to the sexual object, but as *detachment* of the libido from this sexual object, which instead becomes an object of *admiration*. One who admires becomes capable of spreading his or her admiration to the entire world, which is ultimately the sole protection against despair – against that loss of that reason for hope in which reason always consists.

It is through its intimate, native inscription in these collective individuation processes that are idioms that the psychic individual can participate in collective individuations of every kind, always renewing and individuating themselves – exceeding, altering and othering themselves, even if they need to cultivate the feeling of *existing in the other by identifying with that same* which bears this primordial narcissism that also opens onto the myself-other.

Were there no primordial narcissism of the *we*, there would be no process of identification. Primary identification, for example, presupposes the ego ideal of the parent, which itself presupposes the super-ego – which itself orders and metastabilizes processes of transindividuation. It is identification, *thereby* necessitating the *we* – in the varied affective relations within which it is woven through so many *relays* – that makes individuation possible: the primordial narcissism of the *we* enables individuation to occur because it conceals within it, in the most ambiguous way possible, the principle of admiration.

When I admire another's culture – the beauty of a city, of a landscape, of a country where I go to live, or that I visit, resonant with the specific accents of the idiom that has taken shape through an organogenesis and an exosomatization whose most immediately visible marker is its architecture – something else also takes place. There is an admiration that, within myself, and as myself-other, the idiom within which I am myself *transitionally* individuated has made *necessary*, and has done so as the encounter with the desire of the other (of my mother, of my father, of my parent – the one who takes care of me, who in so doing adopts me).

Any adult, mature admiration involves the resurfacing of this *child's play* [*enfance de l'art*] that was the first access to the consistent,[49] to this *other plane* that arises from what Winnicott called transitional space, from whence return, anamnesically and constantly, phantoms and spirits, or what Freud also called phantasms, dressed in inexhaustibly new attire, like Proteus, and as the genius of distance.

No *desire* for other lands and no possibility of being there (which is not always desire, and is sometimes surprising) would be possible were it not thus. This is what one can feel by reading Jean-Christophe Bailly's *Le Dépaysement. Voyages en France*:

> Whatever it may be, including when it is only furtive, the link an artist or a writer has to a land or a city maintains itself in a mysterious way: even though, and this is particularly clear for writers, the link may often be the result of chance, something incurred more than it was chosen, nevertheless something remains, flowing through the air.[50]

In *Ion*, Plato has Socrates say that the rhapsodist is like a current who magnetizes the audience like the stone of Heraclea[51] – also called a magnetic stone. This magnetism is that of transitional spaces that are *first and foremost* idioms, whose echoes reverberate step by step – each idiom being the echo chamber of those closest, and, *step by step, of the most distant*, which is the *originary default of origin*.[52]

A culture *in fact* cultivates its future only provided that it is inhabited by reflections that do not forget the primordial idiomatic spirit formed in mutual admiration – mirrors reflecting each other through identifications that cross borders. Hence Jean Renoir's *La Grande illusion*, where one feels borders and identifications everywhere, culminating in those between Boëldieu (Pierre Fresnay) and von Rauffenstein (Erich von Stroheim).[53]

Through the friends one *makes*, it is possible to exit from the primordial idiom so as to extend transitional possibilities beyond childhood, raised in and by the proximity of one's relatives: going above the primordial narcissisms of the *I* and the *we* into the beyond that is every consistence – which does not exist, but which, precisely *as such*, consists.[54]

4

Administration of Savagery, Disruption and Barbarism

19. The barbarians attack

After 9/11 and its terrifying aftermath, and fed by the second Iraq war, countless catastrophes have followed, all of which seem in retrospect to amount to a chain reaction of planetary proportions and immeasurable complexity.

In this tragic period for the whole planet, the 'economic crisis' that began in 2008 marked a turning point, like the first crack in a sinking ship – for which the Greek people would be the great sacrificial victim whose lament seems to sound the death knell for all of Europe. This totally irrational sacrifice is an exorcism that pretends to ward off evils that have been accumulating for decades – even though it is bound to worsen the effects to the extreme.

Starting in 2014, in the oil zone of the Tigris and Euphrates valleys, where since 1990 two wars against an international coalition (in addition to the Iran-Iraq war) led to the total disintegration of the Iraqi state, a group of former partners of Al-Qaeda, having become competitors to this network and emerging from the manipulations of American intelligence services during the Afghanistan war, baptized themselves Islamic State in Iraq and the Levant and proclaimed in the so-called Sham region a 'caliphate', claiming possession both of what was once ancient Mesopotamia (and which from the end of the First World War until 1932 was occupied by British troops) and the Syrian regions that had engaged in civil war after the impact of the 'Arab Spring'.

Syria, which had been the subject of a highly contradictory set of discourses in the West, then became a phantasmatic region channelling a 'radicalization' through which, in January 2015 in France, barbarism took a new turn.[1]

According to Ignatius Leverrier's summary, Abu Bakr Naji, one of the key strategists behind the war led by Daesh, has argued in *Management of Savagery: The Most Critical Stage Through Which the Islamic Nation Will Pass*[2] for the need to create 'chaotic territories':

> Provoking an outburst of violence in Muslim countries, jihadists will contribute to the *exhaustion of state structures* and to establishing a situation of chaos and savagery. People lose trust in their governments, governments which, overwhelmed, will respond to violence only with greater violence. Jihadis must seize the chaotic situation they have caused and gain popular support by imposing themselves as the only alternative. By restoring security, by providing social services, by distributing food and medicine, and by taking over territorial administration, they will manage the chaos, conforming to a Hobbesian schema of state construction. As 'chaotic territories' are extended, the regions administered by the jihadis will multiply, forming the kernel of their future caliphate. Convinced or not, the people will accept this Islamic governance.[3]

This 'administration of savagery' can thus be considered *disruptive*, and it is worth comparing it to the vocabulary of those, such as the founders of The Family, who present themselves as 'new barbarians', and who themselves 'attack'.

The Family, 'a structure for coaching and financing startups',[4] says that no sector should be spared from disruption. On a website called 'Les barbares attaquent', they set out, through a series of texts and recorded lectures, strategies for conquest (that is, if we take them at their word, for overthrowing civilization) in the following sectors:

> Agriculture – Insurance – Automotive – Business Services – Consulting – Social Dialogue – CIOs – Publishing – Education – National Education – Employment – Energy – Finance – Financing – Building and Construction – Hollywood – Immigration – Real Estate – Luxury Goods – Media – Family Policy – Employment Policy – Public Environmental Policies – Human Resources – Retail – Health – Telecommunications – Regions – Textiles – Tourism – Transport and Logistics – Public Transport.[5]

According to *L'Obs*, the word 'barbarian', as taken up by the founders of The Family, was inspired by the reaction of a young startup entrepreneur named Antoine Brachet against a 'ranking' by the Institut Choiseul, a 'liberal think tank' for those most 'successful' in business.[6] In opposition to this 'elitization', Brachet created another ranking, published on a Facebook page entitled '100 Barbarians', where one can find, notably, Nicolas Colin, co-founder of The Family. It seems safe to assume that Alessandro Baricco's short work, *The Barbarians: An Essay on the Mutation of Culture*, also inspired Brachet.[7]

This list of 100 barbarians, which would therefore purport to be a ranking of the 'best', shows that 'those who succeed [...] are the radical innovators'. Here, *radicalization*, which constitutes the *criterion* of 'success', does not refer to the admirers of Daesh, manipulated by strategists practising the method of chaos, perhaps inspired by the *shock doctrine*;[8] it describes the state of mind of the zealots of disruption, whom we discover to be not the only candidates in the 'digital business' 'success story'.[9] There are also scientists who, as Evgeny Morozov puts it, want to 'hack' the academy, others who want to 'hack' the state, activists who defend the ideas of Pierre Rabhi, and so on.

This new shock strategy that is disruption, according to Sophie Fay, a journalist at *L'Obs*, is, for the '100 barbarians', a matter of 'getting France moving', confronted with the digital disruption to which it would be necessary to 'adapt'.[10] These 'radical barbarian innovators' are above all promoters of new 'scientific and technical opportunities [who] do not care about the usual conventions'.[11] Science and technology are the weapons of the new shock strategy. This is so because, according to Pierre Pezziardi, who is one of their number, 'innovation is a disturbance of the established order'. Is this indeed the case, and what is being referred to here as *order*? And what, then, would be its relation to the *disorder* that comes to disturb it?

Contrary to initial appearances, Bertrand Gille argued from a very different perspective than do the new barbarians: if innovation indeed disturbs an 'established order', it is *successful* only if it establishes *a new order*, or in other words another *metastability*[12] – that is, new circuits of transindividuation – and not a state of shock and *permanent chaos*, to be manipulated for their own benefit by strategists advocating permanent and unlimited innovation, as so many perpetual coups d'état, or constant activity designed to 'exhaust state structures', as the Daesh ideologue put it.

In the 'Prolegomena' to his *History of Techniques*, Gille shows, by referring to François Perroux, that innovation is a process of convergence towards a new metastable equilibrium not just of the technical system within which it occurs, and from which it stems, but also with the social systems that it initially 'disadjusts'.[13]

If such a *metastable readjustment* failed to occur, the technical system would inevitably destroy these social systems, even though, without such social structures, technical individuation itself could not continue and could no longer be revitalized via social individuation, any more than through psychic individuation – since, the one having been destroyed, the destruction of the other would inevitably follow. It is on this fundamental point that, with respect to innovation, the concepts of Simondon and Gille are incompatible with those of neoliberal libertarians.

What we are witnessing today is precisely the destruction of social systems by the technical system, and this is what is referred to as disruption, which the new barbarians claim to exploit, but at the risk of destroying psychic individuals themselves – that is, of making them *both incapable*[14] *and mad*, while promoting their replacement by more controllable automatisms.

But this disruption does not produce dramatic upheaval just for the social systems: it has the same effect on the biosphere in all its dimensions, notably in its dimensions as a climatic, geographical, demographic and biological system.

20. Nihilism, disruption, madness

Consumer capitalism – whose effects in the United States were described by Adorno and Horkheimer at the end of the Second World War – has destroyed the libidinal economy and, *through that*, has installed a 'new kind of barbarism'. It is now trying to compensate for the extreme disenchantment to which exhaustion of the social systems has given rise *by radicalizing itself* – by becoming *purely, simply and absolutely computational*, imposing automated understanding on every kind of activity via the algorithms of social reticulation, which outstrips and overtakes every critique of reason.[15] Reason finds itself systemically short-circuited. *The reality of disruption is the loss of reason.*[16]

Purely and simply computational capitalism is as such the effective accomplishment and perfect completion of nihilism. Nihilism is the process that solidifies what is now called the Anthropocene. In the epoch of disruption proclaimed by the new barbarians, the Anthropocene is reaching its final stage – what, in an article published in *Nature* entitled 'Approaching a State Shift in Earth's Biosphere', twenty-two scientists have called the 'shift'.[17] It is this state of affairs that constitutes Florian's *horizon without expectations*.

Doomed to sink into a blind automatism closed in on itself, aiming like the barbarians of Daesh at the 'exhaustion of state structures' and all forms of public power (that is, all forms of power exposed to contradictory and rational public debate), this purely, simply and absolutely computational capitalism, this radicalized capitalism, reactively engenders radicalizations of every kind, yet it can produce only an *extreme rise of entropy* on planet earth, and with it provoke a global despair bearing the seeds of all manner of madness. Disruption, having become a strategy not just of shock but of chaos, is an extraordinary accelerator of the 'shift', and is in this way itself literally madness.

Imposing itself as permanent disadjustment, never allowing time for a readjustment of the social systems, installing an unsustainable absence of epoch that is necessarily also an absence of reasons for living, thereby ruining processes of psychic and collective individuation, disruption radicalizes the reversal of all values that is nihilism.

This radicalization can lead only to self-destruction: it exhausts the societies that it exploits, and it necessarily exhausts itself along with them – and at short order. The reversal of all values by nihilism requires their transvaluation via a leap into an economy founded on valuing and promoting negentropy and conceived as a reformulation of the 'great health' required by this pharmacology – beyond what Nietzsche called active nihilism.[18]

Hence the conception by Ars Industrialis of a true economy of contribution – which involves the creation of a contributory income remunerating the creation of negentropic value. For this, knowledge in all its forms (of living, doing and conceptualizing) must be rethought under the banner of a neganthropology, so that we may enter the Neganthropocene.

Disruption installs as state of fact an automatic society founded on the generalization of those logical automatons that are algorithms, outstripping and overtaking biological, psychic and social automatisms. But this installation is not an establishing [*instauration*]: in its current stage of *immaturity*, disruption destroys all social bonds, that is, it destroys what Aristotle called *philia*, which alone *contains* madness.

This state of fact cannot last long: it will end either in an exit from disruption as absence of epoch, or in the end of humankind – as *non-inhuman* kind – by an entropic draining that radicalizes psycho-social disintegration, and that is bound to unleash madness *at all levels* of disindividuation[19] by closing off every possibility of forming *critical protentions*.

Critical protentions constitute what Kant called ends. The faculty of reason that forms such protentions is what automatic understanding overtakes: it operates between one and four million times more quickly.

21. Noesis and hallucination

The ghost of madness constantly haunts non-inhuman beings as their inevitable ὕβρις: insofar as they constitute what Canguilhem describes as the technical form of life, they contain an instability. The disadjustment to which the appearance of a new *pharmakon* always gives rise is the historical reflection of this instability, and it is what the knowledge of how to live, do and conceptualize formed within social systems takes care of: these forms of knowledge are always therapies for the ὕβρις contained in the pharmacology of each particular epoch.

The exosomatic genesis of the non-inhuman being becomes organo-genesis by generating, on the basis of the new artificial organs emerging from exosomatization, the new psychic organizations and new social organizations characteristic of the epoch. *Such geneses are possible only on the basis of the phantasmatic and hallucinatory condition of all protentions*, both psychic and collective: the noetic soul *hallucinates* the plane of consistency, which does not exist, and for which transitional space maintains access for the *infans* via transitional objects – which are the first instances of the *pharmakon* in the life of the child.

The madness of ὕβρις always occurs through a confusion of the planes of subsistence, existence and consistence. In barbarism, whether it occurs within or outside civilization, it is the ghost of this madness that possesses the 'barbarians' as their ὕβρις, including when it is manifested in the kind of 'end of an era' behaviour that accompanies all great collapses.

This is possible only because, in the pharmacological condition, the *possibility* of madness is the *condition* of reason, and, so to speak, its reason. Madness, which always remains in the background of reason as its possibility, is both the negentropic and the entropic condition of *Neganthropos*. The Anthropocene is ὕβρις as the entropic inversion of reason in the sense described by Adorno and Horkheimer – an inversion in the course of which the *Aufklärung* turns into a 'new kind of barbarism'.

What this implies is that, if there is a history of madness, as Michel Foucault endeavoured to show, it is fundamentally tied to the history of the *pharmakon* – or what Jacques Derrida also called the supplement. Chapter 9 will return to these themes, passing through Michaël Foessel reading Blaise Pascal, and by revisiting the debate between Foucault and Derrida concerning the status of madness in Descartes.

Before engaging with this question of madness and its history, we will return to disruption as a *radicalization of the pharmacological question, whereby the madness of ὕβρις has been the condition of neganthropo-genesis ever since the default of origin* – where the default of origin, or the default of *identity* hallucinating *identifications*, is a primordial condition of instability and metastability, and where the unstable becomes metastable only via circuits of transindividuation that condition its individuation. Failing which, it becomes mad.

22. Outside the law: the *epokhē* of disruption and domination by chaos

In a short text to which I will return shortly, Alain Juppé spoke of the need for 'reform': 'reform' is the obsession of politicians who call themselves 'reformers'. If 'reform' is necessary, it is because 'modernity' is characterized by constant disadjustment.

Disruption does not burden itself with reform: it dissolves it and replaces it with a *state of fact* that renders the very notion of law obsolete. Chris Anderson's proclamations of the 'end of theory'[20] are utterly comparable with the words of the new barbarians: we do not burden ourselves with law or right [*droit*]. Such is their 'credo' – or their miscreant disbelief [*mécréance*]: they *do not believe* in the *difference of fact and law*, that is, the need to *make* this difference, and to constantly remake it, to take care of it, to cultivate and protect it, however illusory it may seem with respect to facts, however hallucinatory may be the protentions whose sharing makes possible this difference *in the making*.[21]

It is a matter, through the ranking of the '100 barbarians', and in their own terms, of 'spotting the leaders who are truly capable of changing France'.[22]

This idea of change, which in this case has nothing reformist about it, is exclusively concerned with economic initiative, deliberately and strategically short-circuiting any political *deliberation* – on the grounds that it is too slow, hence totally ineffective, and better replaced by the calculations of the market.

It is a matter, in other words, of *radicalizing the conservative revolution* – which was itself a radical critique of social-democratic reformism and the 'Fordist-Keynesian compromise'. It is equally a matter of *submitting all material, formal and final causes to the efficient cause that would be disruption in its self-sufficiency*, that is, *without any other purpose than efficiency itself*. It is a matter of outstripping and overtaking the law and its ends through the efficiency of facts.

This *reign of the state of fact* leads to the liquidation of public power: what the barbarians attack is the *legitimacy* of the public thing – inasmuch as it is in principle not appropriable by private initiatives. The attack of the barbarians is a claim, if not of pure illegality, at least of the *vanity of law*, against which disruption enables 'unblocking France' by multiplying legal vacuums – and thereby creating chaos. It is in this way that accomplished nihilism realizes the 'new kind of barbarism'.

It is a matter of standing outside the law by situating oneself as being *prior* to it, of creating a situation in which it *always* arrives too late, if not of going 'outside the law' in the usual sense. Disruption amounts to nothing other than the ultra-liberal, so-called 'libertarian' programme, which claims to absorb the social and the political into the technological and economic by crushing them: when technology is computational, it makes it possible to *algorithmically dissolve the social*, to reduce it to the calculable in an economy that has itself become purely, simply and absolutely computational.

What follows from this state of fact is the *dissolution of the state of law* itself: this is the real programme of the new barbarians, and it is why they present themselves *as* barbarians. In the name of the purely computational economy, disruption continues to extend in fact the domains of lawlessness [*non-droit*], creating legal vacuums by the very speed at which 'radical innovation' operates. This is, indeed, a radicalization of the way the real is placed outside the law, operating through the creation of a competition of speed.

What the new barbarians denounce in the state of law is something unreal, a juridical fiction that they claim to be illegitimate, an illusion and an economic aberration that invalidates the technologies of reputation – which, in the form of the data economy, and by consolidating the protentions automatically extracted from the digital retentions of us all, produces a purely computational 'general will'. On this basis, they declare that the 'social pact' formalized in law is now obsolete.

What they ignore is that the illusion and the juridical fiction are necessary, in the sense that, obviously, injustice reigns in the reality of facts. The law is indeed an illusion, and this illusion makes it possible, through those hallucinatory collective protentions that are also called principles, *to potentially load reality with neganthropic possibilities* to come, *against the facts*, that is, *against entropy* – to charge it up with what, within entropic becoming, comes to constitute itself in a negentropic way *against every expectation reducible to a calculation.*

It is as this state of fact – the fact of permanent and ever more rapid techno-economic evolution installing a state of lawlessness that is also a generalized state of exception, wherein surveillance and repression (including drones and lethal weapons that organologically transform the administration of death[23]) constantly proliferate – that disruption concretizes the 'new kind of barbarism' foreshadowed in 1944, and it does so by radicalizing it, that is, by taking it to the extreme.

Does this mean that computational and reticular technology, which the new 'barbarians' claim embodies everything that is best about 'hacking' public power, is structurally opposed to the constitution of a new *res publica*, that is, to the birth of a new era of civilization and law?

The conviction of Ars Industrialis is, on the contrary, that the digital *pharmakon* – which, through the speed at which it functions, makes it possible for calculation to *destroy the improbable, that is, desire, affection, attachment, identification, singularity, individuation and the feeling of existing* psychically and thus collectively, which are the conditions of any neganthropy, that is, the conditions of any positively protentional hallucination – this digital *pharmakon*, destroyer of neganthropy, is also the bearer of a new epoch of psychic and collective individuation, that is, of a new neganthropy, constituting and harbouring a new form of epokhality.

From this second stage of the *disruptive* doubly epokhal redoubling, there must emerge a new condition of doubly epokhal redoubling. This requires a complete reconsideration of what is involved in the concept of the 'public' inasmuch as it is woven through processes of transindividuation. A new public thing is required to open up a new age of deliberation, by drawing on the sources of reticulated digital *mēkhanē*, which is first and foremost an organ of publication.

This organ of collective capacitation, that is, of the formation and sharing of protentions characteristic of what we call the public good, must, through a contributory economy, cultivate the knowledge of life, work and conceptualization. And it must cultivate these as *knowledge and power that produce a bifurcation beyond the disruptive chaos.*

This requires that we effect a great bifurcation in the Anthropocene, leaping into the Neganthropocene, and saving the history of life from irreversibly crushing itself. Life forms that have evolved must not be allowed just to turn into dust, which is what looms in the epoch of the new mass extinction of species – the sixth, according to scientists from 'Stanford, Princeton and Berkeley, among others', an event that, according to Gerardo Ceballos of the Autonomous University of Mexico, has every chance of leading to the disappearance of the human species itself.[24]

Such would be the outcome of the kind of barbarism that the new barbarians, who would like to dissolve neganthropic law into the entropy of computational facts, embody as its contemporary realization.

23. Conquest or salvation?

Contrary to their claims, these new barbarians are *in no way radical innovators*. They want to subject European civilization to a disruption conceived and driven outside France and Europe, largely according to the interests and business models of the Big Four, and by entering into a world dominated by what management theory refers to as ecosystems.

All these global-scale companies cultivate such ecosystems, which are conceived according to a form of social Darwinism, where the selection is accomplished through economic competition, itself understood as a struggle for life that eliminates the weak. This perspective – which is not that of Darwin but rather of his cousin, Francis Galton – takes no account of what Nicholas Georgescu-Roegen called exosomatization, inasmuch as the latter radically changes the conditions of life, and does so long before there is any possibility of the kind of 'enhancement' conceived by the transhumanists.

In the exosomatization characteristic of neganthropological evolution, organogenesis is expressed in the constant reorganization of the psycho-somatic organs of psychic individuals, which is demanded by the appearance of the new artificial organs in which this exosomatization consists. Psychic individuals are compelled to undergo long apprentice-ships during which they interiorize collective secondary retentions, that is, knowledge.

On the basis of this knowledge, which they inherit but then transform, individuals and the groups they form produce protentions that are each time new – resulting in collective retentions and protentions via the formation of social organizations. It is this knowledge and the artificial organs that put it to work that here become decisive for the future [*avenir*] insofar as the latter is not reducible to becoming [*devenir*]. This is why the struggle for life, which is the condition of what Schrödinger called negative entropy, or negentropy, can no longer be the criterion for the formation of the societies that constitute neganthropic evolution.

In the technical form of life, the struggle against entropy becomes neganthropic precisely in that it suspends the pressure of natural selection. The latter is replaced by selection criteria operating between retentions and protentions and in the service of a struggle for existence and consistences via knowledge, that is, for the metastabilization of a neganthropological order that is also a relative – that is, metastable – disorder.

Knowledge, here, means the power to produce bifurcations that do not destroy access to those previous bifurcations that are the collective secondary retentions whose accumulation forms knowledge. Preserved and reproduced through the transmission of knowledge, these earlier bifurcations, which grant meaning to bifurcations to come, constitute the ultimate meaning of what, in *Meno*, Socrates called *anamnesis*.[25]

The radicalization of innovation claimed by the new barbarians on the contrary brings to completion the process of proletarianization that in the nineteenth century became the condition of the Anthropocene – that is, of industrial capitalism – and that at the beginning of the twenty-first century has been extended to all human activities through full and general algorithmic automation. The latter is absorbing and eliminating all forms of knowledge, sterilizing them, and thereby dissipating the neganthropic potentials accumulated by civilizations and destroying the ends around which the latter were constituted, leaving non-inhuman beings in despair.

Non-inhuman beings tend then to become inhuman. Ruining the neganthropic dynamic, what the new barbarians claim to be a radicalization of innovation is in fact a continuation of a process that began with the conservative revolution in the early 1980s, a process that aimed to liquidate the regulation of disadjustment by public power.

With digital disruption, this ultra-liberal and now libertarian enterprise *fundamentally impoverishes* economies and cultures (in particular in Europe) by destroying social structures, and knowledge along with them. This in turn leads to the disintegration of the psychic apparatus and to the unleashing of the ravaging and furious drive-based behaviours of barbarism and destructive madness.

Were we to submit to the injunctions of the 'digital Attilas' who would like to be *our* barbarians,[26] we would be condemned to accept that, with the radicalization of innovation imposed by disruption, terrorist radicalization can only progress – a barbarian 'submission' (that of the volunteers for jihad, most of whom have no Muslim culture[27]) opposing another 'submission' just as barbarous (which in no way represents Western culture: it is on the contrary and precisely its destruction, and it claims to be such).

Faced with this, Europe – along with the rest of the world, and with what, in this world, forms the internation[28] – must reinvent the technology it has let drift away from it, as Valéry said.[29] Europe has forgotten its meaning and its stakes: it must, with the rest of the world and the internation that forms therein, *reinvent a state of law that is also a new age of knowledge and of the organology it presupposes and that it engenders.*

If, in the violence that has been proliferating and self-generating since the beginning of the twenty-first century, what is at work everywhere is disruption, then this is what we must begin to think – and to think care-fully [*panser*][30] – not in order to 'conquer the world', but *to save it.*

5

Outside the Law: Saint-Michel and the Dragon

24. Anthropology of disruption

It is unlikely that Bertrand Gille ever envisaged the possibility that has come to pass in the age of disruption to which the barbarians lay claim. Conversely, disruption is almost a direct consequence of what André Leroi-Gourhan did envisage in *Gesture and Speech*, as a possibility that would also amount to the end of humanity:

> The infiltration of urban time [is] now spreading to all moments of the day to suit the rhythm of radio and television broadcasts. A superhumanized space and time would correspond to the ideally synchronous functioning of all individuals, each specialized in his or her own function and space. Human society would [...] recover the organization of the most perfect animal societies, those in which the individual exists only as a cell.[1]

What I have tried to show is that the reticulation of automatisms *tends* towards the concretization of such possibilities, through the production of traces functioning as 'pheromones' in 'digital anthills', where social systems are replaced by global companies that own and operate the algorithms that ensure the hyper-synchronization of what, by that very fact, is prevented from developing into an internation.[2]

Hyper-synchronization *prevents the power of calculation from being socially individuated by the incalculable*. This is precisely what the disruption claimed by the barbarians produces, where the social systems, which are never able to 'catch up to', 'appropriate' or adopt the technical system (which would be to metastabilize an age of transindividuation), are instead disintegrated (in the strict sense[3]) and crushed, reduced to dust, that is, dissolved into entropic becoming.[4] Transindividuation is replaced by processes of transdividuation[5] that are under corporate control, by corporations that are in turn controlled by shareholders who 'manage' them according to a single criterion: the increase of dividends

– at the cost of psychic and collective disindividuation, at the cost of madness.

Every human society is obviously a social network that is both founded on traces and produces traces – that is, tertiary retentions, which generate primary and secondary retentions. What lies at the origin of *Western* civilization is *literate* tertiary retention, which requires citizens to thoroughly reshape their mnesic and anamnesic capabilities, which is also to say their protentional capabilities.

The successive epochs of the Western era (of monotheism, for which the philosophy of Plato paved the way, and which becomes the Christian era) are all, whether recent or distant, thoroughly conditioned by processes of synchronization and diachronization that result from the literate culture of *citizens*, or the *faithful*, or the *subjects* of what amounts to the epochs of the book, which are themselves the eras of monotheism – Judaism, Christianity, Islam. The Renaissance occurs during this age dominated by Christianity in the form of printed books and the 'age of madness', which undoubtedly stems from the great disadjustment caused by printed tertiary retention, from which would emerge primitive capitalism as described by Max Weber.[6]

Analogue traces appeared in the twentieth century with photography, phonography and cinema. The circulation of these analogue traces via broadcast networks produced what Leroi-Gourhan described as the programme industries, giving rise to a process of synchronization that would have been completely inconceivable in the epoch of handwritten or printed literate tertiary retention: it was in terms of this possibility of synchronization, which they understood also as standardization, that Adorno and Horkheimer anticipated a 'new kind of barbarism' – of which 9/11, as a programme and spectacle broadcast into a billion television sets around the world, was a symptom, as was the Nanterre massacre.

The *standardization of symbols* by the culture industry in fact amounts to a de-symbolization, where the senders of what are, precisely, programmes, and no longer symbols, are functionally distinguished from the receivers. Here already, it is a case of economic organizations taking control of the *imagination*, that is, of the *primordial source of protentions*, as well as of the formation of collective secondary protentions, and, ultimately, of individual and collective dreams.

Nevertheless, a new step is taken, one that undoubtedly amounts to a break, with the formation of what Leroi-Gourhan foresaw in 1965, and which he referred to as a 'magnetic library' [*magnétothèque*]:

> There can be no doubt that for several thousand years, quite independently of its role as keeper of the collective memory, writing has by dint of its one-dimensionality provided the analytical instrument indispensable to philosophical and scientific thinking. The preservation of thought can now be envisaged otherwise than in books, which will not for long possess the

advantages of quick and easy manageability. Preselected and instantane-
ously reconstituted information will soon be delivered by a huge magnetic
library with electronic selection.[7]

By somehow projecting outwards from the tendencies expressed in the
analysis of the contours derived from the palaeo-anthropological and
historical documents that he synthesized as a prehistorian, ethnologist
and observer of the contemporary world, Leroi-Gourhan succeeded in
producing an analysis of astounding perspicacity. What he describes
amounts to the advent of digital tertiary retention that would occur after
that of the programme industries founded on analogue tertiary retention,
giving concrete effect to the hypothesis he advanced in his analysis of the
ultimate consequences of this synchronization:

> Because of the development of its body and brain, through the exteri-
> orization of tools and of the memory, the human species seemed to have
> escaped the fate of the polyparium or the ant. But freedom of the individual
> may only be a stage; the domestication of time and space may entail the
> total subjugation of every particle of the supraindividual organism.[8]

There is no doubt that what we are presently living through corresponds
to what for Leroi-Gourhan fifty years ago was a futuristic prediction.
If I am to believe what was once confided to me by Jean-Paul Demoule
during a seminar I organized on Leroi-Gourhan's work,[9] Leroi-Gourhan,
who had taught Demoule, at the end of his life discouraged his students
from reading his own speculations.

In fact, these extrapolations, which express tendencies, are a kind
of 'thought experiment', but, in this experiment, a certain number of
elements are diminished or ignored. These elements amount to those
counter-tendencies that this book is an attempt to highlight:

- On the one hand, Leroi-Gourhan's extrapolations neglect the question
  of the threefold individuation that conditions technical evolution,
  which, itself a process of individuation, requires and makes possible
  collective individuation, which itself requires the psychic individuation that it also makes possible: this threefold individuation amounts
  to a play of transductive relations between three terms, where it is
  difficult to see how any one of the three could be eliminated.
- It is difficult to see because, on the other hand, the condition of
  noetic individuation, whether psychic, collective or technical, is the
  production of neganthropic bifurcations that always translate into
  stages of the process of exteriorization, itself emerging from stages of
  the process of interiorization in which the second stage of the epokhal
  redoubling consists, and does so as the formation of new circuits of
  transindividuation (for a new epoch).

- These amount, then, to the questions: (1) of disadjustment, which corresponds, in the transductive threefold process of (psychic, collective and technical) individuation, to what Simondon called the phase shift [*déphasage*] of the individual through the process of its individuation; and (2) of readjustment, which corresponds to what Simondon described as a bipolar metastabilization forming between collective individuation and psychic individuals. These questions are simply ignored by Leroi-Gourhan, but we can still say that *he formulates the hypothesis of disruption*, as that which eliminates psychic and collective individuations in favour of a techno-economic individuation that our barbarians call radical innovation – behind which looms transhumanism.
- Leroi-Gourhan did not raise the question of the Anthropocene and its entropic consequences, such that they constitute a limit – and he thus brushed aside the question of ὕβρις implied by this Anthropocene.

25. Neganthropology of disruption

The 'storytelling' that accompanies and legitimates disruptive reticulation amounts to the claim that disruption inverts 'top down' power, by permitting individuals to exercise theirs from the 'bottom up'. The disruptors proclaim their individualism – glorifying the ability of individuals to constantly express themselves and interact with the network – which they present as freedom's triumph over the rules, regulations and frameworks of all those orders in which society consists, and which ensure the adjustment between the social systems and the technical system.

But Leroi-Gourhan showed in advance that this reticulation would be anything but some new conquest, triumph or rediscovery of freedom over and against the rules that would supposedly have suppressed it. The 'magnetic library', as we have just seen, is the very thing that suggests that the 'freedom of the individual may only be a stage; the domestication of time and space may entail the total subjugation of every particle of the supraindividual organism'.[10] The barbarian disruptors posit as a fundamental principle that only psychic individuals should 'arbitrate choice' – that is, *produce clicks*.[11] This individualism, which is in reality a destruction of individuals through the herd-like characteristics that prevail, claims to be purely 'bottom up', that is, freed from all the constraints of the 'top down', thereby emancipating itself in a 'revolutionary' way from 'reform'. And this 'reform' is, indeed, totally discredited in the face of this new stage that is digital grammatization, and that is being concretized as automatic and reticulated society: the indigence of national or European public power is in this respect shocking and undeniable.

But what is ignored by the ideology of disruption, driven by the new barbarians who claim to be importing the ultra-liberal vision of the silicon libertarians,[12] is that *freedom is constituted as a transductive relation*, that is, as the *play of the set of constraints shared* precisely as the *rules of this game*. In the case of *disruptive companies*, contrary to what the latter claim, these rules and this play exist, but *'psychic individuals' are no longer the players: they have become pawns* – turned into the powerlessness of a public who watches but no longer arbitrates anything as these games play out, understanding next to nothing of their stakes.

The 'conquest of freedom' through what Leroi-Gourhan called the 'liberation of memory' by the process of exteriorization has always consisted in the law and the fact that psychic individuals can take part in individuation only insofar as the latter is always at once psychic, collective and technical – and by introducing a *diachronic heterogeneity* formed by the singularity of desire, which itself transforms *collective* retentions and protentions into *psychic* retentions and protentions bearing neganthropic *bifurcations*.

This exteriorization that constitutes the principle of 'liberation' is always *also* an 'alienation': it leads to *an offsetting of neganthropic possibilities to exosomatized organs* that amounts to a kind of *dependence*, which is the basis of proletarianization as the loss of knowledge (of how to live, work and conceptualize), *that is, as entropy*. This is nothing new. And it is so as to constitute, on the basis of this exteriorization, which is firstly a loss of knowledge, *new forms of knowledge* – that is, new forms of neganthropy, as the diversity of knowledge, culture and social organizations – that a readjustment must occur, through which it becomes possible for therapeutic circuits to form, circuits of those *pharmaka* in which all exteriorizations consist.

As the Anthropocene approaches its limits, disruption – which consists in *accelerating the process of exteriorization by short-circuiting and annihilating the question of psychic and collective interiorization that is individuation* – establishes the reign of proletarianization by *placing all those 'disrupted' sectors outside the law*.

In so doing, the principle of disruption consists in *hastening the Anthropocene's approach towards its limits.* By violently accelerating this event, disruption risks triggering in each of us the feeling that we are rushing headlong into an abyss, which would be to produce a purely negative collective protention – bringing with it the most barbarous behaviour imaginable.

Faced with this possibility, which in the end Leroi-Gourhan never foresaw, it is imperative to reconsider exosomatization and organogenesis from a pharmacological perspective, where the question is always that of turning the poison into a remedy, that is, of making that which produces a local increase of entropy turn into its opposite, so that it becomes a new neganthropic and neganthropological factor – one that would for this precise reason be epokhal.

26. Providential disruption and the 'wall of time': the reign of dread

As Ars Industrialis has said from its inception, such a transformation requires a new critique of political economy – the economy having become purely, simply and absolutely *anti-political* with the advent of purely, simply and absolutely computational capitalism: the elimination of social systems is the liquidation of political collective individuation, coming at the cost of the destruction of psychic individuals – of their psychic apparatus, which constitutes their reasons to exist in view of consistences, beyond mere subsistence.

The disruption promoted by the new barbarians is the reign of the law of the jungle such as this is dreamed of by the most reactionary conservatives. This air-conditioned jungle – as Duke Ellington called it – establishes a *reign of dread* that is also a new stage of the 'shock doctrine', and it does so as the *attack of the barbarians*, which is, in effect, that is, in the facts, an *organization of chaos deemed to be creative*. And this chaos would also be that to which the advocates of 'Sharia' lay claim, *in awaiting a divine law* that would restore order to this exhaustion of public law, that is, of profane realities.

For the barbarians, who, conforming to the prescriptions of Abu Bakr Naji, await the divine law that will restore order to the chaos that they themselves organize for this very purpose, the *public* thing, which is profane in nature and therefore secular, has become totally illegitimate – the public space of Öffentlichkeit having been *privatized* by what Adorno and Horkheimer describe as a new kind of barbarism.

As for the new barbarians, despite the 2008 collapse and the actions of the European troika that went on to sacrifice Greece, they continue to celebrate the invisible hand as the Providence of God that would be the omnipotent market in the competition of all against all, which is obviously bound to turn into a 'war of all against all'. This Providence that underlies the market proclaims the omnipotence of calculation, which is nevertheless heading straight for the wall of the Anthropocene, while at the same time giving Europe back its old demons – now heavily armed.

Society is constituted through the transindividual construction of social systems. For that to occur, the technical system must be metastabilized – it must maintain a general *form* while *supporting the deviations* that make it *dynamic*. Such a process of internal phase shifting, which is what any dynamic system amounts to, always conditions psychic and collective individuation.

The digital technical system currently individuates by radical (that is, disruptive) innovation, driven by the speculators that shareholders have become. Today, then, the technical system is engaged in a constant and unlimited process of mutation, which, as we will see with Peter

Sloterdijk and Jean-Baptiste Fressoz, amounts to the most extreme stage of a process of disinhibition that began to unfold in the fifteenth century. The result is that, at the same time:

- this radical innovation prevents any metastabilization with the other systems that constitute the social *body*, destroying in advance any capacity on their part to adopt the technical system, or to control, relatively speaking, its effects;
- 'radical' innovation continuously increases the colossal accumulation of capital derived from placing 'disrupted' sectors outside the law – acquiring cash value at the expense both of psychic individuals (who are impoverished also on the plane of subsistence) and of those collective individuals forming public power (which also in this sense becomes public powerlessness).

The *metastabilization* that would render to reason the *time to synthesize* and hence to psychically and collectively '*protentionalize*' the lightning-quick analyses produced by automated understanding, so as to transform them into social ends, is thus the very thing that disruption prevents. It does so by submitting speculative investments in radical innovation to a single criterion: the increase at any cost on the return on investment of those who are therefore no longer investors, but speculators.

Those who speculate on radical innovation, its chaotic effects and the gains they expect to make thereby bet against the social systems, just as in other times they might have bet on being able to make a profit from famine or poverty. In this case, the bet against the social systems consists in hastening the players (the speculators themselves) as well as the pawns (us) headlong into the wall that is the blocked horizon of the Anthropocene, that is, sending it to its limit, which is approaching at high speed.

This blocked horizon is what I once referred to as the 'wall of time':[13] this is the moment when a 'shift' holds sway, that is, a potential for highly chaotic bifurcations, of the kind referred to in the *Nature* article, 'Approaching a State Shift in Earth's Biosphere'.[14] As an imminent event, this must be anticipated, projected and *wanted through a radical trans-formation of radical innovation* – in this way becoming the true *social and open innovation* that lies at the heart of what we call the economy of contribution.

27. Φιλία, différance and ὕβρις

For thirty years, I have tried to build upon the concept of disadjustment, by taking it from Gille and articulating it with Simondon's theory of psychic and collective individuation. I will return to it when I introduce the question of φιλία (*philia*) – that is, of what it is that binds together the members of a community, which is also what is torn apart when a civilization degenerates, giving way to barbarism, and which is related to what Roberto Esposito calls *delinquere*, or what I am here calling ὕβρις.[15]

What Gille describes as *adjustment*, which I describe in Simondonian terms as *metastabilization through transindividuation*, is the condition of formation of what, as φιλία, also conditions this transindividuation. The circular causality this involves leads us to propose that φιλία is that *transindividuation through which what Simondon calls the transindividual is formed*, which is the condition of meaning as sense [*sens*], that is, of *reason conceived not as ratio, but as affect*.

It is in φιλία that the transindividual is given. It is generated by processes of transindividuation and vice versa – transindividuation (and, with it, φιλία) being itself conditioned by *tertiary retention, such that potentials for synchronization and diachronization come to be distributed* in what amounts to a general processuality that is both an organology and a pharmacology.

In the case of disruption, the circuits of transindividuation constituted through φιλία, and vice versa, are short-circuited and replaced by what I have described in the first volume of *Automatic Society* as processes of 'transdividuation' – taking up the concept of the 'dividual' put forward by Félix Guattari.

In his discussion of what he called a 'magnetic library', Leroi-Gourhan showed extraordinary foresight, both in terms of analysing the stakes involved in the culture industries and anticipating the rise of digital networks.[16] He imagined the 'dividualization' of a 'humanity' that would have become totally dependent on the prostheses with which all human knowledge – of living, doing and conceptualizing – would be exteriorized, but a humanity who would then have become *inhuman in a strict sense*, that is, *devoid of any neganthropological capability*.

Such is the intolerable perspective of the 'wall of time'.

There is, evidently, a kind of paradox involved in Leroi-Gourhan's concept of exteriorization, inasmuch as it leads to its own disappearance – where there would no longer be anything to exteriorize, metastability giving way to stability, unless we envisage a war of prostheses among themselves, such as through the intermediary of 'killer robots', where individuals, who will have become cells of a supra-organism, would play no part other than as victims, or where, too, a 'process of interiorization' may be initiated, consisting in effecting, through 'NBIC convergence',[17] the *re-engineering* of life in totality, and, in the first place, through the engineering of 'enhanced' human life.[18]

From an anthropological perspective that has become neganthropological, such scenarios are the least rational: they are what provide the least reasons for living, subjecting the whole of life to this entropic fate that Freud also called Thanatos, as a tendency contained in the technical form of life itself. This is why it is the bearer of that madness concealed in every loss of reason.

What we know from *Beyond the Pleasure Principle*[19] is that technical life *is* noetic life, and, in that, desiring life,[20] constituting the battle-ground of a struggle between two tendencies that Freud calls Eros and Thanatos – which are the life drives and the death drives. What remains deficient in these extraordinary speculations, however, is any account of the fact that non-inhuman life is technical life in Canguilhem's sense.

The emergence of technical life engendered, especially starting from the Upper Palaeolithic, an intense proliferation of these inorganic yet organized forms that are tertiary retentions, and that, from the beginning of hominization some three million years or so ago, constitute as such (as memory[21]) an exosomatization in which *the relation between death and life, that is, between entropy and negentropy, is replayed from beginning to end in the artificial organs that are produced by desire as a dream capable of realizing itself.*

Such a libidinal economy is constituted on the noetically and exoso-matically inflected basis of what Derrida called différance, where *life defers its entropic disappearance via the play of mnesic traces* of all kinds: biological, psychic, social and technical.

Reason then becomes the question of the *constantly challenged conditions of the play of such a différance*, when, noetically, it is conditioned by a tertiary retention that is also a *pharmakon*, that is, an agent of *both* entropy and negentropy, this fundamental ambiguity[22] constituting ὕβρις as such, that is, that which, contained within the jar of Pandora as the condition of all protention, and *freed from all restraint*, can only *precipitate the end that it was a matter of deferring.*

28. Absent from every bouquet

I discovered André Leroi-Gourhan's *Gesture and Speech* by reading Jacques Derrida's *Of Grammatology*. This *grammatology*, which from the outset Derrida declared could present itself only as and in a 'monstrosity', is built upon the foundation of two of Derrida's other texts, written firstly as a response to an invitation by Jean Piel for the journal *Critique*, one on Rousseau's *Essay on the Origin of Languages*, the other a review of the two volumes by Leroi-Gourhan published as *Le geste et la parole* (*Technique et langage* and *La mémoire et les rythmes*), along with two other works on the history of writing.[23]

I read these books between 1979 and 1980. At that time, I lived at Saint-Michel prison in Toulouse, where I had been incarcerated since June 1978, after having been arrested for armed robbery. I was not exactly a barbarian, but a delinquent, that is, an 'outlaw' [*hors-la-loi*]. Sentenced to eight years, I served just over three in the Toulouse prison, followed by a little under two years in the Muret detention centre.

For me, these years were, not 'redemptive', but salutary: they saved me from a dark fate (without 'redeeming' [*racheter*] me from any slavery whatsoever[24]). When I was arrested, my body and mind were seriously run down. I was rapidly descending down a dangerous slope, which terminated in this madness: robbing banks – in a manner that could hardly be called 'professional'.

Reading Leroi-Gourhan's anthropology, I encountered his description of the process that Derrida would himself refer to as a 'monstrosity' – the analysis of which would lead directly to the hypothesis of a coming disruptive state of exosomatic organogenesis – around two years after my arrest. I was at that time 'under review', as they say in prison. From time to time I would be 'extracted' from my cell to the courthouse, where, with my lawyer, I would go to see the judge who was reviewing the file on the basis of which the criminal court would rule on my case. It was while I was observing the work of republican justice, between my arrest and my trial, that I would discover how the Greek notion of ἀλήθεια (*alētheia*, truth) is tied to the notions of δίκη (*dikē*, justice) and κάλλος (*kallos*, beauty).

I remained in prison until February 1983. During that time, I *converted* – not to a religion, but to a noetic disposition derived from a philosophical method: that of the 'conversion of the gaze' conceived at the beginning of the twentieth century by Edmund Husserl, based on what he called the *phenomenological reduction* through which it becomes possible to get past the 'natural attitude'. Husserl describes this phenomenological reduction in his first great work, *Logical Investigations*, published in 1901. At that time, he called it the eidetic reduction, referring to the Greek question of the *eidos* (εἶδος).

Eidetic reduction consists in identifying the εἶδος of a phenomenon through the imaginary variation of this phenomenon. The *eidos* of the phenomenon is the invariant nucleus[25] that, however much the appearance of the phenomenon may vary (for example, the phenomenon of the flower, which can present itself as a poppy or a daisy, as a button closed in upon itself or as a pistil surrounded by its corolla of petals, and so on), is *maintained* as *that which the phenomenon each time aims at*, in each of its instances (as this or that flower, for example).

Hence eidetic reduction consists in *reducing the flower to its ideal traits*, which form what Mallarmé called 'the one absent from every bouquet':

> I say: a flower! and, outside the oblivion to which my voice relegates any contour, as something other than the calyx, there arises, musically, as the same, sweet idea, the one absent from every bouquet.[26]

The Husserlian eidetic reduction attempts to access the conditions of experience and thought – that is, of what, since Kant, has been called the transcendental. But it requires and implies a *phenomenological* reduction that passes through a *radicalization of Cartesian doubt*: for the *undoubtedly possible* accessing of the transcendental dimension of what presents itself firstly as empirical experience, I must be able to *suspend belief in the world around me*, in order to vary phenomena *solely on the basis of what necessarily remains*, of what remains *irreducible*, and *that I cannot eliminate without eliminating myself*, as that *ego* that, for Husserl, constitutes the transcendental condition of the world, which he will consequently call the transcendental ego, the subject of what he will call egology.[27]

This *exceeding of the ordinary*, which enables access to the experience of the extra-ordinary called the transcendental, where the *extra-ordinary is the hidden and forgotten condition of the ordinary*, consists in practising the phenomenological *epokhē*, that is, the suspension of belief in the world.

I myself practised just such a phenomenological ἐποχή, but initially without being aware of it, in the silent void of my cell, where, by accident, I literally dis-covered the extra-ordinary layer that conditioned what had hitherto been the ordinary course of my existence. This cell, which thus delivered me over to an experience of the extra-ordinary, became a laboratory in which – thanks to books, which in this silent void had not disappeared, lingering in my desert like the vestiges of a world now absent – I discovered the works of Husserl.

My reading of Husserl passed through Mallarmé, who, when he writes, 'I say: a flower!', provokes in the silence of reading an event that is not just 'innerworldly' in the sense that Heidegger will give to this word (that is, originating from the world) – because it *makes* world. The Mallarméan text *opens* a world, for example, that of flowers: as a language, not of communication between two speakers, but of poetry inasmuch as it makes appear 'the one absent from every bouquet', the *poetic milieu that idioms form* for those who 'dwell poetically upon this earth'[28] conditions the emergence of a world that presents itself to them as a constant succession of phenomena.

Later, and indeed quite quickly, I went on to encounter, in passing through Derrida, the question of writing,[29] such that, as trace, it itself conditioned my experience of poetic language as the extra-ordinary condition of possibility of the ordinary course of noetic life – that is, thoughtful, spiritual *and* intellectual life, which has, since Aristotle, defined the condition of possibility of non-inhuman beings. But it also defines their 'condition of impossibility', that is, *the possibility of their inhumanity*, and does so as *pharmakon*.

Through this *reflection on writing* that presented itself to me first and foremost as an *experience and experiment of reading*,[30] which for me it continues to be, I came to the conclusion that, beyond writing in the Derridian sense, technics, of which writing is a specific case (constituting a hypomnesic tertiary retention[31]), conditions the formation of the prein-dividual funds of all epokhal experience.

In passing through Heidegger, I then generalized the question of the *epokhē* to that of *history* inasmuch as it is composed of a succession of *epochs*, in which technics regularly provokes, as changes of technical systems, upheavals that generate noetic activity on the basis of which, as the 'doubly epokhal redoubling', an epoch properly speaking is constituted.

Today, thirty-three years after my release from prison, I try to think and to think care-fully [*de penser et de panser*][32] about what makes Florian suffer, and I believe that it is the *epoch of the absence of epoch that, precisely, erases the possibility of acceding to the possibility of considering the one absent from every bouquet, that is, that effaces, annuls and annihilates the possibility of acceding to 'consistence'* – of flowers, gardens, cities, of life on earth, of others, of the κόσμος.

This *systemic obstruction of consistences*, provoked, for example, by the automatic generation of protentions, prevents the projection of singular desires, which always aim at consistences. This systemic obstruction, which disruption installs everywhere, is bound to lead to an ever more murderous madness. It is this that invades the world, today in one way, tomorrow in another, and, in so doing, leads to the world's becoming befouled [*immonde*] – that is, strictly barbarous.

29. My prison studies and the *epokhē* of my life

During the three years I spent in Saint-Michel[33] prison in Toulouse, then, I discovered that a prison cell can, under certain conditions, become a phenomenological laboratory where the practice of the reduction enables one to make genuine discoveries about oneself, and where this is possible insofar as this 'self' harbours what, in 2003, when I wrote *Passer à l'acte*, I called myself-an-other [*moi-l'autre*], which opens its arms to the world, including and even especially when it is absent, which is very rare.[34]

After some months of almost total prison silence,[35] that is, *without speech*, still not knowing who Husserl was, I encountered the 'epokhal' *virtue* of my cell, where I almost always lived alone (at the cost of some struggles with the prison administration). Nevertheless, in the silence in which nothing is uttered, speech never just disappears, and so this consti-tuted a new experience of language, *quite close to madness*, but one that gave access to the extra-ordinary, in the first place as highly singular mental phenomena, which in some way effect a spontaneous conversion of the gaze upon the world become absent – as the 'one absent from every bouquet' – but, at the same time, and at the same stroke, all the more consistent.

At the end of weeks and months, progressively penetrating into this experience of the extra-ordinary, the experience of being imprisoned ceased to be painful and became, instead, an adventure, which I sometimes truly experienced as good fortune: there really were moments of joy. It was the fortune of my life – as the *epokhē* of my life, but also, in this *epokhē* that was a 'conversion' of my life, in the discovery of what, thanks to Joë Bousquet as read by Gilles Deleuze, I began to practise as the *meletē* of quasi-causality.[36]

It was on the basis of this experience, through which I became the 'quasi-cause' of my 'punishment', through which it became my chance, and as my great epokhal experience and experiment, that I came to understand 'stoic logic' as the trans-formation of necessity into virtue, ἀρετή – I had to turn necessity into virtue – and that I began to think the *défaut* [fault, default] as being that for which we must *do what is necessary* [*ce dont il faut faire ce qu'il faut*].

This condition – being by default – soon seemed to me to be that of the exosomatic being spoken about in the myth of Prometheus and Epimetheus. Ἀρετή (*aretē*), which we translate as virtue (from the Latin *virtus*), and sometimes by excellence, then becomes the passage from the potential (*dunamis* – δύναμις) of the *default* to the act (*energeia* – ἐνέργεια) of *that which is necessary*. This can no longer be an ontology, as it was in Aristotle, but amounts rather to *an ethics in the Greek sense of this word*: the discipline of an ἦθος (ēthos) founded on αἰδώς (*aidōs*) and in view of δίκη (*dikē*). I will return to this question in the final part of this work.[37]

Much later, that is, over the last ten years, I went on to discover that Michel Foucault himself had investigated these questions a few months before his death, which he knew was coming, and, some twenty years after his work on the history of madness, had engaged in the study of techniques of the self, government of self and others, care for the self, the hermeneutics of the self and the courage of truth.

This journey, which continued after I left my laboratory and still continues today, thanks to many friends, including Gérard Granel, Jacques Derrida and Jean-François Lyotard, this journey to the limits of the world and the worldless [*immonde*] was possible only by working, and by working hard *so as not to go mad*.

This was possible from the very first days thanks to the constant and unconditional help given to me by Gérard Granel. Granel had been an occasional companion of mine in my nightlife prior to prison. He became a friend by guiding me in the darkness within which I was groping about, within memory that was becoming my milieu: my memory of the absent world, and the memory contained in the books that accumulated in my cell, with which he could supply me because the examining magistrate had granted us the right to use the visiting room reserved for lawyers. I learned to read these sometimes very difficult books with the help of the magnificent professors who taught at the Université de Toulouse-Le Mirail, whose courses I could follow thanks to remote-education programmes.

The principal materials of the laboratory were, on the one hand, my psychic retentions, and, on the other, the protentions that haunted my dreams – all this being sustained, revived, channelled, filtered, and so 'reduced' in a sense that was not strictly Husserlian, because what I began to observe was that all this was transformed and ultimately transindividuated by what I would later call tertiary retentions.

The literal tertiary retentions (written à *la lettre*) that allowed me to orient myself in what would otherwise have remained a noetic desert were:

1. the novels found in the prison library;
2. the books given to me by Gérard;
3. the letters I received from my family and a few friends;
4. the mimeographed courses I received from my remote-education professors;
5. *the notes that I took*, in the margins of books and courses, and in notebooks, some of which resulted in copies of philosophical works, and others of which still feed into my books today – as so many tertiary retentions converting themselves into tertiary *protentions*, and which will continue, I hope, to nourish me as long as I live, Inshallah.

I have previously discussed some aspects of this in *Acting Out*.

30. The existential propadeutic of noetic salvation

My prison studies began not with philosophy, but with linguistics. Early in my adolescent years, between the ages of thirteen and fourteen, I had the luck to meet Alain Bideau, who at that time was friends with my brother Dominique. Alain had himself had an important encounter, at the Lycée Paul-Éluard in Saint-Denis, some years prior to our friendship, with Jean Marcenac: Marcenac was his philosophy teacher, and also a poet who had come out of surrealism, a former communist member of the resistance and an editor of *L'Humanité*.

Through Dominique and Alain, but also thanks to the French Communist Party press, which they themselves read, and which I began to read through imitation and 'secondary identification', I was aware of discussion of Jean-Paul Sartre, structuralism, Roland Barthes and many other thinkers – in particular Derrida and Saussure. And I purchased some of these books, as well as Plato's *Republic* – which I acquired from a great bookstore that supplied literature and essays to the 'grand ensemble' of Sarcelles, where I had lived since the age of seven.[38]

During my pre-prison life, I had never succeeded in reading any of these works – having, moreover, left high school at the end of the *classe de seconde*[39] in the spring of 1968 – with the exception of the beginning of Saussure's *Course in General Linguistics*, which I had begun in the early 1970s, and which I found quite exciting. I had also begun to read many twentieth-century novels. Prior to that, my mother had encouraged me to read nineteenth-century novels – something for which she herself set the example. When I was sixteen, I read a little Lenin and Marx, just prior to May 1968. Ten years later, I was in prison.

Having made a detailed study of my copy of Saussure, which Gérard had sent me in what was thus to become an extra-ordinary *cell of the conversion of the ordinary*, I then also read Gérard Genette, Tzvetan Todorov, Roland Barthes, *Théorie de l'ensemble* – published by the *Tel Quel* group,[40] who also published a journal under the direction of Philippe Sollers and Julia Kristeva – Roman Jakobson's linguistics, the poetics of the Russian formalists, and *Revolution in Poetic Language*, which Kristeva dedicated to Stéphane Mallarmé.[41]

At the same time, every morning at dawn I began to read a poem, or some prose, by Mallarmé, having acquired the La Pléiade edition of his work, as well as his published correspondence, collated by Henri Mondor. After a short time, I decided to enrol in philosophy at the Université du Mirail, on Granel's advice and with his assistance. To do so, I took and passed the university's special entrance examination: I was not a *bachelier*.[42] The chance to sit this special exam had been introduced in 1968.

From Saussure to *Tel Quel*, I came to read 'Différance',[43] *Dissemination*[44] and *Writing and Difference*[45] – after having read *Of Grammatology*, which, in addition to leading me to Leroi-Gourhan's *Gesture and Speech*, led me to Heidegger's *Being and Time*,[46] and then to many other works by this pupil of Husserl, and in particular to *An Introduction to Metaphysics*.[47]

After reading and rereading two short texts, Husserl's 'Origin of Geometry' and Derrida's introduction to his French translation of Husserl's text, I began to simultaneously, passionately and systematically study the Greeks, and firstly Plato and Aristotle, as well as everything by Husserl that had been translated into French – being unable to read German.

From there, I began to make my way through what was then called the 'history of metaphysics', which I studied systematically, and by going backwards in time, passing through Hegel, Kant and Descartes – Nietzsche constituting for me a special case, belonging in some way to my present, while the great thinkers of modern 'metaphysics' already belonged to another age. On this path, I finally discovered Foucault and Deleuze.

I began to think that what we were then living through (between the 1970s and 1980s), arising fundamentally from this age that we call the 'present', was already, perhaps, *no longer presenting itself* quite *as* a present, which is also to say, as a *given* [*don*] – as what Jean-Luc Marion calls givenness [*donation*].[48] It seemed to me that these were the vital, existential and political stakes of the 'deconstruction of metaphysics', at the time of what, in 1944, Adorno and Horkheimer had described as a new form of barbarism, while in 1969 Maurice Blanchot investigated the 'change of epoch' in a text that remains misunderstood: 'On a Change of Epoch: The Exigency of Return'.[49]

Now that the trial and ordeal of the absence of epoch imposes itself *as such*, I have come to believe that the *withdrawal of the present as given* was already the issue at stake in the pronouncement by Hegel, then Marx, and especially by Nietzsche, of the 'death of God'.

In the course of these peregrinations, and without ever forgetting what had so struck me in Leroi-Gourhan's analysis – that man is the technical form of life (and, much later, I would find this formulated in another way in Georges Canguilhem), producing an exosomatic organogenesis that, through that very fact, generates a type of memory that does not exist in other forms of life – I gradually forged what would become my doctoral thesis, and the constant motive of my work.

31. The 'end of the Book', 'Mémoires du futur' and the 'change of epoch'

This thesis was supervised by Derrida at the École des hautes études en sciences sociales in 1992. It posited that what is sometimes called 'phenomenological' time, that is, time lived in the specific mode of what Aristotle called the noetic soul, appears conditioned by, and hence constituted by, technical exteriorization, in return forming a process of interiorization, and, in this, a process that transforms the one who exteriorizes and 'exosomatizes' his or her existence. It was not until 1986 that I would discover, during a discussion with François Laruelle, that Simondon is the philosopher who allows us to think this transformative interiorization as a *process of individuation*, a process always re-exteriorized, and, through that, *transindividuated*.

According to this thesis, it is through this loop – which passes through exosomatization, and which, as organogenesis, *trans-forms*, through the artificial organs that it generates, somatic and psychic organs and social organizations – and only through this loop, that *noesis properly speaking*, that is, thinking, is constituted. This is what led me to write in *Symbolic Misery, Volume 2* that noesis is a technesis.[50]

Such a point of view inevitably leads, however, to what is not just an enigma or a mystery, but the *very limit of the thinkable* [*pensable*] – and, according to L'Impansable, that which cannot be healed [*pansé*], treated, overcome, 'saved'.

For, following anthropological, historical and archaeological studies, but also contemporaneous technological studies of what was then in the course of being established as the *analysis industry*, an industry based on using computers to treat automated data, what Maurice Blanchot called uprooting [*arrachement*], change of epoch and impersonal powers,[51] arising from and towards the 'terrifyingly ancient',[52] was bound to continue as the *end of the Book*,[53] and all the way up to the inconceivable point of what, now, in our day, thirty-eight years after the beginning of that epoch of my life that was the *epokhē* of my existence, we are calling *disruption*.

Eighteen months after my release, Thierry Gaudin, then director of the Centre de prospective et d'évaluation of the Ministry of Research, part of whose team was following the seminar I had been giving at the Collège international de philosophie since 1984, entrusted me with the task of studying the theoretical and practical consequences of what we were at that time calling 'new technologies'.

In this seminar, I tried to articulate the materials of the 'change of epoch' and the 'end of the Book', on which Blanchot meditated, with Derrida's main theses concerning what he called the archi-trace and the supplement – for which he drew from zoo-anthropological perspectives that lead, as in Leroi-Gourhan, to the most extreme stages of exosomatization (of which the 'Derridians', for the most part, have understood next to nothing).

I submitted my report to the Centre de prospective et d'évaluation in 1985.[54] It was then presented at the Château de Vincennes to some fifty people, among whom were artists, writers and scientists I'd interviewed during the inquiry that accompanied this work, as well as philosophers from the Collège international de philosophie, including Jean-François Lyotard, who was at that time its president, Pierre-Jean Labarrière, who was *rapporteur* for my studies, and Jacques Derrida.

This day – which turned into a kind of symposium – was to have numerous consequences for the continuation of my adventures, since, a short time later, the Centre Pompidou invited me, in the context of its tenth anniversary and for the Bibliothèque publique d'information, to design an exhibition, which opened two years later in October 1987, entitled *Mémoires du futur*.

Throughout the years that followed, and until today, I have tried to think this '*impansable*' by always examining the 'change of epoch' induced by the 'impersonal powers' of writing, that is, 'technics in all its forms',[55] above all through reticular digital writing, and by relating it, in more or less intermittent and uncertain ways, both to what in a 1949 lecture Heidegger called *Gestell*, and to the great question of nihilism opened by Nietzsche in 1882.

In the last few years, these two notions, *Gestell* and nihilism, have come to dominate my work. *Gestell* is a way of reflecting on what I am here calling disruption. And nihilism is what leads, today, faced with this ordeal of *Gestell* – and as purely, simply and absolutely computational capitalism – to the question of the madness that unleashes, and that is unleashed by, barbarism.

More recently still, I have initiated an effort to clarify these perspectives, after reading a key article by Rudolf Boehm[56] that had hitherto escaped me, in conjunction with reading *The Shock of the Anthropocene*.[57] I have thus begun to reinterpret the texts of Nietzsche and Heidegger from a perspective of what I am here referring to as neganthropology.[58] It was at that point that I encountered Florian's *parrhēsia* – his παρρησία.

32. Release from prison: on another madness

At the beginning of February 1983, I learned that I was 'to be released': the sentencing judge had signed the order for my conditional release. In prison, I had made friends with a man named Jean-Luc (who became a reader of Bousquet), and it turned out that he and I were to be released on almost the same day. We were to find ourselves thrown back into the world that had thrown us into prison. How would we cope 'on the outside'?

Both of us had the good fortune to have family and friends who had not abandoned us. We had a fair idea where we would sleep when we got out, and of course we were excited by the new perspective from which we would need to *relearn* how to live.

On the Saturday preceding this return to 'freedom', as intoxicating as it was terrifying, sitting on the lawn of the Muret detention centre,[59] I began to warn Jean-Luc of the great difficulties we should expect.

The return to freedom was going to be very hard, not just because we had *unlearned* the elementary gestures of the current world,[60] not just because we were returning to the world like ghosts from confinement, if not as the living dead, but because nihilism, whose accomplishment was accelerating with what Derrida presented as monstrosity, was only going to continue to worsen.

The world to which we were about to return was *much tougher* than the one in which we had lived prior to entering prison, and we needed to know it, that is, accept it (which in no way means to submit to it, but rather to make this necessity into a virtue – or what Nietzsche called active nihilism).

As a general rule, and as difficult as it may nevertheless be, it is easier to arrive in prison than it is to leave. More than that, however, I told Jean-Luc that *all that noetic life had learned since the beginning of hominization*, which we call *knowledge* and *culture*, all this was being *crushed* by what I began to describe to him as a *great rolling machine*.

What I today call disruption is the concretization of this discourse – inasmuch as it unleashes not just a new form of barbarism, but a new form of madness. In early February 1983, I told Jean-Luc that in the future, in our freedom regained, we would need to remember the *struggle against madness* that was our relationship to philosophy and literature during incarceration – which is the carceral absence of the world and its epochs, if not, strictly speaking, the absence of the epoch of the world.

33. Filial experience of veridiction

Ten years later, in 1993, I lived in Oise, thirty kilometres from the Université de technologie de Compiègne, where I had been teaching since 1988, partly thanks to the success of *Mémoires du futur*, which had involved engineering students from Compiègne, and partly because Patrice Vermeren, whom I had met at the Collège international de philosophie, recommended me to Gérard Pogorel, who had freed up a position there.

One evening that year, I was with my children, Barbara and Julien. They were by then already adults: twenty-three and nineteen years old. We had had a party the previous day, as we occasionally did, together with a few of their friends and some of mine. We had indulged in a few excesses, and we found ourselves in that vaguely 'epokhal' state which results from the 'disturbance of the senses' that any real party always somehow aims at reaching – something described in a singular way by Alan Sillitoe in *Saturday Night and Sunday Morning*.[61]

The morning haze on Sundays of this kind sometimes reveals ghosts and hauntings – revenances of the subterranean life of the mind. Out of the blue, Barbara asked me why it was that I was always sombre and silent. I was surprised and embarrassed. The question challenged me as a father to take upon myself, to assume, the absence of a rule when faced with the problem of answering genuine questions, which those whom we engender alone can address, and know how to address, to their parents, establishing the intimate and familial genre of veridiction.[62] Without taking the time to reflect, I tried to offer an unreserved answer to this question that was asked without reserve.

I told Barbara and Julien that now that they were no longer children, even though they were still my children, I had to speak to them on another register, that of their becoming adult. And I told them that, in fact, Barbara was right: I was often sombre, and often silent, not because prison had cast me into gloom in the way that 'ex-cons' are often portrayed in movies, but because, on the one hand, I had encountered *gravity in general*, as what confers upon the world its weight, which is also its price, including and especially in its absence, and, on the other hand, because I had learned that we are heading towards a great trial and ordeal concerning what would be the gravity of our epoch – or of its absence.

I then explained to them what I believed then, and what I still believe, to be my duty. My duty is not *to repress or deny the conviction I acquired in my improvised phenomenological laboratory*: we are heading along a path towards the worst, towards a turning point, a *shift*, one about which we cannot be sure whether it can be overcome – and this moment is not just coming but constantly accelerating its approach.

I also told them that, confronted with this indeterminate but certain deadline, I had resolved to live in such a way that:

- I would never give up continuing to study, document, observe, analyse and critique this becoming in an attempt to find ways of overcoming the ordeal, which is to say, of producing a future;
- I would posit in principle that sustaining such a critique requires being *able to envisage* that the ordeal might *not* be overcome, and hence to assume the radical indeterminacy of the future as *irreducible to becoming*;
- if the ordeal *is not* overcome, if it continues along its path, and if it, nevertheless, leaves behind it heirs who resemble ourselves, because they are *in one way or another* our *descendants*, then only the *traces* left behind by those who would have *struggled* with this ordeal, if enough remained,[63] and did so right up to the end, and by doing *all they could* to overcome it, to *change* its course, to *reverse* its 'negativity' by initiating circuits of transindividuation capable of giving rise to a new epoch, founded on quasi-causality,[64] *only those*, surviving perhaps through these traces, would be *credible, legitimate, fruitful and fertile* for this descendance after the shift;
- I would try to conform my existence to these convictions, and it was *in that light* that they should interpret my dark, taciturn and grave character – not 'pessimistic', any more than 'optimistic', but resolute, Inshallah.

34. The cowardice of optimism and pessimism

I am often taken to task on the basis of holding two opposing discourses. Some ask me why I am so 'pessimistic'. Others ask how I can be so 'optimistic'. The first lot refer to the virulent and radical nature of the critique I am constantly making of what is happening today – in the wake of Adorno, Horkheimer, Polanyi, in some respects Blanchot, and so on. The others refer to the proposals that, since 2005, I along with Ars Industrialis[65] have presented as *a future made possible and necessary by disruption* – which has led us to assert, as a condition of 'active nihilism', the necessity of what we call a positive pharmacology, based on an organology inscribed in what we have more recently begun to call a neganthropology.[66]

To these questions concerning the pessimism and optimism that some purport to see in my work, I reply that, ever since I have tried to reconstitute my former prison laboratory in everyday life, a quotidian existence regained less as freedom than as *generalized demoralization*, this demoralization has sent many of my contemporaries deeper and deeper into madness, in particular across the following sequence: 1990 (first 'Gulf War'); 1993 (onset of 'digitalization'); 2001 (9/11 and second war against Iraq); 2008 (financial crisis).

Contemporary demoralized madness is first expressed in the modes of denial and disavowal,[67] which constitute ordinary, everyday madness in the twenty-first century,[68] but also in more exceptional and clearly devastating modes. It is in these contexts that, after having been liberated, and having practised philosophy in a world that seemed to me to be, and to be for most of my fellows, a *lost* world, I never gave up *struggling against* optimism *and against* pessimism, which are two forms of ordinary madness, two forms of denial.

It is obviously necessary to specify what is meant by what I am here calling 'ordinary madness' – given that madness is always in some way extra-ordinary, and it is firstly in that that it has a relation to 'reason' when the latter turns to its 'conditions of possibility', which, too, are *always* extra-ordinary. Foucault's *History of Madness* turns around these questions, which will also form the heart of Part Two of this work.[69]

I here call 'ordinary' a madness, usually collective, that consists in *denying the obvious* by taking to an extreme level the denial that lies at the heart of the relationship to the end that Heidegger calls being-for-death – or being-towards-death, *Sein zum Tode* – which I have named, here, archi-protention.[70]

In such circumstances, *optimism* and *pessimism* are *unworthy, indecent and cowardly*. As forms of denying the obvious [*dénégation de l'évidence*], and as the disavowal of the situation that this evidence puts on display, these are also 'soft' (ordinary) forms of madness: *we have to be crazy* to continue to *live as if nothing is happening*, while at the same time knowing, *but without wanting to know*, what is underway everywhere. This includes, first and foremost, the fact that the official Paris conference of the United Nations in November and December 2015 needed to provide in principle responses and a programme of action without delay (but failed to do so).

Like optimism, *pessimism* consists in living *as if nothing has happened*. The pessimist uses his negative *fatalism* as an excuse *to do nothing*: everything has *already* played out – it's all *too late*. Optimism, too, *does* nothing, because it postulates that, ultimately, *nothing that happens* is anything but completely *ordinary*.

Optimism and pessimism therefore tend to *reinforce one another by enclosing themselves in denial*, which, as we know, can lead to the unfurling of what psychoanalysis describes as rationalizations, that is, lies perfectly constructed so as to conceal reality. Pessimism and optimism both deny the obvious, namely, the *imperative to be, neither pessimistic nor optimistic, but courageous*.

This criminal denial occurs with the support of an immense apparatus dedicated to encouraging pessimists and optimists in their denials of every possibility and necessity of taking action – an apparatus that consists, on the one hand, in the technologies and corporations that liquidate the will, and, on the other hand, in the disruptive systems that establish public impotence.

Optimism and pessimism deny and disavow to such an extent that they lead, like antechambers of madness, to pathological radicalization processes that in turn lead to the reinforcement of denial and disavowal – by installing terror and the panicked reactions to it that always bring out the worst.

It is to this that I, along with my friends, try to oppose the individual and collective ability to generate reasonable and rational statements, confronted with the 'shift'.

Pessimism and optimism are emotional states – states of the soul. To think rationally, that is, to think the conditions of possibility of acts capable of opening a future always yet to come beyond becoming, is to transform these states of the soul into critical resolutions, forged in public debate. And it is to do so against every process of denial and every process of precipitation, in which, ultimately, always consists that temptation to the worst that is the drives unbound.

Part Two

Madness, Anthropocene, Disruption

6

Who Am I? Hauntings, Spirits, Delusions

35. I am Malcolm X

On 2 May 2015, Louise Fessard, a journalist for the Mediapart website, reported a telephone conversation she had had with D., a 'young Frenchman who seems to be looking for media exposure', a Catholic who 'later converted to the Muslim religion', until, while in prison for drug trafficking, he tipped over 'into a sectarian and apocalyptic logic, advocating armed struggle against "unbelievers" in Syria. [...] D. has only one project when he gets out of detention: "To emigrate, to fight overseas for Allah and not return".'[1] This is obviously a suicidal discourse, related to the 10,000 successful suicides each year in France and the 200,000 attempts, together amounting to so many souls wounded by the absence of epoch. It is also the discourse of a 'repeat offender':

> When he was first incarcerated, D. was about twenty years old. He has received several convictions for drug trafficking and violence, and has spent time in a dozen or so prisons. Coming from a Catholic Caribbean family without economic hardship, D. grew up in a regular suburban household. His universe fell apart when his parents separated and he wound up in a small apartment. 'I've committed sins and not by half, because when I do something, I go all out', he said. 'I've smoked, I've been a dealer. I'm forsaking all this for Allah.'[2]

He was first approached by the 'brothers' of Tabligh, a preaching movement 'born in India and that arrived in France in the 1960s'. 'These ascetics crisscross the French suburbs in order to "awaken a dormant faith, put to sleep by comfort, lust or economic gloom, in a society of over-consumption where money rules".'[3] Then, in prison, he met an inmate 'who had wanted to go to Syria, but was reported by a friend. He introduced me online to other brothers in Syria. I was able to speak with them on Facebook.'[4]

This is somewhat reminiscent of what happened to Malcolm Little, who became Malcolm X,

> also known as El-Hajj Malik El-Shabazz, a defender of African American rights, assassinated on 21 February 1965 [...] [who] participated in drug-dealing, gambling, racketeering and burglary. [...] He was arrested in Detroit in 1946 for larceny and breaking and entering [...]. He was sentenced to ten years (serving seven) in the Charlestown State Prison in Massachusetts. [...] Malcolm was addicted to cocaine, which he had begun to consume when he was in the underworld.[5]

In prison, Malcolm Little discovered reading, as he writes to a friend from his cell:

> I'm just completing my fourth year of an 8 to 10 year term in prison ... but these four years of seclusion have proven to be the most enlightening years of my 24 years upon this earth ... and I feel this 'gift of Time' was Allah's reward to me as His way of saving me from the certain destruction for which I was heading.[6]

It was under the influence of the Nation of Islam, a movement then led by Elijah Muhammad,[7] that Malcolm began to read intensely. He was not accompanied in his prison studies by a generous and demanding mind like that of Gérard Granel, but he struggled against the *relentless segregation* to which African Americans were subject, in Alabama, Detroit and Chicago.

Nation of Islam, an anti-white heterodox Muslim movement, was advocating the creation of a black state in the southern United States, and asserting that white people are demons. Four of Malcolm's uncles 'were killed by whites, including one who was lynched'.

> In 1931, his father was found dead, run over by a streetcar. Malcolm said that the cause of death had been questioned by the black community. He himself later refused to accept it, arguing that his family had often been the target of the Black Legion, a white supremacist group affiliated with the Ku Klux Klan whom his father had accused of setting fire to their house in 1929. At that time, Michigan had 70,000 KKK members, five times more than Mississippi during the same period.[8]

I myself first became aware of Malcolm X from the cover of an Archie Shepp record entitled *Poem for Malcolm*. It was recorded in 1969, at almost the same time as *Yasmina, a Black Woman*, with the musicians of the Art Ensemble of Chicago. After that, Noah Howard, Bobby Few and Muhammad Ali, whom I met with and visited in Paris in 1969 and 1970, talked to me about him. Eventually, in 1971, I learned about his life and his influence in a book by Philippe Carles and Jean-Louis Comolli, *Free Jazz/Black Power*.[9]

Much later, I saw Spike Lee's film, *Malcolm X*.[10] At the end of the film, Spike Lee has a Harlem schoolteacher address her students by saying 'I'm Malcolm X', and she makes them say aloud and together:

'I'm Malcolm X.'

It would be interesting to undertake a detailed analysis, seventy-one years after Adorno and Horkheimer heralded the advent of a new form of barbarism, of what a statement such as 'I'm Malcolm X' could mean in 1965 in Harlem, compare it with what was said in Paris on 11 January 2015, '*Je suis Charlie*', and ask what occurred during this intervening half-century that has seen the epoch of the absence of epoch take hold, along with its new forms of madness, symptoms of barbarism on the rise.

I myself, no doubt, and without ever having thought of it prior to writing it here and now, I myself said to myself, in the early 1970s, in a bar where I made the night owls of Toulouse listen to and discover black music:

I'm Malcolm X.

I said this to myself by idealizing Malcolm X and identifying myself with him, and, of course, by deluding myself.

36. Who are we?

I, who am not black, was inhabited by the spirit of which Malcolm X was an incarnation, which is the spirit of *non-submission*, a spirit that *affirms another, extra-ordinary plane*, which John Coltrane sought ever more rigorously in each new album, and of which sublime African-American music, beginning with the blues, was a manifestation at once tragic and joyous, fantastic and colossal.[11]

Besides the education I received from my family, my friends, as well as my political education and my attraction to books – which provided a sentimental education in what used to be called, in a sense that is now almost totally forgotten, the humanities and humanism[12] – the passion for jazz played a significant role in my psychosocial individuation, starting from the age of fourteen.

Jazz created a collective: it was a machine to make friends through listening to the improbable. I was introduced to it when I happened to listen to two albums[13] belonging to my uncle Jacky Stiegler, my father's brother, and later by my brother Dominique, and then by Saada N'Diaye, a wonderful Malian who became my best friend. In 1977, ten years after this discovery, and so as to be able to listen to jazz and allow others to do so, I opened a bar in Toulouse, 'L'Écume des jours', a nocturnal musical joint closed down by the police shortly before my arrest.

Like Malcolm X in the 1940s, and like myself in the 1970s, D., according to the statements reported by Louise Fessard, observes and declares that in prison, where he has been *radicalized*, 'there's more time to learn, more time to talk about the real issues in life. To focus on more important subjects.'[14]

When I met Derrida in February 1983, after being released from prison, he asked me if there was anything in particular I wanted. I told him that I would like to have a meeting with someone from the government in order to urge them to do something so that time spent in prison should turn out not to have been a complete waste – so that it should not be irremediably lost, along with reasons for living and hoping, which prisoners commonly lack, given that in order to survive, they are often left with little choice but to reoffend.

Derrida arranged a meeting with Jean-Pierre Colin, a law professor from Reims but also a theatre figure, who had been seconded to the office of the culture minister, Jack Lang. During the interview, I proposed to this charming, affable man that we should establish the conditions required so that real studies could be undertaken in prison, in various fields, and by starting, firstly, with sport, then circus arts, the theatre, and therefore reading, writing and so on – and through the development of a genuine curriculum.

In the detention centre from which I'd just been released, I had seen some extraordinary athletes, some of whom had been incarcerated when they were quite young, and had already accumulated close to twenty years inside. They had become great sportsmen, but, in addition to the fact that they had nowhere to play except in prison, they also ran up against the difficulty of not having the opportunity of turning their athlete's body into a circus performer's body, or an actor's, and so on, through which the world would begin to *beckon to them, to make sense [faire signe]*: to signify, to transindividuate – from the microcosm that is the body and that in this way opens itself to the κόσμος, and, in its *admiration*, opens up the macrocosm that is society within the κόσμος.

Jean-Pierre Colin listened attentively, but nothing ever happened. After meeting Jean-Pierre Dalbéra from the Ministry of Culture, who had decided to support the research proposal that Thierry Gaudin offered me, and very busy with this new affair that would make such a difference to my life, it ended up that I myself forgot about my own project to support prison studies based on processes of ἐποχή and conversion of the natural attitude, and in the encounter with the extraordinary towards which prisoners are not infrequently inclined, as both Malcolm Little and D. observed, each in their own way.

37. State of emergency and philosophy

I have often made proposals to the worlds that I have passed through or in which I have lived, and I have done so in order to *change* these worlds. The world is what changes, and it is what must change. I conceive of life in noetic existence as the assumption of this *fatum*, of this fate that is a fact, and it is a question of transforming this fact into a law and a right. Noetic, then, means *neganthropic*. And we know, today, that the reason for this is entropic becoming and what it produces as negentropic future, that is, as evolution, and therefore as organogenesis, which, some two or three million years ago, becomes exosomatic, which is to say pharmacological.

After reading Simondon in 1986, I began to use the term 'individuation' to refer to that change of the world – whose rhythms are themselves constantly changing – resulting from technical individuation.

I conceive of philosophy, which has become my principal activity, as merely a way of creating the conditions for proposals to be elaborated and conceived so as to change the world in law, and not just in fact. Today, in contrast to the tragic age of the ancient Greeks, this question and this imperative arise in the context of the ordeal of nihilism. For Nietzsche, the roots of this nihilism lie deep in Socrates, but in my view this is completely wrong:[15] it is not Socrates but Plato who lays the foundations of nihilism, precisely by reversing, point by point, that which constitutes Socrates' *parrhēsia* – which is tragic through and through. This is what I have tried to show in the courses of pharmakon. fr.

How the world changes has always been conditioned by this world insofar as it is composed, above all, of relations and reticulations *between tertiary retentions*, and, *through them*, between psychic individuals and collective individuations. This retentional condition is formed through what Heidegger describes as a system of 'references',[16] that is, of significance,[17] and in a sense close to what Simondon describes (but on a completely different basis) as the formation of the transindividual.

Today, this system of reference constituting the transindividual is conditioned by the digital *pharmakon*, such that, *in the absence of a politics worthy of the name*, it turns disindividuation into an industry. The price to be paid for industrializing disindividuation is that transindividuation regresses to the stage of trans*dividuation*, prefigured, specifically in terms of the mass control of secondary retentions, by Adorno and Horkheimer's analysis of a new form of barbarism.

Philosophy is not just an academic discipline: it is firstly a way of living – as has always been insisted upon, and which Foucault, at the end of his life, made the motive of his research, doing so, undoubtedly, as a way of dying and of preparing to die. This is how, in particular, he interpreted *Crito* and *Phaedo*[18] – we will come back to this question.[19]

In addition, however, and prior to this, philosophy is a relation to collective individuation, when, by forging a new relationship to law, a new relationship that arises from the appearance of literal tertiary retention, this collective individuation is transformed into *politeia* – into citizenship, a way of living politically, according to a *positive* law that, henceforth governing all social relations in this sense, comes to see in *customary collective individuation* a *regressive* form, that is, something *barbarous and unreasonable*. In the eyes of Greeks of the fifth century, such are those they call οι βάρβαροι: barbarians.

This, then, is how I understand the philosophical task, as a distinction of law within facts and beyond facts, as occurred in my phenomenological 'conversion' in prison, and while awaiting my trial. And, because I understand the task in *this* way, I began – starting from 11 September 2001, a day in whose events I saw signs of a *precipitation towards the worst*, which I had been fearing, a fear that I had confided to Barbara Stiegler and Julien Stiegler in 1993 – to myself write only in an absolutely direct, visible, legible and primary relation to questions of political economy: by politicizing phenomenological questions.

I am convinced that it is *for want of an economic and political proposal capable of projecting beyond the Anthropocene* that barbarian behaviour multiplies – including in everyday life and including behaviour referred to as 'uncivil', which does not just involve those young people that Jean-Pierre Chevènement thought fit to call 'savages'. In addition to a number of public figures in positions of (political or economic) power, it is sometimes 'the aged' who become uncivil – especially with respect to the younger generations, whom they then see *only* as such 'savages', and *of whom they take no care*, being more concerned about their journeys through the 'third age' than about their descendants, condemned as the latter may be to unemployment.

Our sole possibility of struggle is through an economic and political proposal capable of projecting beyond the Anthropocene: to struggle with *what remains of civilization* in the epoch of the absence of epoch, and against barbarities that are as internal as they are external. This is what Marie Peltier succeeded in showing after the attack carried out on a Thalys train on 21 August 2015:[20] 'A society does not become gangrenous on the outside if it is not firstly gangrenous within. Barbarism breaks out when the soil becomes propitious, "fertile", when solidarity recedes, evaporates or weakens.'[21] This state of emergency led me to put on hold (but not to leave dormant) the writing of volumes four to six of *Technics and Time*,[22] in favour of writings aimed at elaborating a new critique of political economy, including when it presents itself as an inherently *anti*-political economy – which is the case with the libertarian ideology of disruption.

38. Economy and politics

The reason I refer here to political economy is because I have not forgotten what I learned when I passed through circles and times when Marx was read, and, in the first place, in the French Communist Party in the late 1960s – after spending time in the Trotskyist movement, which I had frequented in 1967–68, up until the May riots.

As never before, I am convinced that today, and ever since the Renaissance, politics must be thought in direct relation to economics. It was the Renaissance that laid the foundations of what would become globalized industrial capitalism *as the exploitation of tertiary retentions*. It did so in the wake of the printing of sacred books, as well as the printing of those books that constitute the preindividual funds of the humanities, and, finally, the creation of account books, which, with the rise of *monetary tertiary retention*,[23] accompany the transformation of *logos* into *ratio*.

It is because politics must today, and as never before, be thought in direct relation to economics that I was led to write books that are directly and simultaneously addressed to the political and the economic worlds. *The Re-Enchantment of the World*, co-authored with Ars Industrialis, was addressed to Laurence Parisot, who at the time had just joined MEDEF.[24] Some imbeciles interpreted this choice of addressee as a submission. 'To Love, to Love Me, to Love Us' was dedicated to those who voted for the National Front. Ditto.

In 2007, as a way of intervening in the French presidential campaign, I published *La Télécratie contre la démocratie*, which bore the subtitle, *Lettre ouverte aux représentants politiques*. At one and the same time, this book denounced the populism of the two main candidates, Nicolas Sarkozy and Ségolène Royal, analysed the consequences and lessons to be drawn from 21 April 2002, especially for the left, and put forward positive proposals for the future, irrespective of the outcome. I was invited by the host of the talk show 'Ce soir (ou jamais!)', Frédéric Taddeï, to appear on the programme. The following morning, I received a call from Jack Lang, who suggested I meet with him at the Socialist Party headquarters, on Rue de Solférino.

Lang was at that time head of Ségolène Royal's campaign. He wanted to discuss the themes of this campaign. I explained to him the main points of what, since *Disbelief and Discredit*, I had analysed as a major evolution of capitalism, containing the seeds of an equally major crisis, one that needs to be anticipated, and on the basis of which all economic and political axioms would need to be reconstructed.

This was less than six months before the so-called 'subprime' crisis, and just over a year before the collapse of Lehman Brothers, which would trigger an enormous recession whose consequences continue to affect Europe more than ever – like no other industrialized continent. I

spoke with Lang about the 2005 Ars Industrialis manifesto, and about
our analysis of the imminent crisis of consumer capitalism.

Lang replied that we should avoid using the word 'capitalism', and
that these were questions that belonged to the past. As for 'telecracy',
he wanted to use it – for Ségolène Royal's campaign. What he wanted
wasn't concepts, but slogans. He saw himself as part of the 'Mitterrand
generation', and did so through a 'communicative' denial[25] that would
lead directly to what would become the Strauss-Kahn generation.[26] The
latter figure would, five years later, just days before his arrest, be 'on the
campaign trail', riding around in a Porsche[27] with his PR handlers, his
'communicants'.

The full force of the disastrous effects of this *submission by resignation* [*soumission par la demission*] would be felt only with the election
of François Hollande, for whom the Sofitel affair was a windfall,
becoming the 'left candidate' in 2012 at the end of a calamitous 'primary
election'. In February of that year, I, together with Ars Industrialis,
published an article on the *Télérama* website that introduced a blog that
we had conceived, written and filmed with about fifteen participants,
members of Ars Industrialis and invited guests, such as Patrick Bouchain
and Philippe Meirieu. *Télérama* asked us how the election campaign
struck us.

This article, 'Capables et incapables', which was placed on the blog's
home page[28] – and which put the question of what Amartya Sen calls
capability at the core of our proposals for an economy of contribution –
argued that the real stakes of the 2012 election were the 2017 election,
and that the *quinquennium between 2012 and 2017 would turn on
the following issue*. After 2008 and its shattering effects, especially in
Europe:

- either what I am here calling disruption (a term we did not use at the
  time, even if we did describe its effects) will lead to a *new political
  economy, founded on recapabilization* [*recapacitation*], that is, on
  de-proletarianization;
- or disruption will *radicalize incapabilization* [*incapacitation*], and
  generalize proletarianization[29] – as 'algorithmic governmentality'
  (again, a term we did not use at that time, even if we were attempting
  to describe its effects), founded on a full and general automation that
  will rapidly become insolvent.

As we pointed out in our 2010 manifesto, we did not conceive of this
telluric episode of global capitalism – the globalization of capital being
a major agent of the Anthropocene – as a mere financial crisis, but as a
radical mutation of industrial societies. And we suggested that if Europe,
confronted with this mutation, fails to propose for the world to come
some path other than the Silicon Valley model, then it will become a

colony, firstly of the United States, then of the 'emerging countries'. European populations are shifting towards various far right positions, and this is the path towards which France is moving as it heads into 2017, and it is even in the vanguard of this movement, with the help of Michel Houellebecq (but this last point was something we could not have predicted).

We argued that we must take stock of the immense problems exposed in 2008 (problems that were already at stake in our 2005 manifesto) in order to install, in France and towards Europe, a completely new politics. And we argued that unless this is done, the new candidate of the left after the Sofitel affair may well end up thrusting the National Front into the forefront of the 2017 presidential campaign, and the far right may well end up with a presence in government – if not the presidency of the French Republic.

In 2013, we published *Pharmacologie du Front national*, including an Ars Industrialis lexicon compiled by Victor Petit. In March 2015, after the government's dismal failure in the 2014 municipal and European elections, Manuel Valls, who became prime minister, gave a speech in Limoges in which he declared the following:

> The National Front is on the way to becoming the number one party in France, not, this time, in the European election, like last spring, but here, in the departmental elections. And the department is a French symbol. [...] So, what? Will we stand idly by, watching as this National Front threat becomes a reality? Will we 'take note', as an inevitability, that this will soon happen? [...] And I ask: where are the intellectuals, where are the great consciences of this country, the men and women of culture, who, too, must enter the fray?[30]

Now, we await the 2017 election, without François Rebsamen, labour minister in the Valls government, who resigned in anticipation of a coming unemployment disaster – having done nothing to address the immense problems highlighted by Oxford University, MIT, the Bruegel think tank in Europe and the Roland Berger consultancy in France. Nor did the prime minister do anything to address these problems, instead discarding his responsibilities by shifting them onto 'intellectuals', and nor did the president of the Republic.

39. Stories of contemporary madness

On 26 June 2015, an attempted attack on an industrial gas production plant in Saint-Quentin-Fallavier took place, following a murder whose perpetrator was named Yassin Salhi and whose victim was Hervé Cornara:

Yassin Salhi seems to embody the new form of hybrid terrorist who uses the barbaric methods of an Islamist torturer to exact personal revenge, involving a combination of relationship crisis and workplace conflict. Confronted by the police, Salhi eventually admitted the murder of Hervé Cornara, commercial director of the transport company he had joined in March.

His plan was worked out in 48 hours, after he was reprimanded by his boss for a history of dropping pallets.[31]

Salhi's case is typical of the powerful effects of combining psychic disindividuation with collective disindividuation. This joint destruction of the primordial narcissisms of the *I* and the *we* is increasingly leading to all manner of 'acting out' [*passages à l'acte*], from insult to suicide, and it can end in the kinds of homicidal madness we see today – of which Dylann Roof in Charleston and Yassin Salhi in Saint-Quentin-Fallavier are cases.

Concerning Salhi:

the police believe they are dealing with what they call a 'hybrid case', a mixture of personal motivations and ideological commitment. [...] This profile of violent perpetrators, mixing psychological fragility and a claim of identity, is being encountered by police with increasing frequency. [...] A senior intelligence officer worries: '[...] we are seeing the emergence of atypical personalities, sometimes at the limits of psychiatry'. He thinks in particular of Mehdi Nemmouche, who was arrested on 30 May 2014 in Marseille, suspected of having killed four people at the Jewish museum in Brussels six days earlier.[32]

This is followed by a list of similar cases claimed as 'jihad':

The list of 'borderline' individuals does not end there. On 20 December 2014, in Joué-lès-Tours (Indre-et-Loire), a 20-year-old man stabbed three policemen while shouting 'Allahu akbar!', before being killed. Daesh has claimed responsibility for the assault, but it is unclear whether they are doing so opportunistically. The next day, in Dijon (Côte-d'Or), a 40-year-old driver slammed his car into a crowd (wounding ten). He claims to have been acting in the name of jihad. In fact, however, since 2001 he has been admitted to a psychiatric unit 157 times.[33]

It was a few days later that Sébastien Sarron, the 'Nantes *chauffard*',[34] deliberately drove his car into the crowd at a Christmas market, but without claiming affiliation to 'jihad'. Hence *L'Express* does not include him in its list – nor does it include Andreas Lubitz. And yet, like Mohamed Atta and his accomplices in 2001, Sarron and Lubitz utilized road or aerial vehicles as weapons of war. But, in the cases of Sarron and

Lubitz, with what 'war' were they involved? What is it they were battling against? Or, what battle is it that *inhabited and possessed* them?

It is not, of course, the battle between Good and Evil, as Bush Junior claimed – this would be too simple. It is the battle of Thanatos and Eros *unbound*: unbound because they have been deprived of the *bonds created by identification, idealization and sublimation*, through which they form an *economy*, that is, *a dynamic system within which tendencies compose through binding themselves to one another*. When they become unbound, however, these tendencies, Thanatos and Eros, which are *contained* in and by this economy[35] as the two faces of ἐλπίς, are the expression of the two consequences of ὕβρις. These tendencies, unbound and decomposed, inevitably generate *the violent madness of murder, rape, or individual or collective suicide*, and sometimes they do so as the hallucination of a 'sacrifice' – projecting and thereby concretizing a *plane of negative consistences* via *negative forms of idealization, identification and sublimation of every kind*.[36]

What we must now understand – 'now' meaning 'here' in the contemporary disruptive period, and as the hastening of the Anthropocene towards its end – is that, unless we manage a full accounting of the challenges we are describing, acts of madness will continue to multiply in an ever more spectacular way. It is therefore imperative that we open a new perspective.

We will see in the following chapters how all this extends what Peter Sloterdijk has described, in *In the World Interior of Capital*, as a five-fold secular process of disinhibition lying at the root of capitalism and globalization.[37]

Disruption essentially consists in outstripping and overtaking social organizations, and, through that, in short-circuiting collective individuation and transindividuation. Disruption, then, is *based* on the destruction of every psychosocial structure that enables the construction of such an economy – an economy that is simultaneously and indissolubly psychic and social, which means that changing the social also changes the psychic. Disruption, of which the actual barbarians – whether 'new' or not – are symptoms as well as actors, is bound to generalize and radicalize disinhibition, by dissociating Eros and Thanatos. This is already what was at stake in what Adorno and Horkheimer saw coming as early as 1944 with the culture industry.

Reticulated society has accelerated the disruptive processes that were already harboured in the new form of barbarism induced by the reversal of the *Aufklärung* (which, from her own perspective, Hannah Arendt describes as the crisis of culture and the crisis of education[38]), and it has done so by taking this reversal to the stage of *radical rationalization*, thereby provoking *extreme disenchantment* in the Weberian sense. In the case of the new barbarians, heirs of the buccaneers and pirates whose history is examined by Sloterdijk, this radicalization stems from the purely computational exploitation of the traces of individuals and groups who have been radically disindividuated and radically harmed.

All this combines with the despair and hopelessness engendered by the collapse of a thousand promises: in France, those of François Hollande; in the Near East and the Middle East, those of the 'Arab Spring'; among the 'digital natives', those of a new era, of which the NSA, algorithmic governmentality and The Family are the profound and hidden reality. This reality has been laid bare by the revelations of Edward Snowden, by the conflicts around 'uberization', as well as by the analysis of Evgeny Morozov – and it has been laid bare as 'digital disenchantment', or what we have in the past referred to as 'net blues'.

If, as we have seen, the radical and deliberately disruptive digital rationalization propagated by the French new barbarians and their libertarian allies systemically and automatically deprives individuals and groups of their *protentions*, the latter are nevertheless always bound to resurface. But when they do so, they will have totally left their transindividuated pathways. This is why they are increasingly often expressed not in arguments, but through every manner of 'acting out' [*passages à l'acte*] – including as the burst of automatic weapon fire, heard throughout the whole world.

40. Cultures, expectation, madness

The 'cultural' spheres that form as regions and eras of transindividuation via processes of collective individuation are concretely expressed as that which must be *cultivated*: as *that of which care needs to be taken*. As such, they are *indeed* cultures. Such cultures cultivate their neganthropy by *containing* ὕβρις, which means both retaining it [*retenant*, holding it back] and '*protaining*' it [*protenant*, holding it forth].

A culture always, and each time in its singular way, that is, neganthropically:

- retains *what lies at the very bottom of psychic and collective retentions* as the *default of origin* that engenders ὕβρις, and which *is* ὕβρις (of which Prometheus, thief of fire, and the worship dedicated to Zeus as token of recognition of this shame, αἰδώς, are expressions);
- protains, if I may put it like this, what lies *on the horizon of all psychic and collective protentions*, and as their vanishing point: protention *requires* the madness that the imagination bears within it, which *hallucinates* this vanishing point – and the imagination is the protentional faculty, which is also to say the phantasmatic faculty and the faculty of the drives.

A culture takes care of that which, so as to be constantly cultivated, is made an object of constant attention – such as the fire in the hearth watched over by Hestia. Without such worship, the evils contained in the jar of Pandora as so many versions of ὕβρις would escape en masse.

Among these evils – which, when they are contained, are goods (marking the organological condition insofar as it is also pharmacological, and constituting its δύναμις, that is, its potential, its dynamic) – there is ἐλπίς, the archi-protentional[39] spectre of hopes and fears that underlies all forms of madness, from despair to 'vain hopes',[40] that is, hopes held in ignorance of danger.

As *expectation*, ἐλπίς makes possible (and translates into) *attention* – understood as solicitude, cultivating organological and pharmacological forms of knowledge specific to each culture. Culture therefore forms the singular life-knowledge [*savoir-vivre*] specific to a given region and epoch insofar as they are devoted to some or other artificialities arising from ὕβρις. It is always a matter of knowing how to live as such and such: as an Italian, as an Egyptian, as Japanese,[41] according to the epoch and the place where one was born and that one knows. One cannot live without such life-knowledge, which is learned without knowing it, but which lets us know the world [*monde*] and everything that contrasts it with the worldless and the befouled [*immonde*].

To a greater or lesser extent, a culture participates with other cultures, by cultivating those *transcultural fields through which human knowledge is formed* – beyond particular forms of life-knowledge, which belong to 'particularisms' and local idiomaticities. Knowledge that is in this sense transcultural includes the work-knowledge and artisanship of guilds (beyond localities, and this is already the case for art) and the *conceptual, formal or spiritual knowledge* that forms academies, colleges, schools, brotherhoods, churches and all forms of *ecclesia* (the ἐκκλησία is the assembled community) as political and/or religious communities – from the polis to the synagogue and the Ummah, and passing through all forms of power and all institutions through which the diversity of social systems is synthesized in being localized and yet also deterritorialized.

As the power and knowledge of unification, as culture of diversification, as *individuation*, an epoch constitutes the 'spirit' as *Zeitgeist*. In Western Europe, for a very long time, this spirit was, first and foremost, a wholly other possibility accorded by religion, which took as obvious fact the necessary fiction of dogmatic revelation, up to and including Kant.

Before religion, at the dawn of Europe, there was piety – that of the Greeks of Greek tragedy, about whom I doubt it could be said that they practised a religion properly speaking. But they were *mystics*, that is, initiates into the mysteries at Eleusis, through which they celebrated Demeter and Persephone at the return of spring, or, in other words, through which they celebrated the negentropic explosion of life.

All this concerns the Mediterranean West, for which Egypt and the Great Empires were the condition of possibility, including through the establishment of monotheism in the Kingdom of David. Cults of the dead and the divine dynasties of the pharaohs and empires formed other revenances, that is, other ways of being of *spirits* who bring into question the *unity* of the spirit (here we should reread what Freud had to say about Egypt, Amun, Moses and the birth of monotheism[42]).

Before all this, *supernature* [*surnature*], which seems to have taken shape in the Upper Palaeolithic, and probably beforehand, is the horizon of the revenances that the instruments of the shaman have the power to summon – and it seems to have been maintained from the cultures of the 'dream time' and 'dreaming' to the societies of Siberia and North America.

As discussed in Chapter 2, with the new form of barbarism foreseen by Adorno and Horkheimer in Hollywood and early television – all this forming, along with radio and the mass media in general, the industry of cultural goods, the culture industry – it is both psychic individuals and collective individuals who find themselves disindividuated by the coordinated and systematic wounding of their primordial psychic and collective narcissisms, between which psychic and collective retentions and protentions circulate, form and bind together.

This is why we see, today, *psychiatric cases* (who in the past were called *madmen* or *fools*) who, in a delirious (hallucinatory) way, articulate their psychic poverty [*misère*] with the prevailing affective, symbolic, political and spiritual poverty – and do so against a background of economic poverty that becomes unbearable. The primordial narcissism of the *I* suffers, above all, from the wounds inflicted on the primordial narcissism of the *we*. The madness that can be its result tends *inevitably* to be carried to the level of a delirium of the *we*, which the 'professionals of the struggle against radical Islam' present, as in the case of Yassin Salhi, as a combination of 'psychological fragility' and 'identity crisis'.

It is foolish to say 'I'm Malcolm X', just as it is crazy to say 'I'm Napoleon'. These amount to one or another strain of madness, more or less necessary, just as there are necessary fictions. At the same time, however, lies also exist – that is, fictions – and these may be very harmful and highly reprehensible.

There are different kinds of madness, which, as we shall now see, are sometimes necessary, and which are *possible* only because *noesis* can *pass into actuality* as the ἐντελέχεια (*entelekheia*, fulfilment) of the δύναμις (potential) from which it stems, and can do so *only by taking it to the limit*, and, in so doing, *by passing its borders*, which are those *of the dream realizing itself* – at the risk, each time, of turning into a nightmare. These various forms of madness – inevitable, if not always required, because they stem from this *primordial noetic necessity* that is also the issue in Aristotle's essay on melancholy,[43] where the *pharmakon* is also wine – must be *treated and thought about care-fully* [*panser*].

This is what I am attempting to undertake here, a little madly, in trying not to forget how I have taken care [*pansé*] of myself (as myself-an-other), by tackling head-on, and as an uncompromising conversion, the Delphic injunction, starting from a thinking of my own era:

Γνῶθι σεαυτόν – know thyself.

7

Dreams and Nightmares in the Anthropocene

41. Daydreams – or 'The Milkmaid and the Pot of Milk'

Everyone takes their desires for realities: this is the lesson we learn from Jean de La Fontaine in 'The Milkmaid and the Pot of Milk'.[1] This ordinary disposition towards the extra-ordinary belongs to what in English is called 'wishful thinking'. This expression is translated into French as *'pensée magique'* or *'voeu pieux'*. To think 'magically', to want or to wish 'piously', is to *confuse* one's daydreams with reality. The fable of Perrette is like a little theatre in which just such a confusion is played out.

Any examination of wishful thinking will show that *horizons of expectation* (collective protentions), based more or less phantasmatically on 'magical' beliefs, 'pious' wishes and the confusion of desire with reality, have effects on the course of things, effects that may be either positive or negative.

In the case of Perrette, the negativity of these effects leads to the fall of the pot. In the case of the so-called 'Pygmalion effect', positive protentions promote their effective realization – that is, in this case, and in educational contexts, the transformation of reality, in which it is the learner who transforms himself or herself through what he or she learns.

The course of action triggered by wishful thinking can be *realized positively* because the one who acts positively anticipates the consequences of his or her actions, and so strings them together in a fortunate manner, which is to say, successfully – which in turn is not unconnected to so-called self-fulfilling prophecies.

Nevertheless, the disconnection in the course of action between the anticipations of a 'real' yet to come and the reality of a 'real' that is still there – and that the daydreamer dreams of transforming – can lead and often leads to a different outcome: the realization of the dream can then turn into a nightmare. Her pot of milk broken, Perrette sees her 'calf, cow, pig, brood of chickens' evaporate into thin air, and, returning home, she is beaten by her husband.

The moral that La Fontaine draws from this fable begins by observing that the ordinary relation to the extra-ordinary that is presented through waking reveries haunts 'wise men as well as fools':

> What mind doesn't wander over meadows?
> Who does not build castles in Spain?
> Picrochole, Pyrrhus, the milkmaid, everybody,
> Wise men as well as fools
> Everybody dreams awake, there is nothing sweeter.

And he evokes what we can also read in Pascal:

> I am elected King, my people love me;
> Crowns are raining on my head.
> Something happens, and I come back to myself;
> I am John Smith as before.[2]

When I was a child, my mother told me that I ought not take my desires for reality – a *confusion* that amounts to a kind of delirium, and sometimes leads to madness. She encouraged me to be *reasonable*, that is, *wise*. The child enters 'the age of reason' when he or she is supposed to know how to distinguish what is *reasonable* and what is not (as peoples, too, are supposed to have common sense [*bon sens*] and popular wisdom [*la sagesse des nations*] – mostly proving to be a disappointment – which Simone de Beauvoir contrasts with existence conceived as 'philosophical choice'[3]).

The problem is that this distinction shifts over time, which is why people sometimes say that adults who lived in past times were still children. And, while it is true that the transition from *childhood* to the *age of reason* requires the disenchantment of those objects that Winnicott calls transitional, these remain, in the fields of culture and the life of the spirit, *exceptions*, sharing out the division between 'what is reasonable and what is not' – and it is as such that they are called 'extraordinary'. This is why we are all delusional – as Montaigne and then Pascal will say.[4]

The world that surrounds childhood constitutes what Winnicott calls transitional space, which forms between the mother and the child, and which is composed of transitional objects, that is, objects possessing imaginary attributes that the child believes in because his or her mother, and more generally his or her educators, act 'as if' they too believed in those attributes, which do not exist – but which consist.

What is maintained between the mother and her child, which does not exist, is the transitional object that *constitutes* the mother and child in that it links and *attaches* them to one another through a wonderful relationship: a relation of love, of *amour fou* [...] this link through which these two beings become incommensurable and infinite for one another, [...] allowing a place for that which [...] *consists* precisely to the *immeasurable extent* [*dans la mesure et la démesure*] *that it does not exist* – because the only things that exist are finite things.

    This consistence, more than anything else, and before anything else, is what a mother protects when she protects her child [through] the knowledge the mother has of the extra-ordinary character of the object – and that Winnicott calls transitional precisely in order to designate this extra-ordinariness.[5]

The mother is for Winnicott a 'good enough mother' when she knows both:

• how to stimulate and maintain this enchantment in her child, how to cultivate it, and therefore how to introduce the child into what, in the world, consists – *because* it is *extra-ordinary*;
• how to *detach* her child from this enchantment, to introduce it into the disenchantment that 'reality' usually turns out to be, teaching him or her that the objects of reality are finite, that they exist only to this extent, disappointing as this must be in the eyes of a child, which makes these objects as such ordinary – a detachment cultivated by the mother, who, in this way, leads her child onto the necessary path of becoming an adult.

As Winnicott himself points out, however, the adult world preserves transitional spaces: they are *those that are taken care of by culture in all its forms*, what culture *cultivates*, from popular festivals to the sciences and passing through the arts – thereby cultivating this plane of the extra-ordinary that we have already mentioned, operating and culti-vating, at the same time, *conversions* of ordinary men and women, who themselves thereby become 'cultivated' – *the 'new form of barbarism' being what, according to Adorno and Horkheimer, in the form of the culture industry, makes this culture impossible.*

    Disruption is what radicalizes this impossibility and takes it to the extreme, which in turn triggers compensatory reactions, radicalized behaviours founded on phantasmatic processes of 'conversion' – conversion also being, in the *theory of hysteria*, an emblematic form of delusional fantasy.

    What is at stake here is the function of reason – and its relation to fiction, given the possibility and necessity of fictioning, that is, its relation to delirium and madness. The *as if* (*als ob*) is in fact what opens up the very possibility of the ideas in Kant – thereby constituting the kingdom of ends, that is, Kantian reason insofar as it is formed by the so-called regulative ideas, establishing reason as the regulation of practical imperatives.[6] This will be the subject of major conflict between Kant and Hegel, as was shown by Alexandre Kojève.[7]

42. 'All goes ill!': sleep of reason and waking dreams

At night, of course, we enter into a kind of delirium – which raises the question of the difference between sleep, dream and madness. This, as we will see, goes to the heart of the debate between Foucault and Derrida on Descartes' *Meditations*. Obviously we are also in some way delirious when we daydream, that is, whenever we give free rein to our imagination, which is sometimes reflected in our reasoning – and where the latter is always itself a kind of waking dream.

This is why, when reasoning is formulated too quickly or when it is poorly constructed, we can be accused of deluding ourselves, of rambling, of holding extravagant views. 'Extravagant' means to go off the path. '*Delirare*' means to deviate from the furrow. The dreams, delusions and extravagances of madness form the stakes of Foucault's *History of Madness*[8] and its commentary on the *Meditations* of Descartes – a commentary Derrida will contest.

We cannot go to the extra-ordinary – turn towards it, modify the gaze in its direction, abandon the 'natural attitude', make a 'conversion' – without *leaving* the path, the furrow, of oneself, in order to enter into the self-an-other [*soi-l'autre*]. It is precisely in this way that it is *other* – and that there is the other, including and especially as the 'Big Other'. This is the entire issue involved in Winnicott's extension of infantile transitional spaces to the planes of consistence cultivated by the life of the mind and spirit. And this is why Alain Bergala could appeal to the notion of *child's play* [*enfance de l'art*] to think the art of cinema.

This *going to the other* that is the extra-ordinary is also the meaning of being *transported* by love. Such a transport is a delirium, which is to say a μανία (*mania*), synonym of ἔκστασις (*ekstasis*), which, according to Diotima in the *Symposium*, conditions knowledge [*savoir*], or what Montaigne will call sapience, as what gives flavour [*saveurs*] to the world, and, through that, this *taste for the extra-ordinary* without which life would not be worth the *pain and effort* [*peine*] of being lived.[9]

But we are sometimes deluded in terms of the various forms of 'delirium properly speaking' – which then become the expression of *madness* as *de-lirium*, deviation, deviance, in the sense that Voltaire *opposes* to reason.[10] Conceived in this way, madness will become the object of the science of psychiatry – and it will, as Foucault says, be 'exiled' from everyday life, while at the same time 'ordinary madness' will be denied.

It is this that can no longer be sustained at a time when the world is going manifestly and massively mad, in a sense that recalls the words, still so striking and so relevant, set down by Eustache Deschamps in the fourteenth century:

> We are cowardly, ill-formed and weak
> Aged, envious and evil-spoken.
> I see only fools and sots
> Truly the end is nigh
> All goes ill.[11]

I fear that our disruptive situation is *indubitably and infinitely graver* than was the European or global situation in the late fourteenth century. This is why we will need to read and reread Foucault and his analysis of the history of madness – and it is why we cannot be content to take its flaws as grounds for dismissing the fundamental questions he raises.

Today, we consider that certain discourses are, in this extreme sense, delirious. Seen as strictly pathological, these discourses are referred to in our age as mad, even though in the past such discourses may have belonged to what, from his anthropological perspective, Pierre Legendre would call the 'dogmatics' of their epoch – in other words, the *foundations* of that epoch.

To take dreams for realities and to be able to realize them: such is the human condition, and the organological condition.

> Dream – this night, a tiny grand piano – a toy – on a table – and I marvelled at someone moving in the distance – at my command – I did not know myself how this miracle was performed.

> About this I say that I have not been able to explain to myself the *precision* of dreams. The state of disturbance is understood well, but the at times striking precision?

> *The precise dream: the unreal precision of a problem of geometry or mechanics.*[12]

The non-inhuman being digresses [*extravague*] in the sense that, *realizing* its dreams, it *exosomatizes* itself. This is what Marc Azéma notes on the first page of his *Préhistoire du cinéma*: 'Man has always "dreamed". He shares this faculty with a good many animals. But his brain is a machine for producing images [...] capable of projecting his inner "cinema" outside himself.'[13] In the Upper Palaeolithic, to realize one's dreams is firstly to *paint* them, that is, to enter into the era of *Homo ludens*, as Bataille affirmed at Lascaux. Later, it is to *notate* them: hence Valéry. Or to sketch them in order to make *films*: hence Fellini.[14]

By *realizing* its dreams, the non-inhuman being becomes noetic. But, by the same token, it also becomes *furious in the strict sense*: it puts itself *outside itself* – which is what it means to *go into a fury*. Fury and dementia (*furor* and *insania*) are two different forms of madness, and the first of these can always come to affect the sage himself – according to Cicero and many after him.

This is what I tried to reflect upon in *Technics and Time, 1*, in positing that the human being is prosthetic in the sense that its realizations – which, firstly as weapons and tools, are essential to its survival – are posed before it, which is what πρόθεσις (*prosthesis*) means. And I also tried to reflect in this way by reading *Being and Time* starting from this perspective that sets the *pharmakon* at the heart of the existence of Dasein. This is also to say: by re-posing the questions that Derrida addresses to the Husserlian and Heideggerian phenomenologies, but doing so beyond Derrida himself.

In the epoch of negative protention – which affects Florian much more than Eustache Deschamps was affected by the madness of his epoch (which was, undoubtedly, and, if one may put it like this, more than any other, *an epoch*) – all these questions lead back to the early work of Foucault, and to the *immense misunderstanding* between Foucault and Derrida, which will undoubtedly also affect Deleuze and, through him, Guattari.

The *realization* of its dreams is the *condition* of non-inhumanity as *neganthropogenesis in exosomatization*. This 'realization' *artificializes* reality, that is, transforms it, as Marx and Engels will say. This also means that it de-realizes it, and that it thereby de-realizes those who accomplish this de-realization: it 'disrupts' reality, so to speak. And this is what Marx and Engels will describe as proletarianization, but without ever posing the question of the *pharmakon* as such.

What this ultimately means is that non-inhumanity's realization of its dreams always sets up a phase shift [*déphasage*] and a disadjustment in what it de-realizes, which can in the end *really realize itself*, so to speak, as *Wirklichkeit*, only on the condition that it gives rise to a new epoch of individuation – which is the second stage of the doubly epokhal redoubling.

Today's barbarous strategy of 'radical innovation' *prevents this question from being raised*, a question that is also that of the Soviet tragedy and 'real communism' – and this involves theoretical questions regarding Marx that almost nobody among the 'specialists' seems willing to raise, whether they are tenured, or, especially in France, protecting and keeping a monopoly on the Althusserian legacy: there is a little symbolic cheese to be had in these times of scarcity and misery in all areas of existence.

The realization of noetic dreams can be realized *in its effects*, which is to say can find the strength [*forces*] for its realization, only to the extent [*mesure*] *and* in the excess [*démesure*] that opens up to the extra-ordinariness that is obviously manifest in every dream – according to modalities

that are not those of consciousness, but which, in the course of realization, *encounter* those of consciousness. These include, for example,
the mathematical idealities, whose oneiric precision Valéry admired, and
through which the engineer Jiro Horikoshi projected his dreams, as can
be seen in Hayao Miyazaki's film *Kaze Tachinu* (*The Wind Rises*, 2013).

Any mental or spiritual activity idealizes *that which is not*, as a still
unknown promise and in the being-itself of that which is *not yet*. And
from 'that which *is not*', and which *will never exist*, to 'that which
is *not yet*', there is a gulf that simultaneously separates and connects
science and technics. In this way, idealization traces circuits of transindividuation, of which poems and everything we call 'works' [*oeuvres*]
are origin-points as well as key-points (in Simondon's sense[15]) – that is,
milestones, of which the gods of the Greek pantheon are also figures.[16]
It is just such traces that form and grant the time of the second epokhal
redoubling, thereby constituting an epoch.

This is clearly evident in the lines from Pindar that Valéry used as an
epigraph in *Cimetière marin* (they are also an epigraph for this book):

Μή, φίλα ψυχά, βίον ἀθάνατονσπεῦδε,
τὰν δ᾽ ἔμπρακτον ἄντλει μαχανάν.[17]

In *The Wind Rises* – a very dreamlike animated film inspired, on the
one hand, by the works of Paul Valéry, Thomas Mann and Tatsuo Hori,
and, on the other, by the life and mechanical works of Jiro Horikoshi,
a Japanese engineer who designed the Mitsubishi A6M fighter, known
as the Zero – Hayao Miyazaki took these verses in their obvious sense.

Mechanics is indeed the central theme of this wonderful film, which
thus takes at its word the μηχανή of which Pindar speaks. Horikoshi,
as a child, dreamed of flying, admiring Giovanni Battista Caproni, an
Italian engineer, the two of them being inhabited by dreams of Icarus,[18]
Leonardo da Vinci and Clément Ader,[19] a handful of illustrious names
among many others. Exosomatizing their dreams, all of these figures
trans-formed our worlds – until the final disaster, at the end of the
Second World War, in Japan.

43. Dreaming, making, acting – in the Anthropocene and beyond

Every artist takes his desires and therefore his dreams for realities, as
does every scholar, every citizen, every lover and all those who desire –
every non-inhuman being. This *taking* [*prise*] is always a mystery (that
of noetic 'taking form'[20]) and a detachment [*déprise*] (from the reality
that it is a matter of surpassing), which *really is* only by becoming the
taking form of a process of psychosocial individuation. Hence it is only
by this 'taking' that exosomatization occurs – of which *Le Cimetière*

marin is a case, as is the Mitsubishi Zero, even if this 'as' clearly opens the abyss of a difference, which is the *différance of science as technics*. This *différance* presupposes the power to 'protain' 'that which is not yet', as Valéry says.[21]

The verbs *faire* [to do/make] and *agir* [to act] mean *to take* one's dreams with *enough force* for them to *become real* in the sense that:

- The *crafting* [*facture*] in which *faire* consists is what makes, what fabricates, even if '*faire ses besoins*'[22] (an astonishing formulation, if one considers it) is not exactly to fabricate something – but it is indeed to exteriorize something. And it is this *passage to the outside* that *faire* means in this case.[23]
- The *action* in which *agir* consists transforms a state of the world, and does so through this crafting – which contains within it what is constantly being recalled here with the word ὕβρις. This is why fabrication itself is not sufficient: it requires action as *praxeology*.
- Action, inasmuch as it is always tied to making (to an organology that is also a pharmacology), always consists in the initiation of one or more new circuits of transindividuation. As long as this transindividuation has not been accomplished, making and acting can be perceived as extravagances, more or less harebrained, and 'original' in the sense that we sometimes use this term to describe eccentric personalities – or even those that are pathogenic, asocial, excessive and sometimes insane.

Hallucination can lead to concretization, which is to say firstly to *invention*, then to *innovation* in Gille's sense.[24] This socialization of invention through innovation can be crazy: a thousand stories attest to the ever-imminent danger of invention when there is a failure of the adjustment between, on the one hand, the techno-logical *epokhē*, and, on the other hand, the production of the circuits of transindividuation required by this *epokhē*.

Transindividuation thus amounts to the *becoming reasonable* of what will initially have been mad, or of what initially *seemed* foolish. Jean-Baptiste Fressoz and Christophe Bonneuil discuss how foolish risk-taking was not just possible, but systematized, in order to struggle against those who resisted it, and to 'make reasonable' that which has indeed become public opinion fabricated by techniques for the construction of this opinion, techniques totally dedicated to this goal – until, today, this appearance becomes totally inverted.

Today, we know that this systematic risk-taking was made possible by a fundamental transformation of Christianity, one that also coincides with colonialism, about which Peter Sloterdijk has shown that, from the beginning, it involved a process of disinhibition. All this fundamentally complicates the Foucauldian account of the history of madness in the 'classical age', and shows that this age contained the seeds that would grow into the roots of contemporary madness.

Hallucination, whose status evolves between madness and reason, and which lies at the heart of what Simondon tried to think in *Imagination et invention*, is therefore the *condition* of fabrication, that is, of exosomatization.[25] And with disruption, the process of exosomatization not only accelerates, but comes to be exclusively driven by the market, that is, by the strategies of shareholders, or of what shareholders have (in a structural way) become: speculators.

Suffice to say that without an immense transformation of this immense transformation, humanity courts its own doom, and, along with it, the doom of what Husserl called the earth-ark[26] – transhumanism being the technoscientific ideology that tries to lure us to this doom by claiming to eliminate risk at the very moment it exacerbates it.

In our time, innovation is compelled and adjustment fails to occur. On the contrary, social structures are today being destroyed by a destructive form of innovation, which is not innovation as conceived by Gille but a violence without future, a violence that destroys the future itself. In this absence of epoch to which, *therefore*, disruption amounts, *extreme* possibilities of exosomatization are brought to bear, in particular with the advent of digitalization. But in fact it is *well before that*, and with *capitalism's socio-economic translation of what Descartes called the mathesis universalis*, that the Anthropocene comes to be constituted.

What in the Anthropocene's early stages was perceived as dangerous, or even as madness, and hence rejected by society (a rejection that is now called 'resistance to innovation'), *is later metastabilized as 'reasonable'* through a process of transindividuation known as *modernity*, founded on what was then called *progress*. But in the Anthropocene's current stage, that is, in disruption *such as it is claimed by the new barbarians* and *practised* through their more or less Californian 'business models', what becomes increasingly clear is that all this, when the Anthropocene unfurls its entropic and disruptive consequences, *again becomes unreasonable, in the sense that it appears once more to be such*.

At the end of the Anthropocene become disruptive absence of epoch, progress is dominated and covered over by what the geographer Élisée Reclus called 'regress'.[27]

All goes ill.

44. The deliberate exploitation of toxicity and the systemic carelessness that results

The absence of epoch and its characteristic form of abandonment [*déshérence*] are the 'backlash' or 'backdraft' or 'boomerang effect' of which the postmodern *petits récits* were precursor symptoms – and often dissimulating ones.[28] In and as this *après-coup*, the absence of epoch and the way it leaves us abandoned are experienced as all the more threatening with the emergence of so-called 'transformational' technologies. Not only do they overturn social systems: they also compromise and sometimes destroy biological systems, physiological systems, geographical systems and the ecosystems that form therein, while introducing hitherto unknown physical systems. By reticulating billions of earthlings in real time, all the conditions are thereby met for the production of an explosive global panic that could prove fatal to the biosphere.

Nanostructured materials and nanomachines will disrupt every boundary between what I am here calling physical systems (which are concretions and material concretizations of the universal laws of physical science) and biological, somatic, psychic and social systems. Disruptions of this kind will be possible thanks in particular to the penetrative and invasive possibilities opened up by the nanomaterials arising from 'NBIC convergence', which paves the way for exosomatic and endosomatic regimes of a completely new kind.

'Transformational technologies' *materialize* a transhumanist, hyper-disruptive ideology, which also spreads into synthetic biology, aiming at the immense market concerned with health, suffering and toxicity in the Anthropocene, where, instead of curatively and preventatively taking care of individuals, groups and societies harmed by the disinhibition of risk-taking, the market of exosomatization turns every *misfortune* into an *opportunity*.

This 'hyper-disruption' legitimates and paves the way for an *enormous reactive rejection of any coming exosomatization*: invention and innovation, which are merely the concrete acts through which exosomatization is realized, today rightly incite an increasingly percep-tible mistrust and hostility, including and indeed especially in those who throw themselves headlong into the most disruptive forms of hyper-consumerist capitalism, and who thereby become its victims – beginning with the misery and poverty of the 'digital encounter'.

These radical forms, claimed by the new barbarians, invade and engulf every technology of digital tertiary retention by obstructing any other possible model. They impede anyone striving to recompose socializa-tion through a new form of public power that would combat rather than exploit the radically toxic effects of disruption.

Hence, for example, if the toxic effects of food consumption have led to an explosion of diabetes in the United States, the response from industry is to develop chemotherapeutic medicine, health 3.0 and ever more sophisticated prostheses to replace the limbs amputated as a result of this disease, or to treat blindness.[29]

Toxic food is one case of the destruction of the oldest and most fundamental social systems – the culinary arts valuing the best local food resources (in Asia, we can see the way in which these aesthetic wonders that are Japanese, Chinese, Vietnamese or Thai recipes are threatened by the disastrous standardization of eating habits). But instead of struggling against toxic food, what the perspective of 'growth' sees in these degradations of life by deficiencies of health or economic and ecological aberrations are 'opportunities'. And this is also the kind of thing HSBC conceals while blithely avoiding paying tax by every means possible, in order to produce increased shareholder returns.

In the current stage of capitalism, research, invention, innovation *and therefore exosomatization* form a functional whole that constitutes the essential terrain on which competitive advantages are gained and profits are earned. As a result, exosomatization now coincides *in fact* with the most disruptive advances of capitalism – the most 'voracious', the most 'savage' and the most 'barbarous', founded on ultra-speculative unreason and sparking a thousand forms of barbarism in return. This is capitalism at its *least sustainable and most irrational*.

The question that lies at the heart of any consideration of this situation is the question of exosomatization itself. Exosomatization is not 'natural', that is, derived from the 'laws of nature', nor is it simply 'artificial', at least if, with Leroi-Gourhan and Canguilhem, we consider that the process of exteriorization and the artifactual organogenesis in which it consists delimit the technical form of life. This 'technical form' means the life of man having 'extended his organs by means of tools',[30] thereby setting up his *organological, and no longer organic, 'normativity'*[31] – and, along with it, his 'pathology', which is also a production of sense, that is, of πάθος (*pathos*), which itself stems from original forms of the 'infidelity of the milieu' that result from the normativity derived from this *pathogenesis*.[32]

If we accept the analyses that – from Kapp to Simondon, via Leroi-Gourhan, Canguilhem and many others (including Bertrand Gille, as I continue to emphasize in this book, but also Nicholas Georgescu-Roegen, and, more recently, in anthropology and related fields, Merlin Donald, Kim Sterelny and Michael Tomasello) – see in exosomatic organogenesis the *specific characteristics of the evolution of the human form of life*, the question arises of *knowing who will decide why and how this evolution will continue to unfold*, now that it has become *programmable and controllable* – at least up to that point where the risk it entails becomes excessive, which is to say the point where it becomes *uncontrollable*.

This point at which contemporary exosomatic evolution ceases to be *programmable or controllable* is the point at which the *laws of the improbable and the unpredictable assert themselves but in negative fashion*. This is precisely the *point of absolute madness* that I have attempted to delineate as *that point when the Anthropocene becomes structurally, massively and effectively self-destructive* – bringing to an end its own continuation by finally annihilating itself, through a *disruptive contraction* that becomes *absolutely and irreversibly entropic*.

45. Everything happens, nothing happens

Hence it is that today, in the ordeal of disruption by which the Anthropocene is brought to its *fatal extremes*, the question of exosomatic organogenesis imposes its necessity. This *imposing* question, caught up in the dizzying whirlwind of media logorrhoea, itself subjected, directly or indirectly, to marketing, *has not succeeded in posing itself*, that is, in *posing the question* of the criteria *in law* that would enable this literally apocalyptic state of fact to be overcome.

This is so because:

1. the specific organogenesis involved in exosomatization has still not been properly studied;[33]
2. Western metaphysics as a whole has been elaborated in such a way as to *systematically deny* the relevance and necessity of such a question, and to prevent such a study, which could seem to it only to be madness.[34]

After the collapse of this metaphysics, computational cognitivism, which is the *metaphysics of absolutely computational capitalism* bearing within it the ultra-liberal-cum-libertarian project, becomes and remains the new form of denial, while also becoming the theoretical core of transhumanism.

The continuation of exosomatic organogenesis seems inevitable. Not because innovation demands that we constantly innovate, to struggle against the tendency of the rate of profit to fall (this is how Schumpeter counters Marx, under the name of 'creative destruction'), but because psychic individuation, which cannot operate without contributing to collective individuation, that is, to collective transformation and alteration, is *made both necessary and possible only by technical individuation*. And this necessity and possibility occurs as the *production of metastable states*. Today, however, it has become difficult to see how such states could become stable – a failure that would in effect mean the end of exosomatization.

Leroi-Gourhan envisaged such an end in *Gesture and Speech*, when man, becoming a button-pusher as the hand regresses to being just fingers (the digits), tends in some way towards becoming a kind of jellyfish, an 'anodontic human race living in a prone position and using such forelimbs as it still possesses to push buttons'.[35] Such a suggestion is far from absurd, but it passes through the 'digital anthill', which still has a long way to go, and it could unfold in fact only *on the condition that humanity does not first destroy itself* through its negative externalities[36] or the conflicts these bring. Furthermore, as I have already shown, in formulating this hypothesis Leroi-Gourhan pays no attention to the consequences of disruption in terms of the destruction of social systems, in turn destroying psychic individuation.

As much as its cause, technical individuation is the consequence of the evolution of the social systems themselves, insofar as they individuate themselves, and through that individuate the collective and psychic individuals that grow within them – and vice versa. All this involves a transductive relation between three terms, which co-individuate during this process, whose main phases form the doubly epokhal redoubling, and of which the process of transindividuation is the concrete reality – and which, today, is threatened, short-circuited and disrupted by trans-*dividuation*, as I attempted to describe in the first volume of *Automatic Society*.

Until recently, the geophysical systems forming the 'great equilibriums' typical of the biosphere seemed largely unaffected by exosomatic organogenesis. But in the Anthropocene's recent history, this is no longer the case. If we are to believe Christophe Bonneuil and Jean-Baptiste Fressoz, one can find, *from the very outset* of this new geological era – whose beginning coincides with the advent of indus-trial capitalism and thermodynamic machinism (such as it gives rise to the question of entropy, a fact that, though not discussed by Bonneuil and Fressoz, remains crucial) – concern about the consequences of the Anthropocene, and therefore about its specificity as a new geological period. But this 'awareness of the risks' was concealed and stifled by a socially, politically, economically and ideologically organized form of denial.

This question is fundamental, firstly because it is one that continu-ously arises. At the very moment each of us worries about the 'future of the planet', to the point that the United Nations has made this its major mission after seventy years of efforts to contain the ὕβρις leading to war, at the moment when Florian speaks aloud what so many young people do not want to know and prefer to repress, *even if they dwell on it in whispered tones without being able to think it* – which they cannot do because, 'disrupted', the generation above them have been *disqualified*, and hence are unable to offer them circuits of transindividuation to which they could conjoin – at this moment, this denial is now more violent and more striking than ever: *nothing happens, except for the lament that rings out, ever more unrelentingly in the threatening silence, about the fact that, indeed, nothing ever happens.*

The intergenerational or rather de-generational abandonment that leaves both young people and their ascendants isolated and silenced, repressing what therefore only gnaws at them all the more, generates specific and dangerous forms of psychic suffering that can end only by manifesting themselves in the most unexpected and disturbing ways. But the politics of 'who gives a shit?' [*je-m'en-foutisme*], which understands nothing, couldn't care less about any of this – a fact for which it is ever-increasingly despised.

46. Legal and theoretical vacuums

In this context, discourses arise, some more rigorously argued than others, aiming to take stock of this situation – including the present work. Among them, two are of particular interest:

- the discourse of transhumanism, which is a way of denying the impasses of the Anthropocene – on the physical and environmental planes but also on the psychic, economic, political and social planes;
- the discourse conducted by Christophe Bonneuil and Jean-Baptiste Fressoz themselves, as historians, undertaken in order to deconstruct, in the manner of historians, the narratives that accompany what they call the 'Anthropocene event' – an enterprise that was previously begun by Fressoz in *L'Apocalypse joyeuse*,[37] to which we will return.

Just as disruption in general occupies the *legal vacuums* that it more or less deliberately creates through exosomatization, so too transhumanism occupies the *theoretical vacuums* left by epistemology, philosophy, anthropology and all those disciplines (including history) that in one way or another contribute to the current exosomatization without analysing it or critiquing its disruptive dimensions (information theory, economics, design, social science in general, engineering sciences, physical and biological sciences, and the humanities).

As for the major work undertaken in *The Shock of the Anthropocene*, it rightly advocates for environmental studies to clarify the stakes of this new anthropic geological era, and to locate points where it is possible to take action. But in so doing, it neglects the specific questions raised by the doubly epokhal redoubling, and the question of the transindividuation that must operate through it – this being what, at the end of the Anthropocene and with the rise of trans*dividuation*, fails to occur.

It is for this reason that Bonneuil and Fressoz come to the conclusion that we must now *adapt* to the Anthropocene: 'The Anthropocene is a point of no return. We must therefore learn how to *survive*.'[38] We *absolutely disagree* with such a renunciation, and, on the contrary, we believe that the Anthropocene is *intolerable*, and that it is imperative that it be overcome – in order to lead to what we call the Neganthropocene.

It is impossible to combat transhumanist discourse if we do not re-politicize it – if we do not succeed in raising the *precise* question of disruption as a disadjustment between the technical system, social systems and biophysical systems, and, consequently, the question of the need for a new public power.

As for the discourse of the historians of the Anthropocene, it must be conceived not just in view of environmental studies, but in relation to what the Digital Studies Network, created by the Institut de recherche et d'innovation (IRI), practises under the name of 'digital studies',[39] precisely as it relates to the systematic and transdisciplinary examination of the conditions and consequences of exosomatic organogenesis after the advent of what Leroi-Gourhan describes in terms of the process of exteriorization characteristic of hominization.

In this book, we will not explore these questions as such: they will be the subject of *La Société automatique 2. L'avenir du savoir*.[40] But what must be emphasized here is that *the issue they entail is that of the relationship between reason and madness*:

- It is *crazy* to keep running straight towards a brick wall, and to deny the passage to the limits[41] provoked by digital innovation, freed from all regulation and any *prescription transindividuated via the social systems*, which must, contrary to this, be unified into a *new public power*.
- It is completely unreasonable to deny the inevitable continuation of exosomatic organogenesis as the condition of an individuation that is always at once psychic, collective and technical.

The question of tomorrow is that of the *reconstitution of this threefold individuation* – outside of which there will no longer be any noetic individuals, or psychic individuals, or collective individuals, but just an appalling *denoetization*. The reconstitution of *noetic* threefold individuation can be founded only on a new epoch of knowledge, a new macroeconomic organization and an ecological and social politics tackling these questions head-on through a new form of public power. It is a question of survival.

47. Technologically integrated totalitarianism and madness as neganthropological possibility

Now more than ever, and as Simone de Beauvoir already suggested,[42] the improbable contradicts 'popular wisdom' [*la sagesse des nations*] – and common sense in general, just as 'reason', when it is submitted to rationalization, becomes, in the age of disruption, the *computational and probabilistic unreason* that now constitutes the basis of purely and simply computational capitalism. The challenge [*défi*] confronting humanity requires the advent of an im-probable bifurcation in the law of capital that has become computational unreason.[43] Such a bifurcation is improbable precisely in that, being neganthropological, it defies [*défie*] the probabilities.[44]

Transhumanism, as a discourse on exosomatization and endosomatization *calculated* to break with neganthropological psychic and collective individuation, is a technologically conceived ideology that is forged and concretized in order to assure a form of domination. Domination was, moreover, the initial and central question posed by *The German Ideology* in its consideration of humanity in terms of exosomatization

– in the wake of a current of thinking initiated by Herder against the transcendental (that is, idealist) conception of reason.

With transhumanism, it is a matter of imposing *total domination*, which amounts to a form of totalization, if not totalitarianism (in the sense in which this was described in the twentieth century, by, for example, Hannah Arendt): a *technologically integrated and hence full* [*intégrale*] *totalization*.

To combat this totalization, we must first consider the actual, effective conditions of the exosomatic organogenesis characteristic of hominization itself. By ignoring and denying this initial fact, rational thought is incapable of understanding the historical sense of transhumanism, and is thus incapable of fighting against it qua elimination of every question[45] – that is, qua *organological fulfilment of nihilism*.

Transhumanism is a negation of neganthropological constraint: it posits in principle the *obsolescence* of *collective* individuation, of social systems and of noetic *deliberation* in general – and *political* deliberation in particular. As artificial intelligence becomes correlationist, that is, turns into the 'data economy', or connectionist ('deep learning'), transhumanism articulates it with endosomatization and transformational technologies, and this in turn leads to the articulation of 'health 3.0' with synthetic biology and nanotechnology. In so doing, transhumanism denies the entropic constraints of organogenesis, in order to make calculation the *sole and total* principle of its enterprise, which is fundamentally commercial: it is, here, the *market* that *alone* orients and directs exosomatization and endosomatization, and does so exclusively according to its own interests.

To subject transhumanist ideology to philosophical critique (rather than just denounce its ends) by analysing its *possibilities* as the realization of dreams, but also its *impossibilities* as the concretization of nightmares, is firstly to inscribe these possibilities and impossibilities – these impasses – back into the general conditions of organogenesis and exosomatization, insofar as these conditions form the dynamic principles of neganthropogenesis.

In neganthropogenesis, the hallucination of what does not exist is paramount: that which *does not yet* exist can be realized from out of a field of consistent possibilities – consistences that themselves do not exist, and that will *never* ek-sist, but which constitute the idealities *nourishing* reason as kingdom of ends and as function. To think in this way is *to exceed the opposition of idealism and materialism*.

It is the *originally* hallucinatory and as such *ideal* character of exosomatization that enables the disadjustment that leads to unreason in all its forms. At the origin of what Simondon called a technical lineage, there is a fabrication, which must thus firstly have been hallucinated, then realized, and, finally, involved in various phases of the *process of adoption*[46] by the technical system *and* by the social systems within which this invention occurs. Such a fabrication can obviously destroy the one who fabricates it: this is, above all, what ὕβρις means.

'*La technique*', here, refers to the technique of the comedian, the musician and the artist, as well as the scholar, the jurist, the doctor, the craftsman, the engineer, the data scientist and so on. Such neganthropological bifurcations can occur only through *chains of fulfilled possibilities*, possibilities that *at first* seem to be *impossibilities*, and which, when reversed, become *quasi-causes*.

These reversals of *(de)faults* into *that which is necessary* [*ce qu'il faut*] (that is, into new bases of *necessity*) are always essentially organological and pharmacological. If fabrication can destroy the one who fabricates, and even, beyond that, the world within which this fabrication occurs, it is because organogenesis is always a pharmacogenesis – whereby the *therapeutics* can *always* go astray or begin to fail.

At the origin of a technical lineage – that is, an exosomatic organogenesis considered from the perspective of its complete social dimension, and hence as a social practice constituting a knowledge (of how to live, do or conceptualize) that is always also a therapeutics – there lies the *reversal of an impossibility into a possibility*, producing the reality of a *negentropic bifurcation* that is not just unpredictable but improbable, *always initially appearing, therefore, as crazy or foolish*, because neganthropological, and always capable of destroying the one who fabricates it, or the world in which it has been fabricated.

This neganthropology is *above all*, then, a pharmacology, in which the ἄνθρωπος (*Anthropos*) to which, ordinarily, *Neganthropos* amounts can lose itself – by losing its reason, that is, its taste for living, and, therefore, its very life: not just that of its individuals, but that of the neganthropic living species itself – *as a living species that has become anoetic, being made to serve its own automated exosomatic residues*, like an aphid bred to be exploited by a colony of robotic ants (a stage following that whose possibility I discussed in 'Allegory of the Anthill'[47] on the basis of Leroi-Gourhan's reflections on this point, previously mentioned,[48] and which has already been partially realized via 'digital pheromones'[49]).

48. Madness, reality and truth

The possibility of losing reason, that is, of going mad, is the *fatum* of the tragic condition in which the ancient Greeks found the horizon of their piety – where the possibility of ὕβρις, which stems from the conflict between Zeus and Prometheus, is the mortal condition. Hermes is the *interpreter* of this ὕβρις, that is, of the limits it imposes, as αἰδώς (*aidōs*) and as δίκη (*dikē*). When the latter are not respected, they are punished by Nemesis – who inspired Ivan Illich and his concept of counterproductivity.

Today, we are living through the return of the question of ὕβρις as the ordeal of this *extremity* that is the Anthropocene, of which the age of disruption is itself the radicalization: such is the context of contemporary delirium – at once as transhumanist delirium, extreme madness of computational rationalization and disruptive extremity of the 'new form of barbarism'.

The problem of madness is not that it is 'unreasonable' in the sense that it would be 'false' – in the sense where the false would be what could not become real. Hence the madness of Hitler, for example, did *become* real. The madness of the transhumanists, too, is in the course of becoming real, with the help of precursors that are more or less masked, more or less conscious, more or less unconscious.

The rationalization denounced by Adorno and Horkheimer has also become real, and it is of the same order – even if, of course, it is a madness that has so far been less abominable, less barbaric and less criminal, at least in the West, than Hitler's, who in his 1934 plebiscite received 89.93 per cent of the vote, Adorno having fled to England, before taking up Horkheimer's invitation to come to the United States.

In the Near East and the Middle East, the devastation of Iraq in the early 1990s, which is the starting point for the present calamities and the exodus of millions of people, was clearly abominable: the barbarous acts of criminals led to the worst extremities of crime – which is also another word for ὕβρις – and up to a point not so distantly removed from the homicidal madness of the Nazis. From the Abu Ghraib prison to the gratuitous killings of civilians by American mercenaries (which was just as horrific as the Oradour-sur-Glane massacre carried out by the SS), and via Guantanamo, which continues to operate right up until this very day, the crimes of the West have prepared the ground for the barbarity of Daesh and the madness of 'converts'. Meanwhile, at the time of writing, Bradley Manning (now Chelsea Manning) continues to sit in prison.[50]

Reality can turn mad only because madness can become real. And it can do so only because genius always stems from a 'grain of madness': 'No genius has ever existed without a grain of madness.'[51] As a possible consequence of exosomatization, the *future always* itself consists in the *realization of a delirium* – of a pathology, a being-affected and, in that, of a feeling (πάθος) – and therefore of a sense that, realized, can become normal. It is only from within the horizon of ὕβρις that the question of the relationship between madness and truth arises, in such a way that the *parrhēsia* of Solon can offer the Athenians the prediction that the madness they attribute to him – they accuse him of 'going mad', *mainesthai* – will eventually prevail as truth.

As a *bifurcation that we should, would and could effect today* in what we call truth, the latter, ἀλήθεια (*alētheia*), is *that which reinforces neganthropy* as the *preservation, cultivation and development of buried anamnesic capabilities*, through which, when a bifurcation occurs, a truth *appears*, becomes *a-lēthēs*, dis-covered, un-concealed – or what Socrates calls *anamnesis* and Husserl *Reactivierung*, founded on what Roberto Esposito describes as *delinquere*.[52]

As kingdom of ends, reason is not just analytical, but synthetic. This means that even if, in its sleep, it can engender monsters, *reason must dream* – and *realize* its dreams. And today, in this case, it can and must realize its dreams, that is, its ends, against the nightmares being prepared by the libertarians, disruptors and transhumanists. We must, then, be *much crazier* than these disruptors – whether they are so unconsciously or deliberately – who provoke an outburst of barbaric violence without precedent. We must realize a 'madness' of another kind – by provoking a *much more unexpected and unanticipated* bifurcation than those anticipated by all these madmen. We must produce a great negentropic bifurcation of the Neganthropocene.

There are several kinds of madness, as Erasmus points out in *The Praise of Folly*.[53] One kind of madness, which Montaigne calls creative, remains, for him, the condition of wisdom: 'Is there anyone who does not know how imperceptible are the divisions separating madness from the spiritual alacrity of a soul set free or from actions arising from supreme and extraordinary virtue?'[54] With regard to what, in the Anthropocene, has become 'rational', but which, as *rationalization*, engenders the new form of barbarism that now concretizes itself as disruption, reason must *combat the prevention of dreaming*,[55] and it must do so by *realizing* the dream of this *immense, absolutely improbable and unexpected bifurcation*, that bifurcation which, Inshallah, the Neganthropocene will be, and which archi-protention bears within it as neganthropic power. All this must be thought with Heidegger – but Heidegger did not himself think it:[56] he confused it with *Eigentlichkeit*.

Eigentlichkeit is 'authenticity', which contains all those possible compromises with the worst moments of twentieth-century European history. That is why we must read Heidegger.

49. Hubris and *boulēsis*

The possibility that allows dreams to be taken for realities (and sometimes to be realized in fact) stems from a type of delirium that Diotima discusses with Socrates: the μανία (*mania*) of *Eros*, such that, related to enthusiasm, and as expectation (ἐλπίς) projecting and protaining the extra-ordinary consistences *for which the one who is loved is the screen*, this delirium *constitutes* noetic φιλία (*philia*) – φιλία that includes φιλότης (*philotēs*), that is, friendship.

Friendship (φιλότης) is itself the condition of παρρησία (*parrhēsia*), which is always risky, as Foucault emphasized many times during his last seminar:

Parrhēsia, the act of truth, requires [...] a challenge to the bond between the two interlocutors [...]: it involves some form of courage, the minimal form of which consists in the parrhesiast taking the risk of breaking and ending the relationship to the other person which was precisely what made his discourse possible. [...] This is very clear in *parrhēsia* as spiritual guidance, for example, which can only exist if there is friendship, and where the employment of truth [...] is precisely in danger of bringing to question and breaking the relationship of friendship which made this discourse of truth possible.[57]

With philosophy and via Socrates, according to Foucault, *parrhēsia* enacts an evolution in the history of the *socialization of truth*, an evolution that *constitutes* the *polis* – and that constitutes it through and through. From the moment of its appearance, it configures the modes of veridiction (themselves forming a regime of truth[58]) that, in Athenian society, condition any βούλησις (*boulēsis*), that is, any form of *will* in psychic and collective individuation, the latter having become political individuation.

The becoming-political of psychic and collective individuation, however, is made possible only because literate tertiary retention, as the generalized practice of the alphabet,[59] rearranges primary retentions and secondary retentions, giving rise to *collective retentions and protentions of a hitherto unknown form* – through this *pharmakon*, writing, the ὕβρις of which will be denounced by Socrates, and which we always practise as the foundation of the humanities (in the Renaissance sense) and the Enlightenment, which arose after the printing and circulation of books and with the *circulation of mail* that *thereby* formed the *circuits* that would come to constitute the 'Republic of Letters'.

The new arrangement of collective retentions and protentions that appeared in Greece with the alphabet is constituted by the βουλή (*boulē*). It is this institution that forms the ἐκκλησία (*ekklēsia*) as such. Ὕβρις is *excessive will* [*volonté démesurée*], that is, beyond all measure, μέτρον (*metron*), which, as the framework formed by knowledge, governs a *wise* relationship to *pharmaka*, and, more generally, to *tekhnē*. This wise relationship is called *sophia* and *sophrosunē*, where the μέτρον is *what provides the measure*, that is, *contains the excess* [*démesure*] (ὕβρις).

This will comes to occupy the centre of modern thought by transforming itself, that is, by transforming its relationship to pharmaka – and, more precisely, to written hypomnesic tertiary retention. This is played out with Descartes – not in the *Meditations*, in relation to which Foucault and Derrida will come to blows, but in the *Rules for the Direction of the Mind*: these rules establish the hypomnesic, organological and pharmacological basis of the *mathesis universalis* that will be generalized in the *Discourse* and *Meditations*, laying the foundations of modernity.

50. Will, disinhibition and denial

With his *Rules*, Descartes opens the way for what, via Leibniz's 'universal characteristic', will lead to digital tertiary retention and to its ὕβρις – that is, to the disruption characteristic of the end of the Anthropocene, with its specific forms of madness and barbarism.

This *new conception of the will* eventually becomes the basis of the thought of *logos*, and of *logos* as *ratio*, that is, as calculation, hence making the calculating subject the foundation of any veridiction. And, in making calculation the foundation of veridiction, this new conception of the will is what will ultimately give rise to the current stage, where it no longer passes through the βούλη, nor any institution of this kind, until, ultimately, it can dispense with the *cogito* itself. This is precisely what Chris Anderson asserts[60] – and with him, so too does computational cognitive science.

As the rational will of mastery and possession, the modern will finds its foundation in the *mathesis universalis*. But this is possible only because, *while mobilizing tertiary retention and equipping it with instruments* – tertiary retention being the condition of rational attention (see Rule Fifteen[61]), that is, of the *mathesis universalis* itself – *it conceals its facticity and technicity*, that is, ὕβρις. This concealment is essential to the Anthropocene qua industrial realization of the Cartesian dream of being master and possessor of nature.

Setting 'will' *at the heart* of classical moral and political philosophy, and doing so on the basis of the fundamental denial of the madness that constitutes this will, insofar as it ignores the ὕβρις of retention, this realization turns, now, into a nightmare. We will return to the Cartesian moment, and in so doing we will be brought to Foucault's thesis on the modern age – which is the first moment of the modern epoch (or the modern era?) – a period he separates from what he calls the 'classical age' (all these denominations requiring in-depth examination).

By also reading Peter Sloterdijk's *In the World Interior of Capital* and Jean-Baptiste Fressoz's *L'Apocalypse joyeuse*, we will see how the history of modernity can be viewed from a very different angle than that which Foucault chose to adopt: with these works, 'modernity' will prove to consist in a regulating/deregulating process of immense disinhibition.

The *retentional* dimension of madness, that is, of ὕβρις, is what neither Foucault nor Derrida were able to see – neither in the history of madness *in the classical age*, nor in madness and its historicity *in general*.[62] Descartes, while constituting the certainty of the *cogito* through the method of doubt and the evil genius, effects the 'great division' described by Foucault,[63] of which Voltaire, already cited,[64] is a perfect example.

But the most important thing for us, and the thing that Foucault could not see,[65] is that in this Cartesian *denial* [*dénégation*] of madness as condition of reason it is the *contemporary age of madness* that is preparing itself, through the forgetting of the *pharmakon*, as its disavowal [*déni*], leading to that blindness that will be constitutive of the ὕβρις of the modern will *and beyond*.[66]

8

Morality and Disinhibition in Modern Times

51. The exosomatization of the life of the mind, spiritual life as exosomatization, computational unreason

> It is generally helpful if we draw these figures and display them before our external senses. In this way it will be easier for us to keep our mind in a state of attention.[1]

Such is the fifteenth of the *Rules for the Direction of the Mind*, in which Descartes posits in principle that, in order to *direct* his thought, he must *exteriorize* it, and, in so doing, make it an object not just of a retention, but of the *form of attention* that this exteriorization makes possible – and, already, makes possible *as the realization of the dream in which all thinking consists*, as we are about to discover.

With Rule Fifteen, Descartes places attention at the heart of what, for modernity, is his foundational approach, by defining attention as *analytical* capacity. The issue is not representation, as many of those who interpret Descartes as the founder of modern metaphysics have tended to believe, but rather the exteriorization of representation, in the form of points, lines and surfaces that 'fix our attention'[2] – where this also amounts to the fixing of the *already there of our thinking, which thereby reflects upon itself* through the fact of its self-exteriorization. Leibniz will say the same thing in almost the same terms. And Husserl would no doubt have read and reread this rule, as well as followed it.

In Rule Sixteen, in fact, it is no longer a question of attention, but of *retention* – and, more precisely, a matter of tertiary retention becoming, already, a kind of *automaton*:

> As for things which do not require the immediate attention of the mind, however necessary they may be for the conclusion, it is better to represent them by very concise symbols rather than by complete figures. It will

thus be impossible for our memory to go wrong, and our mind will not be distracted by having to retain these while it is taken up with deducing other matters.[3]

Preparing the way for Leibniz's Characteristics as the fulfilment of the *mathesis universalis*, Descartes here describes the conditions in which analysis can and must constitute itself in and by its exteriorization, such that the analytical function of the understanding can then be delegated to this exosomatization of mental life. Memory, this rule tells us, 'is often unreliable',[4] and this is what Derrida, commenting on Husserl, will call 'retentional finitude'.[5] This can, however, be overcome, by 'the practice of writing. Relying on this as an aid, we shall leave absolutely nothing to memory but put down on paper whatever we have to retain, thus allowing the imagination to devote itself freely and completely to the ideas immediately before it.'[6]

Descartes, here, apparently does not recall Socrates' warnings about hypomnesis as the pharmacological condition of anamnesis: he makes no mention of this issue. On the other hand, he does describe that which will inaugurate the coming realization (in the nineteenth century, with Charles Babbage, Ada Lovelace and Herman Hollerith) of a new stage of grammatization, which will be made possible by what, in Descartes' time, will become algebra, which is also the condition of the algorithmic: 'We shall do this by means of very concise symbols, so that after scrutinizing them (in accordance with Rule Nine), we may be able (in accordance with Rule Eleven) to run through all of them with the swiftest sweep of thought and intuit as many as possible at the same time.'[7] These *Rules*, too, will provide the axiomatic outline of a treatise on the algorithmic that will be brought to completion only with Leibniz, 'critic of Descartes'.[8] 'Therefore, whatever is to be viewed as one thing from the point of view of the problem we shall represent by a unique symbol, which can be formed in any way we like.'[9] This way of describing Cartesian analysis lies at the origin of Kant's distinction between understanding and reason. But this analysis is possible only as a new stage of grammatization, which is obviously not unconnected with the rise of movable type in the epoch of printing, for which the shock of the Renaissance paves the way, and which will be concretized as a new technological shock and a new stage of the exosomatization of the mind with the US census of 1890.[10]

Descartes lies between two shocks provoked in the life of the mind by the *fatum* of the doubly epokhal redoubling – Cartesianism and its consequences for modern philosophy tracing circuits of transindividuation that, linking these two epochs, make all this *both possible and impossible*.

All this belongs to the process of grammatization as exosomatization of the life of the mind and spirit, which, set outside itself, interiorizes

itself by already dreaming of the next stage of its self-exosomatization
– a chain reaction that typifies that which opens the possibility of
the Anthropocene, in which self-exosomatization ineluctably becomes
hetero-somatization.

This is what we are currently living through in the *disruptive divorce*
of computational understanding and reason, which, as rationalization,
becomes computational unreason – the (mis)calculated loss of reason.

52. Modern will and disinhibition

Between ortho-graphic grammatization and algorithmic grammatization,
a sequence of operations unfolds: forms of discretization, comparison
and the creation of tables and series – seriations that are mathematical
'reasons', that is, laws of series extracted on the basis of their numer-
ation, and of which the correlationism of 'big data' is an immature
(unreasonable) stage. All this amounts to the *modern organology of
the mind* that, exosomatizing itself, thereby concretizes itself. In *Taking
Care of Youth and the Generations*, I have tried to show how and why
Foucault ignores this, but that he does so in a way that is nonetheless
suggestive, and how he raises the therapeutic question in his final works
by examining techniques of the self.

As suggested by Foucault, but without him thinking it himself as
such, the Cartesian will amounts to the transindividuation of the shock
stemming from the printing of books and the advent of the Reformation[11]
and then the Counter-Reformation, resulting in one transformation
after another, including the introduction of money into everyday life,[12]
account-keeping practices and practices of silent, solitary reading that
cultivate a penchant for the 'solipsistic'.

At that time, however, and contrarily, *globalization*, product of the
madness of Portuguese and Spanish *explorers* who set out to conquer
the ('East' and then 'West') Indies, is already well under way: 'The
Portuguese and Spanish expeditions could never have been undertaken without motivating systems of delusions to justify these leaps
into the unclear and unknown as sensible acts.'[13] In other words, even
were it true that Descartes had excluded madness from reason, and
indeed from unreason (on which Derrida casts doubt) – a gesture that
would according to Foucault be characteristic of the classical age, an
age that organizes and concretizes this exclusion through the 'great
confinement' of the mad and the deviant – nevertheless, in 'Delusion
and Time: On Capitalism and Telepathy', the tenth chapter of *In
the World Interior of Capital* and the chapter from which the above
quotation is taken, Peter Sloterdijk shows that the modernity of the
modern age rests, on the contrary, on the unleashing of a new kind of
madness – and perhaps, already, on the exporting of the West as extra-
territorialized barbarism.

Sloterdijk, who clearly perceives the 'anthropotechnical' condition of humanism and modernity,[14] deplores the fact that 'a philosophically thought-out history of discoveries, terrestrial and maritime alike, has never been considered'.[15] He then shows that this history (whose main outlines he sketches) would be that of an *immense process of disinhibition characteristic of capitalism*, which today has become planetary, after having been – from its inception – the engine of *globalization as this disinhibition*, that is, as this *madness becoming the norm* of which rationalization would be the justification. Here, 'rationalization' should be understood simultaneously in the senses of Weber, Adorno and Freud.

I will attempt to show, by combining Foucault's analysis with that of Sloterdijk, that what the latter describes as a process of disinhibition is made possible by the tensions and contradictions that occur over the course of the successive doubly epokhal redoublings that unfold from the late Middle Ages to the Renaissance, the classical age, the first industrial revolution, the advent of Taylorism (which is also the advent of consumerism, the culture industry and marketing), and, finally, digital technology – agent of the contemporary disruption.

It is over the course of these epochs, and as the consequences of these successive shocks, that the conditions of the Anthropocene as it is described by Bonneuil and Fressoz are set into place. In other words, the Cartesian will and the *mathesis universalis* that it founds (and that founds it) are the absolutely necessary, albeit insufficient, conditions of possibility of the Anthropocene: the analytical discourse of Descartes presupposes tertiary retention that will lead to the question of calculability as the *characteristica universalis*.

The Anthropocene comes into effect only with capitalism qua *economy of disinhibition*, which, as we will see with Sloterdijk, is tightly and paradoxically articulated with this calculability that, before too long, moves towards algorithmic probabilities – which, in our time, themselves bypass statistics as the technology of state power and biopower. These post-statistical probabilities, which complete the psychotechnological organology that the culture industries would already have constituted, then become the exosomatized architecture of a neuropower of which transhumanism 'tells the story'.

53. Disinhibition and discipline as pharmacological consequences of tertiary retention

The articulation of disinhibition (that is, of madness) and calculability characteristic of the classical age and its reason as *mathesis universalis* is, in a way, the antithesis of the processes described by Foucault. Foucault seems to invite the conclusion that these processes lead, on the contrary, to a systematic reinforcement of inhibition, both by 'morality' and by the criminalization of deviance: this is what Foucault describes throughout *History of Madness.*[16] These two phenomena, of course, are in fact anything but contradictory: they feed into each other – and they are the result of a dynamic of tendencies that *are* contradictory and that tertiary retention always sets up, this being precisely what makes the latter a *pharmakon*. We will see that this dynamic of mutually reinforcing tendencies amounts to an *organology of temptation* of which the monotheisms are at-tempts that are, by turns, curative and toxic.[17]

This capitalism and this economy of disinhibition can become industrial only by combining with the great turning point that is the Reformation,[18] where, in dissonant counterpoint to the puritanism that unfolds with Calvin, disinhibition operates through a succession of disadjustments that eventually lead to the current stage of the Anthropocene qua disruption. The latter amounts to *total disinhibition*, and as such, paradoxically, to a kind of totalitarianism, seemingly 'soft' but in reality extremely violent: a violence (only just beginning) that encompasses every level – verbal, moral, sexual, police, economic, delinquent, terrorist and so on.

From this economy of disinhibition (which today becomes a diseconomy, founded on the systemic dissimulation of its insolvency) stems the *crazed will* of noetic and neganthropic beings. With the globalization of this mad will, a generalized anthropization of the world is established – installed on the model of Western capitalism and precipitating an immeasurable increase in the rate of entropy. Today, we must trace the history of this mad will – which requires a *combined rethinking of madness and will*, and with regard to contemporary ὕβρις.

The history of madness is always also the madness of history – in ways and by means that are each time singular, for which *In the World Interior of Capital* provides a wealth of material, where 'will' is expressed on a wide variety of registers that we will examine after recalling and commenting on the broad outlines of Foucault's work.

Behind the acts of madness by which history is fashioned, such as those of the explorers discussed by Sloterdijk, as well as the campaigns of conquest by which 'historical figures' are made – *actualizations* [*passages à l'acte*] that are perhaps the most energetically and most deeply repressed dimension of *Geschichtlichkeit* as *Historie* – behind all this madness, one finds that it always, and always *first of all*, somehow involves tertiary retention in the broad sense: in the sense that *any technique, any technics*, is a tertiary retention, that is, a *pharmakon*.

Bearing technological epokhalities or borne by them, a new tertiary retention always reopens, in one way or another, the dehiscence in which ὕβρις always consists, and within which, alone, the process of disinhibition can occur – that process whose history and economy Sloterdijk investigates, in a way that is both parallel to the Foucauldian enterprise and often implicitly in argument with it.

From the Sloterdijkian perspective, the *certainty* that the foundation of the Cartesian subject is supposed to provide, far from dominating the classical age, in reality just creates a space for risk-taking, for calculations of probability and for insurance mechanisms of all kinds, which *rationalize* the *new ordinary madness* of the conquerors, and which accompany the *immense uncertainties* in which capitalism also consists.

This process of disinhibition requires, in the first place, globalization itself, and, along with it, the 'great adventure', that is, the conquest of the high seas. In this adventure, piracy plays a key role – and this is the case despite the fact that these endeavours are accompanied by evangelizing missions, in a dynamic involving the conquest both of minds and territories.

The industry of cultural goods and its various wireless networks, which today include the data economy and its digital networks, obviously transform the nature of this globalization.

Thirty years after Adorno and Horkheimer, all the consequences of decolonization unfold as the beginning of the 'crisis'. This crisis will profoundly transform the structures and goals of capitalism,[19] until the advent of Hayekian ultra-libertarianism, advocating total disinhibition, glorifying 'hacking' (but in a quite peculiar way), that is, piracy (the hacktivists, many of whom I count as friends, prefer to see themselves as so many Robin Hoods rather than as pirates, in the belief they are realizing the dreams of Hakim Bey at the very moment they are actually serving the cause of Hayek).

54. The tragic and ὕβρις

In one of his most discussed pensées, Pascal writes: 'Men are so inevitably mad that not to be mad would be to give a mad twist to madness.'[20] This is what Foucault recalls in his preface to the first edition of *History of Madness* in 1961.[21] He also quotes Dostoyevsky: 'It is not by locking up one's neighbour that one convinces oneself of one's own good sense.'[22] As we shall see, Sloterdijk, too, refers to Dostoyevsky, whom he presents as being the great thinker of, precisely, disinhibition.[23]

It is starting from these references to Pascal and Dostoyevsky that, at the beginning of his preface, Foucault presents reason in the classical age in its relation to madness as a 'trick that madness plays [...] through which men, in the gesture of sovereign reason that locks up their neighbour, communicate and recognise each other in the merciless language of non-madness'.[24] Let us measure the scope of this statement: *at the origin of reason in the classical age, there would lie a 'trick played by madness'* [*tour de folie*]. The history of madness in the classical age would not just be, therefore, that history during which reason enclosed madness, locked it up, and treated it as unreason: it would also be the *history of this classical reason as a kind of madness*, as a *new form of madness* – and it could be seen as preliminary to a 'new form of barbarism' that will not occur until three centuries later, eventually unfurling itself fully after yet another century, that is, after 11 September 2001, at the dawn of the twenty-first century, which will also be the latest period of the Anthropocene, and as the age of disruption, that is, as absence of epoch.

This perspective, however, opened by Foucault in the preface to the first edition, is not explored any further in the body of the text.

My own thesis is that this 'trick of madness' that would be reason must now be rethought as that which leads to the Anthropocene, to disruption, to transhumanism and to the various forms of barbarism in which all this consists – including with the 'neo-barbarians', as the radicalization of this new form of madness that 'classical reason' would have borne within it, and as its ὕβρις. It is in this way – that is, otherwise than Foucault himself – that, today, we should read *History of Madness*.

What leads to this kind of questioning is that 'man', who today is certainly no longer 'modern' (which does not mean that he would be 'postmodern', or that what is covered by the latter would adequately describe his present situation), 'man today', who is obviously no longer at all classical (but the classical age is in Foucault an *epoch* of the 'modern age', so that the latter, in this regard, would instead constitute an era – inclusive of the first two epochs of industrial capitalism), this man *becomes* mad, even in his everyday life,[25] and he does so through the unfolding of an 'ordinary madness' that is *quite extra-ordinary with respect to the history of madness*, especially as Foucault conceives it.

Contemporary everydayness *becomes crazy* and even *requires madness* while at the same time denying it – we all feel it, outside ourselves *and within ourselves*. This exigency stems, in particular, from disruption, if it is true that madness is an expression of the absence of epoch, that is, the impossibility of producing collective protentions.

We *know* that ὕβρις, which is thus carried to a point that is truly incandescent, could become fatal to the great adventure of hominization, that is, to this technical form of life that appeared some two or three million years ago. We also *know* that the technical form of life, having become, as globalization, generalized and uniform anthropization, ultimately constitutes the geological era referred to as the Anthropocene. But we know it in a mode that *does not recognize it*, by abandoning ourselves to denial.

It is at this moment that the words of Florian appear: he reopens the question of *parrhēsia*, and of its relation to madness – to *our* madness, as the madness *of denial*.

We constantly repress the consequences of this unrecognized knowledge. And we do so through this process of denial that has perhaps become the most characteristic and systematic indication of our age,[26] but which rests on a primordial dimension of what Heidegger calls Dasein – namely, the concealment of 'being towards the end' by what he calls *Besorgen*. But it is starting from this unrecognized knowledge, and from these *primordial givens of what unleashes all forms of madness*, that we are now trying to measure the scope of Foucault's first work.

The preface to *History of Madness*, published in 1961, was removed from the 1972 edition following the critique that Derrida presented in 1963, in 'Cogito and the History of Madness'.[27] Before challenging the Foucauldian interpretation of the section of the *Meditations* already mentioned, Derrida criticizes the way that Foucault, in this preface, passes very quickly over ancient Greece and its relation to madness, while at the same time Foucault claims that Greek Logos, unlike classical *ratio*, *would have had no opposite, no other* – which in this context means that it would not be defined by the *exclusion* of madness.

In making this claim – in a way that was no doubt too brief, as Derrida was right to regret (in what follows, I will explain why this assertion seems to me, too, to be much too brief) – Foucault nevertheless hinted at the theme of ὕβρις:

> The Greeks had a relation to a thing they called ὕβρις (hubris). The relation was not solely one of condemnation: the existence of Thrasymachus, or that of Callicles, is proof enough of that, even if their discourse comes down to us already enveloped in the reassuring dialectics of Socrates. But the Greek Logos had no opposite.[28]

Here we must emphasize that Foucault defined his whole enterprise from a *tragic* perspective – by drawing on Nietzsche. But it seems to me that he does so by utilizing a notion of the tragic that remains inchoate, if not vague and allusive. This is especially so with respect to madness inasmuch as it stems from being originally by default and outside itself, and as such 'furious' in its original sense, a *situation* that constitutes the lot of mortals (this is the very *fatum* of the tragic itself) inasmuch as they are condemned to prostheticity, to facticity, to artificiality, and, in all things, ultimately, to fiction – which is also to say, to ὕβρις.

The speed and almost lightness with which Foucault passes over the status of ὕβρις and the tragic in general, and of their status in ancient Greece in particular, does seem, therefore, rather strange and problematic. But I am not sure that Derrida himself succeeded in taking the measure, precisely, either of Foucault's general intention or of the stakes of ὕβρις. And this is so, not just with regard to the Greeks, but with regard to madness in general, *with regard to Cartesian discourse observed from this angle*, and therefore with regard to classical reason. We will come back to this.

55. On the need to read or reread *History of Madness* in the twenty-first century

The 1961 preface summarizes the theme and presents the goal of Foucault's *History of Madness*, which we can put in the following terms:

1. After the Greeks, who, through ὕβρις, cultivated a relationship to madness that was not in opposition to Logos, and since the Middle Ages, 'European man'

> has had a relation to a thing that is confusedly termed Madness, Dementia or Unreason. It is perhaps to that obscure presence that Western Reason owes something of its depth, as with the threat of hubris, the σωφροσύνη (sophrosyne) of Socratic speechmakers.[29]

2. At the end of the Middle Ages, madness becomes the symptom of a

> sudden unease that appears on the horizon of European culture [...]. Madness and the figure of the madman take on a new importance for the ambiguousness of their role: they are both threat and derision, the vertiginous unreason of the world, and the shallow ridiculousness of men [...] a great unreason which could be blamed on no one in particular but which dragged everyone along in its wake in a sort of tacit agreement. [...] From the fifteenth century onwards, the face of madness never ceased to haunt the imagination of the West.[30]

This being haunted by madness put an end to the Middle Ages in that it 'took over from the seriousness of death. From the knowledge of that fatal necessity that reduces man to dust we pass to a contemptuous contemplation of the nothingness that is existence itself.'[31] It is here that Foucault quotes Eustache Deschamps declaring that 'all goes ill',[32] on which he comments that 'the rise of madness, its creeping invasion [...] indicates that the world is close to its final catastrophe: it is the insanity of men that called it forth and made it inevitable'.[33] This omnipresence of madness, to which Erasmus, Montaigne, Shakespeare, Cervantes and so many other, lesser-known figures, bear witness, right up until Pascal, is what the classical age, through Descartes as well as through a new politics of social exclusion,[34] will repress, efface, exile – by a gesture that is social in the sense that it *fundamentally* reorganizes society.

3. Foucault refers to this gesture as the 'great confinement', for which a 'date can serve as a reference point: 1656, date of the decree founding, in Paris, the Hôpital Général'.[35] From this starting point, institutions of confinement come to replace the leper houses of the Middle Ages and a

correctional system is erected whereby the mad and deviants in general are confounded together as 'asocials':

> To that way of thinking, confinement was merely the spontaneous elimi-
> nation of the 'asocial'. The classical age is taken to have neutralised, with
> sure-footed efficiency – all the more efficacious for being blind – the people
> who, not without hesitation or danger, we now divide between prisons
> and corrective institutions, psychiatric institutions and the psychoanalyst's
> couch.[36]

How can we not conclude, reading these lines after having considered a few cases (among many) where crime and madness are today confounded in a context of internal and external protean barbarism, how can we not conclude that the dossier opened by Foucault, and closed perhaps a little too hastily by Derrida, needs to be reopened – while considering the place that twentieth-century philosophy gave to madness, in particular in France, principally through schizophrenia, for example as 'schizo-analysis'?[37]

4. In this preface that will later be removed, Foucault places his whole approach 'beneath the sun of the great Nietzschean quest, [an approach that] would confront the dialectics of history with the immobile struc-tures of the tragic'.[38] The tragic and its '*immobile* structures' refer here to *limits*, of which it is a matter of undertaking to do the history – and, as it were, the history of a 'tragic structure'.[39] This history *of limits, that is, of the tragic,* is a history of madness itself, inasmuch as the latter is not constituted just by the medical description of a pathology whose clear contours can be identified through psychiatric nosology. This history of limits is a history of madness inasmuch as the latter is a *fundamentally historical* reality, resulting from social transformations and the tensions they generate. There is no 'immutable continuity of a madness already fully armed with its timeless psychological equipment'.[40]

56. The most mad

The great confinement, which in the eyes of Foucault comes to define the classical age, is also the confinement of beggars, that is, of the poor, who are to be classified into two categories:

> the good and bad poor, those of Christ and those of the Devil [where
> the latter must be deprived] of their freedom, a freedom for which they
> had no use other than the glorification of Satan. Confinement was thus
> doubly justified, as a benefit and a chastisement. It is both reward and
> punishment.[41]

Foucault tries to give a history of a set of structures that link the various processes of inclusion and exclusion through which classical society is formed, where it is a question of the history of madness as a multi-stratified, paradoxical, contradictory complex, which the classical age tries to simplify, even though

> the consciousness of madness, at least in European culture, has never been a monolithic fact, undergoing metamorphosis as a homogeneous whole. For Western consciousness, madness has always arisen simultaneously at multiple points, forming a constellation that shifts little by little, transforming its outlines and whose figure perhaps conceals an enigmatic truth. Meaning here is always fractured.[42]

For Foucault, this is the accomplishment of a new Western consciousness of madness in the classical age, a radical separation of reason and unreason,[43] for which Descartes would be the pivot: it is on this point, especially, that Derrida will challenge Foucault. It is, for Foucault, a new regime of *will* that will thus have been established, and not just of truth – where will and truth are not separable:

> [Methodical] doubt [...] is wrapped in the will [...] which [...] is a voluntary wrenching away from the temptations to give in to madness. [The] will to doubt has already excluded the involuntary charms of unreason [...]. Long before the Cogito, there is an archaic implication of the will and of the choice between reason and unreason.[44]

The will is thus constituted as *condition of truth*, and vice versa, on the basis of banishing madness – where thought would therefore never be mad:

> Madness has been banished. While *man* can still go mad, *thought*, as the sovereign exercise carried out by a subject who sets out to seek the truth, can no longer be insensate. A new dividing line has appeared, rendering that experience so familiar to the Renaissance – unreasoning Reason [*Raison déraisonnable*] – impossible. Between Montaigne and Descartes an event has taken place: something that concerns the advent of a *ratio*.[45]

This *ratio* breaks with a thought of madness that is also a thought of thinking in which the possibility of madness would always remain the *condition* of thought – as if thought had, at bottom, and as its object, to think only its own madness, in order to delimit it, and to invest it, and where, by the same token, as Montaigne suggested perhaps more so than Pascal, the objects that thought gives itself are its projections. The whole question of technoscience would need to be reconsidered as such from this *fatum*.

It is this question that Montaigne suggests is raised, before Pascal, by Seneca. And all of them think this way, *before* the separation, at the dawn of the Anthropocene, of philosophy and the sciences.

But what does *ratio* mean? And in what sense does the passage from reason as *logos* to reason as *ratio* impose this exclusion? I have already indicated that this development is fundamentally tied to the fact that calculation and calculability, in passing through the rules that Descartes assigns for the direction of his mind as *analytical* practices of reading (Rule Fifteen) and writing (Rule Sixteen), *thereby* become the condition of all rationality. Rationality is *thereby* directed along the path towards rationalization, which according to Weber is the reality of capitalism, and which according to Adorno and Horkheimer inverts the *Aufklärung*.

What underlies this condition of calculability (as reading and writing) is the question of hypomnesis – that is, of the ὕβρις that the *pharmakon* always contains, as we discover in the epoch of algorithmic governmentality.

Classical morality – soon to become bourgeois morality – is equally shaped by the classical conception of unreason, a morality that becomes moralism in the common sense of the term, and therefore, too, a rejection not just of deviance, whether delinquent or religious, but of transgressions and sexual perversions:

> The gesture of confinement [...] had social, political, religious, economic and moral meanings. In all probability, they concerned certain essential structures of the classical age as a whole.[46]

> [M]adness too was divided up [...] according to the moral attitude it manifested [...] sometimes under the category of beneficence, sometimes repression. Every internee [...] was treated as a *moral subject*.[47]

> [I]n the space of less than half a century [madness] found itself a recluse in the fortress of confinement, bound fast to Reason, to the rules of morality and their monotonous nights.[48]

From the nineteenth century on (but this movement had already commenced in *Rameau's Nephew*[49]), with Jean-Étienne Dominique Esquirol who testifies to the horrors of the great confinement, with literature and poetry that gave a language back to madness as the expression of 'man's hidden truths', and with the rise of psychiatry in the sense it continues to have today, 'the refusal of madness was no longer an ethical exclusion, but a distance that had already been granted. It was no longer necessary for reason to divide itself from madness, but to recognise itself as always anterior to it, even if it does on occasion lose its way within it.'[50] This turning point, however, leads us to call for a complementary analysis of what will occur with respect to madness and its social significance in the twentieth century – not just with psychoanalysis and its critique by schizo-analysis, the latter having not yet appeared at the time Foucault was writing, but with the industry of cultural goods, bearer of a 'new form of barbarism' and no doubt a new form of madness,[51] and with what accompanies it in a functional way, namely, *marketing as the organization of innovation by disinhibition, which becomes hegemonic*, and of which the theory of disruption is only the most recent development, and the *most mad*.

Unlike Deleuze, who fundamentally links the advent of control societies to marketing, that is, to the exploitation of affects via calculability, Foucault pays little attention to the historical fact of marketing.[52] His analysis remains limited in that he fails to see that, in the twentieth century, biopower presupposes this psychopower.

Let us now see how Peter Sloterdijk allows us to think the genesis of the psychotechnologies of the psychopower that is consumerist capitalism, which in turn becomes control society, and which, as neuropower, is now radically reshaping biopower itself in terms of the 'transhumanist' perspective.

57. The Modern Age as the 'propensity for madness'

Sloterdijk's work is inspired by Fyodor Dostoyevsky's *Notes from Underground*, itself the result of the impression left upon the writer by the Crystal Palace. The latter was built in London in 1851 for the 'Great Exhibition of the Works of Industry of All Nations', which Dostoyevsky visited in 1862:

> In his tale *Notes from Underground*, published in 1864 – which was not only *the founding certificate of modern ressentiment psychology*, but also *the first expression of an anti-globalization stance*, assuming such backdating is legitimate – there is a formulation that encapsulates the world-becoming of the world at the incipient end of the globalization age with unsurpassed metaphorical power: I am thinking of the reference to Western civilization as a 'crystal palace'. On his visit to London in 1862, Dostoyevsky had visited the site built for the International Exhibition in South Kensington.[53]

What Sloterdijk undertakes in *In the World Interior of Capital* amounts to a *history of disinhibition*, a history that begins *with* and *as* the process of globalization (firstly as colonization), which is also to say with and as the expansion of capitalism, and to start with, pre-industrial capitalism.

Through the analysis of this process of disinhibition, Sloterdijk uncovers a *propensity for madness* characteristic of modernity as a whole and beyond, of which the classical age was a key stage of development, and *starting from which* capitalism will eventually, in the

United States and in the twentieth century, undergo a fundamental evolution: 'Columbus was an agent of a *pan-European willingness to embrace delusion – though it was only psychotechnically perfected by the USA in the twentieth century* (and re-imported to Europe through the consultancy industry).'⁵⁴ The analysis of the 'consultancy industry' here complements the Adornian perspective on the culture industry: the well-known conflict between Habermas and Sloterdijk should not cause us to forget that Sloterdijk's analysis is also an extension of the claims made by Adorno and Horkheimer in *Dialectic of Enlightenment* – even if the terms of their argument are very different.

Today's 'consultancy industry' should be considered with respect to its fundamental relationship to the absolutely computational capitalism of algorithmic governmentality.⁵⁵ Along with the culture industry and the data economy (the inclusion of the digital in relation to these issues is my own contribution, and is not something that Sloterdijk discusses), this now constitutes a totality that is formed from out of the 'activity culture of modernity'.⁵⁶

This 'activity culture' emerged at the dawn of the Modern Age with *those madmen who are the explorers, pirates and swindlers* [*chevaliers d'industrie*] who, through colonization, but on a much broader scale and within cities, establish an 'organization of disinhibition',⁵⁷ thanks to which the whole ensemble of social structures begins to transform.

In this highly complex and often paradoxical process, the response of Ignatius of Loyola to the Reformation, conducted according to his *Spiritual Exercises* – the latter belonging to what Foucault called techniques of the self⁵⁸ – prefigures, according to Sloterdijk, the development of the psychotechnics that will be essential to globalization, from colonization until today: 'As an explicit attempt at psychotechnical and medial modification, Jesuit subjectivity was driven by the longing to understand the successes of the Protestants better than the Protestants themselves. [...] The first subjects of the Modern Age [...] were [...] the Jesuits.'⁵⁹ Just as spiritual exercises can lead to their opposite, namely, to psychotechnologies in the service of what has been described today as an economy of attention, which is in reality a *destruction* of attention (its *diseconomy*), so too, what seems to constitute the speculative or transcendental sphere of mental or spiritual life in *otium*, in reality works (without knowing it) for the establishment of the hegemony of *negotium*: 'The dominant figure of modernity is thus by no means the excess of reflective inwardness [...]. What becomes manifest in the process is that the task of reflection is to prepare the desired disinhibition.'⁶⁰

There are many ways in which we might interpret such an idea. For myself, it seems to express the dynamic involved in what I refer to as the doubly epokhal redoubling, such that, in the *second moment* of this double redoubling, which is accomplished, for example, as the constitution of those circuits of transindividuation that arose with the

Republic of Letters and, more precisely, with modern philosophy (from Descartes to the Kantian *Aufklärung*), the 'task of reflection' is to trigger the technological and scientific epokhality of the next stage[61] – which has the paradoxical result that the 'owl of Minerva' arrives always too late.

58. From Raskolnikov to disruption, via Schumpeter: mercilessly clearing the way for the territories of disinhibition

As Hegel taught in the nineteenth century – at the moment when exosomatization suddenly accelerated into machinic becoming (the first steam engine arriving in Berlin in 1795), thereby inaugurating the Anthropocene era – *the life of the mind is the life of its exteriorization*. Through exteriorization, the mind enters into a contradiction with itself that Hegel believed to be dialectical, leading to the great synthesis of absolute knowledge through which it would regain peace with itself.[62]

As for us, what we learn from *disruption* is that *this becoming is not dialectical, but tragic*, that is, *pharmacological*. This is what Foucault sought to grasp but without success, having failed to conceive ὕβρις in a tragic manner (just as, oddly, *The Birth of the Clinic*[63] overlooked the industrial pharmacopeia and pharmaceutical-chemistry that turned health into a market – just as Google is doing now with the digital industry – despite the fact that Foucault's teacher, Georges Canguilhem, did indeed raise the question of drugs and of their place within care).

Sloterdijk shows how disinhibition results from the delay and advance that plays out in the exteriorization that we are here calling the doubly epokhal redoubling. It is in this way that disinhibition constitutes the condition of possibility of the Anthropocene and of the passage to limits that has already led this era to a critical turning point (but, as we will see,[64] Sloterdijk does not raise the question of these limits, a question we learn from Foucault but on another register – itself insufficient).

This disinhibition is what leads to the sense of a permission to commit crimes: it is this that became clear to Dostoyevsky at the Crystal Palace, a building through which Britain above all celebrated the great power of its global capitalism, and which could equally be seen as the invention of the leisure park, and, through that, of the entertainment industry, if not 'cultural goods'. After the Great Exhibition, the building was shifted from central London to South Kensington, where it ultimately burned to the ground in 1936.

Hence it is that we shift from *Notes from Underground* to *Crime and Punishment*:

> No one has illuminated the way in which the disinhibiting auto-persuasion of future perpetrator subjects works in individual cases with greater

precision than Fyodor Dostoyevsky in his novel *Crime and Punishment*, written in 1868 – a psychological-moral study that can be read [...] as a handbook of practical philosophy with particular reference to special permission for crimes.[65]

It is starting from this relation to crime (to ὕβρις), and as the extremity or radicalization of disinhibition (in the epoch of those one refers to, in the world of Dostoyevsky, as nihilists), that Sloterdijk conceives innovation and what will become the economic theory of 'creative destruction', all this being conceived as a *theory of progress*, which is also to say of disadjustment:[66] 'Two generations after Raskolnikov, Joseph Schumpeter would state in his theory of economic development that in economic life, functionally speaking, there are ultimately only innovators and imitators.'[67] This state of fact was established, therefore, by the systemic organization of disinhibition in which capitalism consisted at the dawn of the Modern Age, an organization that resulted from new regimes of the doubly epokhal redoubling – all of this, today, seems in retrospect to have reached that critical threshold of the Anthropocene we are calling 'disruption'. From this, however, we must not conceive a 'naïve ontology of progress in which the distance between the vanguard and the main body can consistently be interpreted as the pilot function of those at the forefront'.[68]

In other words, those at the forefront have no use for those who lag behind, for those who arrive 'late': the former are the pirates and criminals who mercilessly clear out territories for disinhibition, without the least regard for what might otherwise have remained of 'civilization' – which, coming always too late, can but fill them with *contempt*:

> In this schema, the headstart of those who are extraordinary is made possible by a vocation to disinhibition that forges ahead solely through active contempt for the restrictive power of morality and convention – hence the thesis of the inevitable criminality of the innovators.[69]

The thinker who will conceive progress and advance on this register, that is, as a philosophy of becoming, and no longer as ontology, is Nietzsche. With '*Thus Spoke Zarathustra* [...] Columbus's deed had arrived in thought'.[70]

59. Risks, probabilities and protentions: reflective madness

This event, which is the advent of nihilism in Nietzsche's sense, leads to the fading away of every narrative of origin, that is, of territorial and historical belonging, and to the pre-eminence of risk and novelty:

'A human of his type exists not from their origin, but rather from their headstart.'[71]

The United States of America, conquered by mad explorers and populated in large part by reclusive types, will become the country of immigrants who themselves become 'pioneers' who will hunt, annihilate or enslave the indigenous inhabitants. In this way, a wholly other form of society will be invented, one that could not have been conceived in any other country. Even the Bolsheviks, who wanted to make a 'clean sweep' of the past, were not able to succeed in erasing from those societies the heritage lying at the 'source' (*provenance, Geschichte*). Heidegger will make this the context of his *An Introduction to Metaphysics*.

The disruption now underway, as a new stage of the organization of disinhibition and an extremization of those tendencies character-istic of the Anthropocene, is at the same time being extended to the entire planet, via digital networks functioning at two thirds light speed. Among its effects is the breakdown of inherited territorial immunities – in the United States and everywhere else – heritages, cultures and social structures originally emerging from their origin [*provenance*], and not from this advance. All this does nothing but prepare the way for an immense counter-reaction, as it triggers a chain reaction of *incalculable* consequences.

This society of risk that according to Sloterdijk appears with the Modern Age has nothing to do with 'risk society' as described by Ulrich Beck, for whom risk appears with the exhaustion of an initial form of modernity and its replacement by what he calls 'reflexive modernization',[72] in a sense close to the 'reflexive modernity' of Anthony Giddens.[73] For Sloterdijk, risk and its socialization are, on the contrary, located at the *beginning* of the Modern Age.

In this way, Sloterdijk is in agreement with the positions espoused by Bonneuil and Fressoz: 'all these new risk-nationalists: the Portuguese, Italians, Spanish, English, Dutch, French and Germans who hoisted their flags on the oceans [...] had learned by 1600 at the latest how to make their risks calculable through diversification'.[74] This culture of risk is what leads to the formation of insurance organizations:

> The new insurances seemed suitable to outwit the sea and its cliffs economically. Humans and property can be in what one calls danger; 'a commodity at sea' (Condorcet), on the other hand, is subject to a risk, that is to say a mathematically describable probability of failure, and calcu-lating solidary communities can be formed to combat this probability.[75]

In other words, in addition to *statistics*, which according to Alain Desrosières[76] is the science of the state in the service of what Foucault called biopower, the calculation of *probabilities*, which is something similar (but which, like algorithmic governmentality, is essentially distinct from it[77]), is what shapes psychopower as the *control, pooling and amortizing of protentions* achieved through the use of probability calculations.

This *probabilization of protention* is what leads very early on in the Modern Age to a kind of *reflective madness*: 'Here the risk society comes about as the alliance of well-insured profit-seekers. It unifies the insane who have thought everything through beforehand.'[78] These are those *rational madmen* who are always so keen to distinguish themselves from 'ordinary madness', the better to maintain their business affairs:

> The blooming of the insurance idea in the middle of the first adventure period of globalized seafaring shows that the great risk-takers were willing to pay a price in order to be taken seriously as reasonable subjects. For them, everything depended on establishing a sufficiently deep divide between themselves and ordinary madmen.[79]

This leads to a differentiation between reason and madness, but by a completely different path than that traced by Foucault: as the Modern Age becomes the classical age, philosophy, like insurance, begins to legitimate these 'insane who have thought everything through', *and vice versa*. 'Such insurance systems as Modern Age philosophy drew their justifications from the imperative to separate reason and madness clearly and unambiguously.'[80] Insurance replaces worship as a means of consolidating a possible future in the chaos of improbabilities. The improbable is replaced by probabilities as the protentional horizon within which improbability is *dissolved*:

> one defines modernization as a progressive replacement of vague symbolic immune structures [...] with exact social and technical security services. [...] Prayer is good, insurance is better: this insight led to the first pragmatically implanted immune technology of modernity.[81]

60. Modernity as a process of reflexive disinhibition

In *L'Apocalypse joyeuse. Une histoire du risque technologique*, Jean-Baptiste Fressoz brings to bear, by starting from the history of inoculation, analytical elements that are fully in keeping with what Sloterdijk describes in terms of disinhibition and the calculation of risk. The introduction to Fressoz's book is entitled 'Les petites désinhibitions modernes', and he shows that it is 'subtle twists of reality and certain *moral dispositions* that, in the eighteenth and nineteenth centuries, send us on the path towards the abyss'.[82] This disinhibition, which is a factor leading to the taking of such risks, is the result 'of the scientific and political production of a certain modernizing unconsciousness'.[83]

According to Fressoz, however, there is a resistance to this development, which has been erased by the orthodox history of 'progress', and the first example he offers is that of Eugène Huzar, author of *La Fin du monde par la science* (1855). Huzar condemns 'an ignorant, short-sighted science, [...] a progress that marches on blindly, with neither criterion nor compass',[84] and he formulates objections to *a posteriori* experimental science that are not totally foreign to the kinds of claims that Husserl will make in *The Crisis of European Sciences*.

Industrial technology emerges from experimental science, and the risks posed to humanity by the development of this technology are *deliberate*. This is what Fressoz shows, just as he shows that such risks are *justified by a science of calculation and probability* that posits that risk-taking is more rational than ignorance of a risk that exists regardless – for example, in the case of variola inoculation, or variolation, which, at the beginning of the eighteenth century, was the origin of vaccination, the theory and practice of which was established by Pasteur in the second half of the nineteenth century.

But what Fressoz shows above all is how this discourse, which met with much opposition, led to the deliberate and artificial manufacture [*fabrication*] of a trust [*confiance*] in innovation that in turn led to these 'small modern disinhibitions': 'Trust was not something self-evident and it was necessary to produce – in a calculated way, at every strategic and conflictual moment of modernity – disinhibiting ignorance and/or disinhibiting knowledge.'[85] From that point onwards, a vast politics is implemented that falls under what I have described in previous chapters as a process of adjustment between the technical system and the social systems, and that constitutes what Gille describes as a politics of innovation: 'For innovation of any significance to occur, it is necessary to circumvent moral reticence, social opposition, impacted interests, suspicions and critiques of its real-world consequences.'[86] This process is accompanied by theoretical productions that legitimate accidents and catastrophes in order to attenuate ethical resistance and constitute a reasoned process of disinhibition that runs parallel to that described by Sloterdijk, and which Fressoz describes in the following terms:

> The word disinhibition condenses two moments of the *passage à l'acte*: that of reflexivity and that of its being disregarded, that of taking danger into account and that of its normalization. Modernity was a process of reflexive disinhibition aiming to 'legitimate the technological *fait accompli*'.[87]

In this way, Fressoz enters a debate with the study of history as to the causalities at work in the emergence of that 'European (or rather British) exceptionality' which lies at the origin of industrial society, and where, according to Fressoz, 'the technological dimension of the historical transformations that took place between 1750 and 1800 was somewhat erased'.[88]

This is already a little like what Lucien Febvre said to Bertrand Gille.[89] But Fressoz's enterprise is built upon an historical gaze (after the fact) quite different from Gille's. The history (*Historie*) of industry is the history of a madness, and a madness of history (*Geschichte*), in the sense of being a history of a *fait accompli* outside of law:

> Historically, technics has never been made the subject of a shared choice. Certain actors actively brought it into being and it then became necessary to regulate them. Contrary to the sociological dream of having mastery over technoscience, of a gentle incline of progress, the history of technics is that of shocks [*coups de force*] and subsequent efforts to normalize them.[90]

The expression 'sociological dream', which is rather odd, is worth noting: through it, the question of the dream enters by the back door, a question to which it will be necessary to return.[91]

Fressoz's remarkable work must, in fact, be inscribed within a wider problematization – that, precisely, of ὕβρις, which obviously does not date from 1750, nor from the time of the Conquistadors or the Renaissance: it lies at the heart of anthropogenesis, which, for this reason, and so as to integrate the questions raised by Fressoz, we must understand as, *from its earliest moments, a neganthropogenesis* – a neganthropogenesis that is an exosomatic organogenesis.

As we have already seen, neganthropic organogenesis *presupposes the dream, whether nocturnal or diurnal*, that is, *the fantasy that can pass into actuality*, which is the agent of ὕβρις par excellence: this is what we are told in the myth of Icarus. Daedalus, father of Icarus, is indeed the 'very type'[92] of the technician who, to satisfy Pasiphaë, wife of Minos, builds a 'wooden cow' that enables her to mate with the bull with which she had fallen in love. So was born the Minotaur, for whom Daedalus would build the labyrinth within which he would keep this hybrid being locked up, whereas, himself locked up with Icarus by Minos, he would see his son lose all sense of measure and so perish, despite the *inhibitory* recommendations given by the father.

In this madness that is Humanity, which 'does not exist at all yet or [...] barely exists',[93] there is a history of ways of *dealing with* ὕβρις [*faire avec* ὕβρις] – and it is not absurd to say that all of history is above all the history of these ways that ὕβρις can be handled, that is, with the possibilities, compossibilities, incompossibilities and impossibilities that it opens up.

Capitalism amounts to one remarkable moment of this history, one that reveals this condition by transgressing the principles of containment, which amounts above all to a process of disinhibition *as such*. This is what, after Sloterdijk, Fressoz shows, the ecological dangers of which were already highlighted by Marx. But over the course of this history, what evolves are *forms of knowledge,* that is, those processes that produce circuits of transindividuation, processes that form between the two stages of the doubly epokhal redoubling.

If in general, and since the dawn of exosomatization, what is manufactured [*fabriqué*] must first have been hallucinated and then *realized,* until eventually it is engaged in the various phases of a process of adoption by the existing technical system, and then by the social systems within which this invention occurs, nevertheless such bifurcations can occur only through series of *fulfilled quasi-causes.*

Such is the case with the transformation of a fault in the Lenoir engine (auto-ignition) that led to the diesel engine (with engineer and businessman Rudolf Diesel in 1893), just as it is when the theory of energy dissipation for heat engines is given a *new theoretical foundation* (as the theory of entropy). In these cases, it is always a question of a necessary 'major improvement' in a more or less extensive field of formal knowledge, where this is provoked by a phenomenotechnics that appears more or less accidentally, and becomes the condition of a *recuperation from the shock* caused by this or that technical epokhality of greater or lesser extent.

What Fressoz shows is that this recuperation has, in the Anthropocene, and ever since the engagement of this process of disinhibition that is capitalism, been confronted with a test and an ordeal of limits. This ordeal is not that of Georges Bataille (although it is clearly related to the question of transgression that lies at the base of Bataille's work) but of Icarus, opening up the problem of compossibilities and incompossibilities that is generated by the appearance of any new *pharmakon,* requiring a new type of care, that is, a new type of knowledge.

In the Anthropocene that emerges from capitalist disinhibition, the 'transindividuators', which is what scholars and other clerics ought to be, end up internalizing the state of fact that consists in a structural lateness, and they do so through *theory, which no longer plays any role other than to legitimate, after the fact – to 'rationalize' – that which makes a state of law impossible.* This is why the analysis of the genesis of the Anthropocene proposed by Fressoz must be continued and completed via epistemological considerations leading to a critique of forms of knowledge in general, including, of course, historical knowledge itself.

For, in positing that this 'fact' has always been imposed by the apparatus organizing disinhibition and the more or less blind trust in risk:

• Fressoz renders practically unthinkable the singularity of the moment that I am here calling disruption;
• he does not problematize the process of phase-shifting that lies at the

origin of any process of psychic and collective individuation inasmuch as it always has an intrinsic relation to technical individuation;

- he does not make it possible to raise in a new way the question of law (and, alongside it, the question of the history of law), or of the dependence of history on notions of law, or of the relations between law and knowledge in general, including in the epoch of industrial science;

- he then comes to posit with Bonneuil that we must 'learn to live in the Anthropocene',[94] and in so doing makes impossible the thought and the will to pass into the Neganthropocene as a new dream of ὕβρις or of what it contains that is necessary, as a source of all neganthropy – as if he had internalized the state of fact that he criticizes while establishing it as unsurpassable.

Nevertheless, the fact remains that *L'Apocalypse joyeuse* is a fundamental work of and in the history of risk-taking, inasmuch as the latter sends us 'on the path towards the abyss' unless there is a *complete reconceptualization of the neganthropological condition*. And it is fundamental, not just because it offers a precise and well-documented historical analysis, but because it brings to light the theological and philosophical foundations[95] that were put to use and recycled so that they might accompany the rise and rationalization of 'market capitalism'.[96]

His analysis shows how these theological and philosophical sources were enlisted or combatted in order to constitute a *new morality* (in particular through Cotton Mather and the Reverend Colman) that welcomed this disinhibition, if not other, more universal forms: 'Pastors changed the relationship between nature and morality by articulating them not according to universal sentiments deposited by God in his creature, but by probabilistic laws that must be discovered in the world.'[97] In France, in the eighteenth century, at the instigation of Charles Marie de La Condamine, a wide-ranging debate took place that prefigured the questions that we are rediscovering today with transformational technologies, in relation to which the transhumanist fable would have us accept its vision of the 'augmented human'. According to Guillaume J. de L'Épine,[98] summarized here by Fressoz,

> to allow inoculation would be to endorse the existence of a third type of medicine, a medicine of 'transformability' or of the 'mutability' of the human body. Doctors would be headed onto a dangerous interventionist slope aiming at the endless improvement of the human body according to the desires of patients.[99]

This morality consisted of a practice both of puritan prohibition and the destruction of public power so as to *deregulate not only the circulation of commodities but also industrial science*, and as such it resembled an early elaboration of the transhumanist discourse of the libertarians.[100]

61. Descartes and the Anthropocene, pirates and money, Sloterdijk and ill-being

According to Sloterdijk, the development of insurance that eliminates the improbable – and that leads towards the 'death of God' – would find its 'inner basis of certainty' in Cartesianism's success in 'modernizing self-evidence' as its reassuring logical ground: 'Perhaps the rationalist branch of continental philosophy that followed on from the emigrant Descartes attempted precisely that: providing a new breed of risk-citizens [...] with an unshakeable logical mainland on which to stand.'[101] This 'foundation', however, this 'basis', inexorably loses its credit thanks to the effects of what it makes possible, namely, *the new 'technical world' that is the Anthropocene*:

> On the market of modern immunity techniques, the insurance system, with its concepts and procedures, has completely won out over philosophical techniques of certainty. [...] Insurance defeats evidence: this statement encapsulates the fate of all philosophy in the technical world.[102]

It was piracy that opened these pathways, by practising atheism in an empirical and factual way: 'In this context, piracy – [...] the foremost manifestation of a naïve globalization criminality – [...] is the first entrepreneurial form of atheism: where God is dead, [...] the unimaginable is indeed possible.'[103]

This leads us, once again, back to the libertarians, who today in France advocate their 'new barbarism', which is an-archist in the sense that it is fundamentally hostile to all public power and all ἀρχή (*arkhē*): 'the moderns conceive of the dangers of libertarian and anarchist disinhibition in terms of piratical atheism'.[104] The question of the relationship between power – inasmuch as it constitutes itself in social structures, which are here generally referred to as *immunitary* structures – and the spheres that, in Sloterdijk's philosophy, constitute there-being, falls within what he names macrospherology. 'I use this term to encompass the reflections with which the theory of intimate spheres (microspherology) is "elevated" to the level of a theory of large immune structures (states, realms, "worlds").'[105] On this point, which leads to the consideration of today's relationship between psychic individuation and collective individuation – which are articulations between micro- and macrospherology – the conclusion arrived at in *In the World Interior of Capital*, a work published two years prior to 2008, does not manage to reach the heights of what precedes it: 'In truth, money has long since proved itself as an operatively successful alternative to God.

This affects the overall context of things today more than a Creator of Heaven and Earth ever could.'[106] This statement ignores the question of archi-protention and of its 'existential' conditions in Heidegger's sense, of which, of course, Sloterdijk is well aware. It is this overestimation of probabilities – which are the condition of the 'operative success' of 'money' – that leads Sloterdijk to virtually ignore the existential question itself.

In Sloterdijk – as for that matter in Simondon – there is no pharmacology: there is none of the sense of the tragic around which Foucault turned in 1961, without the latter ever quite seeing where, precisely, the question of madness in all its forms (in the combined senses of Foucault *and* Sloterdijk) truly lies. This is why Sloterdijk does not (in 2006) feel the rise of this question: the question of what we refer to as a 'new ordinary madness', within that new form of barbarism that in 1944 Adorno and Horkheimer already feared – and which in France, in 2015, becomes that of the new barbarians, and, in the world generally, that of disruption in general, and, in the Middle East, that of Daesh.

The new ordinary madness is what, in issue 413 of the journal *Esprit*, and under the title 'Aux bords de la folie', Marc-Olivier Padis, Jacques Hochmann and Michaël Foessel describe as a form of ill-being [*mal-être*].[107] This ill-being results from what makes existence impossible, whereas Sloterdijk still believes that existential opportunities can emerge from the fact of disinhibition itself, in which he seems ultimately to invest unfailing trust – and there, perhaps, lies his own 'propensity to madness':[108] 'From the [moment, in the crystal palace that is global capitalism, that] a radical de-scarcification of goods [occurred,] a leap [took place] in the pampering history of *Homo sapiens* – a leap that opened up an enormously expanded space of existential opportunities.'[109] In this cynical tone, Sloterdijk celebrates excess – that is, ὕβρις, which is also to say crime – which he relates also to chaos, and by referring to Deleuze and Guattari: 'The wretchedness of the conventional forms of grand narrative by no means lies in the fact that they were too great, but that they were not great enough. [...] For us, "great enough" means "closer to the pole of excess".'[110] It is here that Sloterdijk quotes Deleuze and Guattari: 'And what would *thinking* be if it did not constantly confront chaos?'[111] What 'confronts chaos' [*se mesure au chaos*], however, what finds its measure in chaos, is not just excess [*démesure*]. It is, precisely, chaos as the opportunity to bifurcate. Excess, that is, ὕβρις, is its condition. But this condition is not sufficient: it lacks a therapeutic.

It was around 1990 that Deleuze began to reflect on this question, and it is what lies on the horizon of Guattari's *The Three Ecologies*.[112] But this is another subject, to which we will return in *La Société automatique 2. L'avenir du savoir*.

9

Ordinary Madness, Extraordinary Madnesses

62. On 'the ordinary madness of power'

'Aux bords de la folie'[1] revisits, from various angles, the question of madness, the history of psychiatry (in both the nineteenth and twentieth centuries), the history of confinement and hence the history of the relationships between the asylum and the prison, the history of pharmacology and behaviourism, the history of the representation of madness, and, finally, the question of the madness of power as that of the madness within power – in the Pascalian sense, which provides the occasion for Michaël Foessel to return to the debate between Foucault and Derrida concerning the respective roles of madness and dreaming in the constitution of classical reason.

Before delving into Foessel's article, which is entitled 'La folie ordinaire du pouvoir',[2] we must indicate a few of the salient features of this issue of *Esprit* with respect to our purposes here, and firstly the data provided by Marc-Olivier Padis in 'Derrière la folie, les malaises ordinaires': 'According to the World Health Organization, one out of every four people experiences mental illness during their lives. In fifteen years, the number of children in France treated by child psychiatry has doubled.'[3] Referring to a 2012 report of the 'Conseil économique, social et environnemental' concerning France, and to a book by Jean-Paul Delevoye, *Reprenons-nous!*,[4] Padis speaks of 'psychic fatigue' and 'collective burn-out'. While defending himself against the charge of wanting to make 'a medical diagnosis of our society', Padis nevertheless highlights a profound change in contemporary society with regard to mental illness: 'Psychological concerns have become a mass phenomenon. Hence mental illness can no longer be localized on the remote margins of society. This does not mean that the madman does not retain his unsettling strangeness.'[5] A key question is, precisely, the relationship between mental illness and madness – where it would not be possible to confine madness to mental illness if it is true that, as we have already suggested, and as Gladys Swain argues when she revisits the history of psychiatry in a way that deviates from Foucault,[6] madness is constitutive of the history of thought, and hence equally constitutive of the 'phenomenology of spirit' in the strict Hegelian sense, where spirit can appear only in passing through moments of *division, that is*, of madness, those 'intermittences' of noetic life that would be essential to noesis, in which respect Pinel would provide Hegel with clinical data concerning mental illness.

In 'Les contestations de la psychiatrie', Jacques Hochmann, after recalling the genesis of the 'sectorization' of French psychiatric care,[7] goes on to recall the debates that ensued between institutional psychiatry, anti-psychiatry, psychoanalysis and sector psychiatry, debates that focused on the place of delirium and more generally of psychological suffering in the process of individuation.[8] Hochmann emphasizes the therapeutic scope of Paul Ricoeur's concept of narrative identity, while recalling that what he calls the 'pathologies of ill-being' are not cases of madness.

What, then, are the relations between the extra-ordinary dimension of madness and what, in this special issue, is named – perhaps in contradiction with Hochmann's recommendation – 'ordinary madness'? It is starting from these contemporary questions that, in §§68–69, I will revisit the debate between Foucault and Derrida. And I will do so in order to try to think the *psychosocial stakes of disruption* – inasmuch as it stems from the dream and the nightmare, thereby constituting a factor contributing to the expansion of madness in its various forms, whether 'ordinary', 'rational' or 'extraordinary' – with Sloterdijk and his analysis of disinhibition constituting the principle of the Modern Age and capitalism.

In 'La folie ordinaire du pouvoir', Foessel invokes Pascal and his study of the relationship between power and madness – after briefly referring to the 'Sarkozy case' and to the strategy drawn from it by François Hollande:

> Shortly before the 2007 presidential election, the weekly *Marianne* launched a campaign on the alleged 'madness' of Nicolas Sarkozy. [...] By 2012, the turmoil of a chaotic quinquennium had convinced François Hollande to present himself as a 'normal' president. As if mental stability provided an entrance ticket for the Élysée.[9]

There remains much to say about this last point – and we will come back to it. What designation of 'madness' would the Sarkozy case have fallen under? And as for Hollande, might he not have been classified by 2015 as himself counting among the 'rational madmen'?

Foessel remarks that in 2012, in fact, 'the feeling [that Sarkozy is crazy] goes well beyond partisan divisions: something, decidedly, was not right about this presidency [of Nicolas Sarkozy]'.[10] On the basis of this remark, Foessel extends the question of this instability, this disequilibrium, to the whole of society, and he does so by referring to Foucault: 'Foucault [...] taught us that it is the nature of power to diffuse itself into the whole of society without it becoming possible to isolate some place or character that would totally escape its grasp.'[11] The symptomatology of power as exercised by Sarkozy should, therefore, be inscribed into a diagnosis that is more alarming and 'undoubtedly more characteristic of our epoch, according to which we would all be more or less crazy to the extent that we are alienated by a power that slips into the folds of our psyche'.[12] As for this fundamental dimension of Foucault's contribution, which posits that power is the condition of individuation (and here we should obviously mobilize Simondon with Foucault), it is necessary to point out that the exercise of knowledge as techniques of the self is a remedial practice that Foucault prescribed to himself.

In this contemporary symptomatology of power, *what* power, precisely, does this most concern? Should it still be understood starting from Sarkozy, Hollande and other representatives of national executives who can be found in the present world, or is it not rather a matter of *a wholly other form of power* – one that *would precisely no longer be public power*, and hence no longer political?

63. Ordinary, extraordinary, morality, imagination

The power in question is no longer political, but economic, and the 'representatives of national executives' are its pitiful playthings. The *ordinary* character of contemporary madness results from this state of fact, which is brought to its peak, before any other cause, by disruption, and, more precisely, by the *ordinary madness* that results from the *liquidation of the extra-ordinary by the nihilism* in which capitalism fundamentally consists.

If this is not precisely what Foessel investigates, these are the questions to which his analysis nevertheless leads: 'Is this a madness reserved for the powerful or a general spread of madness? Is it an irrational and ancestral pathology afflicting those who govern us or just ordinary delirium sustained by the extreme rationalization of contemporary ways of life?'[13] Here, once again, we find the question of the new form of barbarism that would be contained in this rationalization – as the process by which the *Aufklärung*, that is, 'classical reason', is inverted into a rationalization that generalizes not just *Dummheit* (stupidity, stupor, stupefaction), but madness, and does so by somehow depriving it of its very extra-ordinariness, by the fact of *counting* with it.

It is a question, then, of *specifying ordinary madness* in a history of madness that extends beyond modernity, but still as the *pursuit of the process of disinhibition* in which capitalist madness consists, despite the fact that, long before the advent of modernity, 'power [...] produces delirium wherever it operates'.[14] It is here that Foessel turns to Pascal, who situates the 'grain of madness' that makes madness *possible* in the imagination – whether it is the madness of those who govern or that of the governed: 'No one before [Pascal] had ever described this desire [of the governed] to obey legitimate power as a hallucination close to madness.'[15]

Now, what is this 'desire to obey'? We, who no longer live in the age of La Boétie or Spinoza, we who come after Freud, we know that power, which is an instance of transindividuation and a synthesis of all transindividuations, power insofar as it is primarily *legitimated* only on the condition of *constituting itself in law* (and here, undoubtedly, it is necessary to do more than just repeat Foucault), stems from a libidinal economy of the drives through which identifications and idealizations constitute a hallucinatory horizon that is indispensable to the formation of a *we* that this power embodies while, inevitably, disembodying it.[16]

Insofar as it is transindividuation, that is, binding, power cannot be thought without conceiving an economy of the drives, of which it is also the diseconomy – that is, the unbinding, and as ὕβρις.

As this (dis)embodiment, power is a matter of desire, of economy, a question of the binding and unbinding of the drives. It is on this basis that we must rethink the will to power – within a process of transindividuation that, from the dawn of the Modern Age, is *also* a process of disinhibition. Here we must examine the latter with respect to two fundamental points. The first is constitutive of *exosomatization in general*, as the care taken of the *pharmaka* that it generates. The second is *historical*, and *characteristic of the Anthropocene*, where exosomatization and disinhibition combine as the fulfilment of nihilism (that is, as the dissolution of use values and practical values into exchange value):

- Psychic and collective individuation presuppose technical individuation, which I now call exosomatization (in so doing following Georgescu-Roegen's economic analysis), and which is the fundamental factor involved in those phase shifts that the Greeks referred to also as ὕβρις – such a statement cannot be confined, in this generality, to the 'logic of the supplement', for the latter is in fact accomplished only through the history of this supplement.
- Disinhibition disembodies individuation and transindividuation by progressively destroying the social structures of super-egoization, to the point of total unbinding, thereby generalizing what Durkheim called anomie; with disruption, this disinhibition replaces transindividuation with transdividuation, that is, with the automatic dividuation of those 'ordinary madmen' that we are all becoming, insofar as we are 'wholly calculated', and there is complete destruction of the very notion of what in an earlier time [*autrefois*] was called 'the moral'

[*morale*] – not in the sense of 'bourgeois morality' or 'Judeo-Christian morality', but in the sense of the Stoics and Montaigne.

What, in fact, does '*autrefois*' refer to here? There are alterities of this *altra volta*, which our epoch tends to relate only to that other time that was the age of 'Christian morality' (whose relations to Mosaic law and guilt should be understood more profoundly): on that other side, that of Montaigne and the Stoics, 'moral philosophy' takes us back to the question of αἰδώς (*aidōs*) and its relationship to δίκη (*dikē*), *hermeneia* and the *pharmakon* – and of what connects them in also constituting the question of ἀρετή (*aretē*), and hence of ἀλήθεια (*alētheia*), and therefore of παρρησία (*parrhēsia*), and all this as what (*es*) *contains* ὕβρις.

Before making our way to these lands of other times, which may still lie *before* us, remaining *to come*, perhaps, and perhaps even as *the only future possible for us beyond* disruption and the Anthropocene – visible and accessible through a very narrow doorway, as thin as the eye of a needle – let us continue reading 'La folie ordinaire du pouvoir', where the key question is the imagination. For Pascal, writes Foessel, 'all men imagine, which leads them to a universal delirium. "I am not speaking of madmen, but of the wisest men" (Pascal).'[17] Here, Pascal extends Montaigne, who extends Seneca. The question is imagination, that is, the dream – and it is around the dream and its status in Descartes' *Meditations on First Philosophy* that Derrida and Foucault will face off against one another.

64. The dream of Descartes and the question of powerlessness

In 1954, seven years before *History of Madness*, Foucault published a long introduction to the French translation of 'Dream and Existence', an article published by Ludwig Binswanger in 1930 in which the question of the dream and its status in psychoanalysis, psychiatry, psychology and philosophy is central.[18] The reflections on imagination that Foucault puts forward in his introduction are at times astonishingly close to those of Simondon[19] – whom he would at that time no doubt have encountered in the courses and seminars of Canguilhem and Merleau-Ponty.

The dream, as I have insisted upon repeatedly here, is the condition of exosomatization, and it is also the seat of the unconscious and the pathway of the expression of the drives. But we cannot reduce it to this nocturnal scene of the unconscious, and we cannot interpret it solely from the perspective of its 'latent' content – this is what Binswanger says, and then Foucault. A fantastic as well as historial – *geschichtlich* – foundation is its condition, composed of those 'affective tonalities' characteristic of individuals, such as these are inscribed into epochs, and, beyond these epochs, into transgenerational existential dimensions.

According to my own analysis – which extends the analysis that Foucault set out before he published *History of Madness*, and, as we shall soon see, where this latter work itself extends, in his reading of Descartes' *Meditations*, his own 'Introduction' to 'Dream and Existence' – we must refer, here, to an *oneiric condition* that *makes* exosomatization *possible*, and that is itself *made possible* by exosomatization, as the *fund* [*fonds*] *of tertiary retention that is the vector of fantasies, hallucinations, collective retentions and protentions of every kind, characteristic of the epochs that are thereby formed*, and *linking* these epochs in the *never* achieved, *always* threatened and *necessarily* threatened unity of *Geschichtlichkeit*.

The questions of madness and dream belong to the broader question of imagination, inasmuch as 'humankind [...], each time that it imagines, finds itself, unawares, close to hallucination'.[20] In the conflict of interpretation that opposes Derrida and Foucault with respect to the *Meditations*, a dispute I will not claim to settle, Derrida's objections, however powerful they may be, do not seem to me to do justice, whatever may be their necessity, to the clarifications Foucault offered in 1972 in 'My Body, This Paper, This Fire'.[21] This text, Foucault's reply to Derrida, clarifies – by opening a question that I believe to be crucial – the scope and significance of the *meditation as dream*, and *as the dream dreamed by Descartes*. The dream that meditation would be will condition and nourish not only the project of 'classical reason' but the process of disinhibition examined by Sloterdijk (see §57).

In Pascal's epoch, 'institutions [...] make people delirious inasmuch as the latter confuse them with reality, and imagine that they must honour the great because they are great and not because they possess the signs of greatness'.[22] But if so, what then is our situation today, that is, in the 'epoch' of 24/7 capitalism that *destroys the common faculty of dreaming?*[23] Does there not appear, on this point, a *solution de continuité*,[24] a 'break in continuity' (that is, a dissolution), between the classical age and ourselves?

Do we not find ourselves in this epoch of the absence of epoch precisely insofar as this description characterizes a certain regime of ordinary madness striking, in an extra-ordinary way, not just the governors and the governed, that is, public powers, but also private powers, who have become immensely powerful by dissolving, through disruption, the difference between public and private? This difference, as we know, is a *major component of the psyche, that is, of the extravagances of all kinds of which it is capable*, and to which each of us can testify through our own dreams.

On this point, Foessel opens doors but he does not cross the threshold – he does not enter the labyrinths onto which they open. After noting that our world is saturated with images 'far more than was Pascal's', he adds that, in our societies, 'the trappings of power can no longer sustain their illusions. Doctors no longer wear "square hats", and judges do so less and less often for fear of seeming ridiculous. It is, rather, an epoch where the powerful are looked upon with what has become a widespread ironic gaze.'[25] But who are the 'powerful' involved here? Would it not be better to refer to the *powerless* – and to the question of *powerlessness*, which, perhaps, tends to incite distrust more than it does irony, if not contempt? And where this is a distrust and contempt in the face of what, perhaps, stems precisely from a new form of madness, as well as of barbarism, triggered by a new age of ὕβρις, which means in particular that power is itself transformed, and that the forms hitherto taken by power find themselves struck with impotence?[26]

As for these forms, which were those of public power, Foessel quotes Pascal, who narrates a fable to a young nobleman in the first of his *Three Discourses on the Condition of the Great*: 'a man is tossed by a storm onto an unknown island, whose inhabitants were having trouble finding their king who had gone missing',[27] and whom he resembled. The king having disappeared, 'all legitimacy has deserted the world. The inhabitants are in want of a king and they desire nothing more than to fill this void with their imagination',[28] by recognizing their king in the shipwreck survivor. 'It is their imagination, and not an act of will or reason, that makes of them a "people".'[29]

We could, however, present things somewhat differently, by positing that their imagination is constituted by tertiary retentions that produce collective secondary retentions and collective secondary protentions, supporting the libidinal *economy* that a society cannot do without – failing which, no longer being able to 'economize' its drives, it sees them unleashed, giving rise to the barbarism that will destroy it. It is this that Freud describes at the beginning of *The Future of an Illusion*.[30]

65. Hyperpower

The king cannot be a king without what Foessel and Pascal describe as his trappings. But these alone are not enough to underpin his power: they require other artifices, whose functions are not simply decorative or sumptuary. These other artifices are supports of processes of transindividuation, through which the retentional and protentional compromises that promise a common future are metastabilized – and they form epochs by unifying collective protentions. Understood from this perspective, *imagination, will and reason* cannot be neatly distinguished, still less isolated from one another.

This is the whole question posed by what Sloterdijk refers to as psychotechnics, which is today implemented by a psychopower that is ever more elaborate, and that is specific to capitalism. The industry of cultural goods heralded by Adorno and Horkheimer bore the germ of a 'new form of barbarism' precisely in that it *is* such a psychopower: as such, it leads to public powerlessness by transferring the symbolic power of political embodiment – of which 'trappings' are the most visible surface, and the most visibly contingent – to commodities, and through this merchandise to the merchants, that is, to private powers.

It is for this reason that Foessel can end his study in a rather sceptical tone, with a note about what came to be called the 'spirit of 11 January':[31]

> Most recently, the French authorities proclaimed the existence of a 'spirit of 11 January' in which national unity would (magically) reside. This was not without an element of collective psychosis from the moment it was forgotten that this 'spirit' is at best only a metaphor: eight-year-old children who did not have the chance to perceive it found themselves at the police station.[32]

Indeed, it is a question of knowing how the classical forms of power, which have become contemporary forms of powerlessness, apprehend the new forms of *hyperpower* granted in particular by the contemporary disruptive situation that constitutes

> another discourse of the Truth: that of the market, where, by right, anyone can claim the status of leader. In the collective imaginary, 'oil kings' take the place of kings draped in purple robes. It is true that proof of their greatness no longer depends on fantasies of birth right: it is generally in some unassuming garage that they claim to have had some commercial or technical idea that was to revolutionize the world.[33]

This narrative could no doubt benefit from some nuance. Long-term investments, led by military public authorities, conceived then concretized by universities endowed with massive resources, and channelled by an industrial politics that is also a wholly other culture of risk – which should be analysed with Sloterdijk and Fressoz – make these kinds of narratives possible, oscillating between fairy tales and marketing strategies and based on planetary-scale 'storytelling'.

The issue, therefore, is to return to the problem of the foundations of legitimate authority, and to understand to what extent [*mesure*] (and in what excess [*démesure*]) they allow a becoming-in-common that does not degenerate into explosions of ordinary and extraordinary madness perpetrated with countless *pharmaka*, which could, in such cases, end up being transformed into fatal means of destruction.

In order to grasp the *possibilities of contemporary delirium* – both ordinary and extraordinary – we must therefore place them in a *new history of madness*, where it is precisely a matter of thinking the immensely varied instances of ὕβρις in its primordial relationship to the *pharmakon*, that is, to tertiary retention, and in particular inasmuch as tertiary retention is what makes the process of disinhibition possible, which occurs at the very moment of coalescence of the conditions that give rise to the state form [*État*].

To explore these questions more deeply, let us turn back to the debate between Derrida and Foucault with respect to the relations between *dream, madness and reason*.

66. Madness, δαίμων, ὕβρις, Derrida (right up) against Foucault

Derrida summarizes Foucault by recalling that the 'great confinement' is a 'political decree' corresponding to the 'Cartesian decree', the latter defined as the 'advent of a ratio' that is intrinsically isolated from all madness, which 'would have been impossible for Montaigne, who was, as we know, haunted by the possibility of being mad, or becoming mad, in the very act of thinking itself'.[34] The 'Cartesian gesture' belongs, for Foucault, to the 'historical (politico-social) structure of which [it] is only a sign'.[35] We shall see how Derrida rejects this point of view, about which he makes the rather unnuanced claim that it stems from a 'structuralist totalitarianism'.[36]

Having recalled this, the essence of the debate between Derrida and Foucault concerns the *relationship between madness and dream*, and above all the *function* of dreams in the experience of doubt:

> the hypothesis of dreams is the radicalization [...] of the hypothesis according to which the senses could *sometimes* deceive me. In dreams, the *totality* of sensory images is illusory. It follows that a certainty invulnerable to dreams would be *a fortiori* invulnerable to *perceptual* illusions of the sensory kind.[37]

The issue is *certainty*, and the (classical, modern, Cartesian) *truth* that forms therein: 'certainties and truth that escape perception [...] are [...] of a nonsensory and nonimaginative origin. They are *simple* and *intelligible* things.'[38] These 'simple and intelligible' things, however, are not themselves dubitable in the dream. All *representations*, on the other hand, may be 'doubtful' or false, because they are 'composite things'.

Because their objects are such 'composite things',

> Physics, Astronomy, Medicine and all other sciences which have as their end the consideration of composite things, are very dubious and uncertain; but [...] Arithmetic, Geometry and other sciences of that kind which only treat of things that are very simple [...] contain some measure of certainty and an element of the indubitable. For whether I am awake or asleep, two and three together always form five, and the square can never have more than four sides.[39]

It is after the experience of the dream, and in the encounter with the indubitable that resists every test of being put into doubt, that, according to Derrida, Descartes will be led to a new extrapolation and radicalization, which is that of madness, but which for Derrida, contrary to the claims of Foucault, is merely the *extension* of the hypothesis of the dream.

Derrida wants to show, contra Foucault, that what is excluded is not, in and of itself, madness – it is *illusion* in general, and above all *sensory* illusion:

> *All* significations or 'ideas' of sensory origin are *excluded* from the realm of truth, *for the same reason as madness* is excluded from it. And there is nothing astonishing about this: madness is only a particular case, and, moreover, not the most serious one, of the sensory illusion which interests Descartes at this point.[40]

When Descartes formulated the hypothesis of 'a more common, more universal experience than that of madness: the experience of sleep and dreams',[41] he envisaged, *at the heart* of the life of the mind, *at the heart* of reason, 'the possibility of an insanity – an epistemological one – much more serious than madness',[42] which will lead, Derrida tells us, to the 'hypothesis of the evil genius':

> Descartes has just admitted that arithmetic, geometry, and simple notions escape the first doubt, and he writes, 'Nevertheless, I have long had fixed in my mind the belief that an all-powerful God existed who can do everything...' [...] [T]he hypothesis of the evil genius will evoke, conjure up, the possibility of a *total madness* [...] that will no longer be a disorder of the body, of the object, [...] but [...] will bring subversion into pure thought and into its purely intelligible objects, into the field of clear and distinct ideas, into the realm of mathematical truths that had escaped natural doubt.[43]

Madness is what, in the thought of Descartes, *originally haunts the very possibility of thinking*: this is what Derrida will attempt to show, counter to the Foucauldian thesis according to which classical thinking is constituted by the exclusion of madness, consigning it to the exterior of thinking in the *Meditations on First Philosophy*, which was published in 1641 and so would correspond to the 'great confinement' of 1656 described in *History of Madness*.

Having established his own, contrary thesis, Derrida then introduces the question of the relationship between *language and madness*, which is also examined by Foucault at the very end of *History of Madness*, where the issue is that of an opposition between *normality* and madness: 'Foucault says: "Madness is the absence of a work [*oeuvre*]." [...] Now, the work begins with the most elementary discourse [...]. The sentence is, by its essence, normal. It bears normality within it, that is, *sense*.'[44] Sense belongs to language, and so that which can manifest itself as non-sense, as madness, is *thus above all historical*.[45]

Having established this point, Derrida goes on to show that the point of the *Meditations* is not to establish what reason is in its relation to madness, but what the Cogito is, for which madness is, on the contrary, a *possibility* – which inhabits and conditions every possibility of thinking, and firstly of thinking rationally. For Descartes, it is a matter of thinking the Cogito (Second Meditation) such as it *is*, and as that which underlies everything that is, which is to say everything that makes sense, and hence, also, everything that thinks:

> The hyperbolical audacity of the Cartesian Cogito, its mad audacity [...] would consist in the return to an original point which no longer belongs to either a *determined* reason or a *determined* unreason, no longer belongs to them as opposition or alternative. Whether I am mad or not, *Cogito, sum*. Madness is therefore, in every sense of the word, only one *case* of thought (*within* thought).[46]

The Cogito is neither enclosed nor enclosing: 'Invulnerable to all determined opposition between reason and unreason, [the Cogito] is the point starting from which the history of the determined forms of this opposition, this opened or broken-off dialogue, can appear as such and be stated.'[47] The possibility of this opening up or breaking off, however, introduces the possibility and even the *necessity* of ὕβρις. The Cogito is, indeed, 'the impenetrable point of certainty [...] where the project of thinking this totality by escaping it is embedded. By escaping it: that is to say, by exceeding the totality, which – within existence – is possible only in the direction of infinity or nothingness.'[48] Such a possibility, which is therefore that of ὕβρις, is also that of madness: 'In this excess of the possible, this excess of law and meaning, over the real, the factual and the existent, this project is mad, and acknowledges madness as its freedom and its very possibility.'[49] It is inhabited by this ὕβρις that manifests itself as 'demonic':

> it is not human, in the sense of anthropological factuality, but is rather metaphysical and demonic: it first awakens to itself in its war with the demon, the evil genius of nonmeaning, by pitting itself against the strength of the evil genius, and by resisting him through reduction of the natural man within itself. In this sense, nothing is less reassuring than the Cogito at its proper and inaugural moment.[50]

The 'demon', however, the 'evil genius', is also the return [*revenance*] of the δαίμων of Socrates, which constitutes the inaugural experience of philosophy as such, and which, in Book VI of the *Republic* (509c), appears with the question of ἀγάθων (*agathon*) as that which 'exceeds the totality of the world' – ἀγάθων, the question of which lies also at the heart of the reading of *Phaedrus* in 'Plato's Pharmacy':

> This project of exceeding the totality of the world, as the totality of what I can think in general, is no more reassuring than the dialectic of Socrates when it, too, overflows the totality of beings, planting us in the light of a hidden sun that is ἐπέκεινα τῆς οὐσίας. And Glaucon was not mistaken when he cried out: 'Lord! What demonic hyperbole, "δαιμονίας ὑπερβολῆς"'.[51]

From one hyperbole to another, from Plato to Descartes, from the tragic 'demonic' to the 'evil genius' of monotheism, Derrida reminds us that at the bottom of thinking there lies the same madness, including and firstly for and in Descartes, and that this is ὕβρις as such: 'Such a ὕβρις keeps itself within the world. Assuming that it is deranged and excessive, it implies the fundamental derangement and excessiveness of the hyperbole that opens and founds the world as such by exceeding it.'[52]

67. Dream, structure, history and totality

It is here that, with an incredible audacity (the article appeared in 1963), and perhaps with a certain excess, with a kind of ὕβρις, Derrida refers to a 'structuralist totalitarianism' that 'would operate' in Foucault. This speaks volumes about the 'poststructuralist' question (and, as we shall soon see, shedding light on the years 1953–55 may be useful for an understanding of this debate on structuralism, and on the tendency towards 'totalization' that it manifests, a tendency which, here, Derrida opposes via Foucault[53]): 'Structuralist totalitarianism would operate, here, an act of confinement of the Cogito of the same type as the violences of the classical age. I am not saying that Foucault's book would be totalitarian [...]; I am saying that it sometimes risks being so in the implementation of the project.'[54] Behind the concerns with regard to Descartes, it is the whole Foucauldian project that is here challenged in advance – illuminating the relationship of Derrida to Foucault, which is a key dispute with respect to the question of history: 'I believe [...] that (in Descartes) everything can be reduced to a determined historical totality except the hyperbolical project.'[55] But what I myself believe is that this challenge [*contestation*], which is *in itself* incontestable, that is, necessary, obviously bringing with it an indispensable vigilance – which is highly characteristic of deconstruction in its attention to the necessities of that which it is nevertheless a matter of deconstructing – this challenge is also contestable *for us* (the readers of Foucault as well as of Derrida) in that *it leads Derrida into missing the heart of Foucault's project in reading Descartes*, which is to say where, in the *Meditations*, it is a question of the *dream* and of its *noetic bearing*.

Foucault had already tried, in 1954, and with Binswanger, to think the dream in its fundamental relation to anthropology, which was, however, *not yet* the anthropology of structuralism. Foucault plays out this question once again in his *History of Madness*, but in the meantime he has encountered the anthropological structuralism that Lévi-Strauss outlined in *Tristes Tropiques* in 1955.

Between the dream on one hand and 'structure' on the other hand, the issue is the play between:

- the first stage of the *epokhē* provoked by every shock in the organo-logical genesis of a new *pharmakon*, and which is *at first* a *problem*;
- and, following that, the second stage of the *epokhē*, which is the transindividuation of the first shock, and which *creates a question*.

This play is that of the dream insofar as it may or may not be realized. This is obviously not how it is *for Foucault* himself, neither in 1961 nor in 1954, but it is how it is *for me*, if not *'for us'* (my readers and I, and *you*), today.

Even if this doubly epokhal redoubling is obviously not the question asked in 'Dream and Existence', the 1930 article that the French public became aware of via Foucault's 1954 introduction (an introduction that is lengthier than the text it introduced, as would be the case for the long introduction to 'The Origin of Geometry' that Derrida would write at around the time *History of Madness* was being published), and even if this doubly epokhal redoubling is also not the question asked in *History of Madness* in 1961, nevertheless the *privilege given to the dream* in *History of Madness* and in its reading of the *Meditations*, which is very clear in Foucault's response to Derrida in 1972 (in the afterword to the second edition), does succeed in highlighting a *porosity between dream and meditation that is clearly fundamental with respect to my own thesis*, which finds its origin in the reading of Binswanger, who on this basis himself proposes elements of an *existential anthropology* that forms the foundation of a *Daseinsanalyse* that would be a *fundamentally new proposition in the history of psychiatry as well as psychoanalysis*.

In his critique of 'structuralist totalitarianism', what Derrida interrogates is the *Foucauldian concept of history. These questions themselves obviously relate to the doubly epokhal redoubling as the historical, factual and accidental condition of transindividuation* (as the test and ordeal of the accident in which technics fundamentally consists qua exosomatic organogenesis that can 'be otherwise than it is', as Aristotle defined technics in its artificiality[56]).

Of this doubly epokhal redoubling, the age of disruption would be one notable stage *as the absence of epoch, that is, the absence of dreams,* where ὕβρις would, as it were, by itself unleash itself, that is, unbind itself from all social systems, which are themselves, *above all, binding systems* [*systèmes de liaison*].

Beneath the *question of* ὕβρις (which also finds expression in the dream, which thereby communicates with madness), and behind the question of its *containment*, there lies the *question of history* as a *succession of transindividuated shocks, forming its epochs*, or its 'ages'.

68. The différance of madness

History, both as movement of collective individuation (*Historie*) and as knowledge of this historicity (*Geschichtlichkeit*), *presupposes* hyperbole, that is, ὕβρις as excess towards nothingness or infinity, and as indeterminacy, but *such that it ceaselessly composes with its determination*, that is, its *normalization*, which is a passage.[57] The principle of history involves a play between contradictory tendencies, the epochs of which are negotiations that mark passages and that *pass* through philosophy:

> historicity in general would be impossible without a history of philosophy, and I believe that the latter would in turn be impossible if we possessed only hyperbole, on the one hand, or, on the other, only determined historical structures, finite *Weltanschauungen*. The historicity proper to philosophy is located and constituted in this passage, in this dialogue between hyperbole and finite structure, between the excess beyond totality and closed totality, in the difference between history and historicity...[58]

This difference, which is a différance, is also the work of what I try to think starting from the Simondonian 'transindividual' as the process of transindividuation

> ...in the difference between history and historicity; that is, in the place where, or rather at the moment when, the Cogito and all that it symbolizes here (madness, derangement, hyperbole, etc.) pronounce and reassure themselves then to fall, necessarily forgetting themselves until their reactivation, their reawakening in another statement of the excess which also later will become another decline and another crisis.[59]

Here, the question of language returns inasmuch as it conceals the possibility of a revenance and contains not only the possibility but the *necessity* of madness – just as the jar of Pandora contained all ills:

> From its very first breath, speech, submitted to this temporal rhythm of crisis and reawakening, is able to open the space for speech only by enclosing madness. This rhythm, moreover, is not an alternation that

additionally would be temporal. It is the movement of temporalization itself as concerns that which unites it to the movement of logos.[60]

Now, this also amounts to the question of the relationship between genesis and structure, which will be the cross that structuralism will have to bear, and a fundamental theme for Derrida, from 1953 (*The Problem of Genesis in Husserl's Philosophy*) to 1968 (*Of Grammatology*) and beyond, passing through '"Genesis and Structure" and Phenomenology', delivered in 1959 and published in 1967 in *Writing and Difference*. An economy is established between these two tendencies that no dialectic can ever 'sublate', that is, overcome:

> crisis or oblivion perhaps is not an accident, but rather the destiny of speaking philosophy – the philosophy that lives only by enclosing madness, but which would die as thought, and by a still worse violence, if a new speech did not at every instant liberate previous madness while enclosing within itself, in its present, the madman of the day.[61]

Hence is manifested an *essential anachronism* of philosophical ὕβρις (of its fundamentally demonic tenor). Contrary to Foucault's claims in *History of Madness*, this amounts to the question of the 'madman within us' as the destiny of the finite thought that will impose itself on the basis of this reading of the *Meditations*: 'the reign of finite thought can be established only on the basis of the more or less disguised confining, humiliating, chaining and mocking of the madman within us, of the madman who can never be but the fool of a logos which is father, master and king'.[62] This *différance of madness* is *above all* an *economy*:

> *At its height*, hyperbole, the absolute opening, the aneconomic expend- iture [*dépense*], is always taken back and taken over into an *economy*. The relationship between reason, madness and death is an economy, a structure of différance whose irreducible originality must be respected. This wanting-to-say-the-demonic-hyperbole is not one want among others.[63]

What is concealed within it is, *indeed*, the very *possibility* and *necessity of wanting*. It is the *will* inasmuch as it is not opposed to the imagination:

> This wanting to say, which is not the antagonism of silence but its condition, is the original profundity of all will in general. Furthermore, nothing would be more incapable of grasping back this will than volun- tarism, for, as finitude and as history, this wanting is also a primary passion. It keeps within itself the trace of a violence.[64]

In other words, wanting stems from ὕβρις, and what we call the will is its derivative – but we have seen (and this is something to which we will return) that ὕβρις stems from an imagination that itself stems from the dream, which will take us back to Foucault, by projecting us *beyond* that which Derrida opposes *in* Foucault.

The will, then, keeps the trace of violence, and this also amounts to the condition and the question of *critique* as the *crisis of reason's madness* when it *truly* reasons, that is, the question of critique *conceived* (as a child is 'conceived') after this crisis, analysed after this synthesis, and vice versa:

> But this crisis in which reason is madder than madness – for reason is non-sense and forgetting – and where madness is more rational than reason, for it is closer to the wellspring of sense, however silent or murmuring, this crisis has always already begun and it is interminable. It suffices to say that, if it is classical, it is not so in the sense of the *classical age* but in the sense of the essentially and eternally classical, albeit historical in a very unusual way.[65]

Derrida concludes his reading of *History of Madness* thus: 'For what Michel Foucault teaches us to think is that there are crises of reason in strange complicity with what the world calls crises of madness.'[66]

10

The Dream of Michel Foucault

69. Dreaming and meditating with and according to Foucault

In his response – which will not arrive until nine years later[1] – Foucault argues that in the *Meditations*, contrary to Derrida's reading, dreaming and madness do not lie on the same plane. After recalling Derrida's argument ('a more common, more universal experience than that of madness', 'the madman is not always wrong about everything', madness 'affected only certain areas of sensory perception, and in a contingent and partial way'), Foucault maintains that 'Descartes does not say that dreaming is "more common and more universal than madness". Nor does he say that madmen are only mad from time to time and on particular points.'[2] In short, the dream is by no means a 'hyperbolical exasperation of the hypothesis of madness'.[3] On the contrary, it constitutes *the condition of possibility of thinking as meditation*, which is always *at the limit of not being able to differentiate*, and it *is* – that is, it thinks, it meditates – only *as* this limit, *at* this limit and only *at the price* of this limit: 'thinking about dreams, when one applies oneself to it, is such that its effect is that of blurring the perceived limits of sleeping and waking for the meditating subject at the very heart of his meditation'.[4] Like the demonic disturbance of Socrates, and like dialogical argument capable of *reactivating the forgotten in the anamnesic experience* in which *alētheia* consists,

> the dream disturbs the subject who thinks it. Applying one's mind to dreams is not an indifferent task: perhaps it is indeed in the first place a self-suggested theme; but it quickly turns out to be a risk to which one is exposed. A risk, for the subject, of being modified; a risk of no longer being at all sure of being awake; a risk of *stupor*, as the Latin text says.[5]

Risk thus appears here, too, in Foucault, but afterwards, and *as coming from the dream*. This cannot but be of interest to us.

It is, then, evident that *Foucault reads Descartes with Binswanger*

– for whom the dream is also the fundamental dimension of Dasein, and as such the starting point of every *Daseinsanalyse*:

> dreams may well modify the meditating subject to this extent, *but they do not prevent him, in the very heart of this stupor, from continuing to meditate, to meditate validly, to see clearly a certain number of things or principles*, in spite of the lack of distinction, however deep, between waking and sleeping.[6]

The dream as hypothesis in the exercise of doubt is for Descartes a *process* consisting of a series of moments,

> a possible, immediately accessible experience [that] is really and actually produced in meditation, according to the following series: *thinking* of the dream, *remembering* the dream, trying to *separate* the dream from waking, *no longer knowing* whether one is dreaming or not, acting voluntarily *as though* one were dreaming. [...] By means of this meditative exercise, thinking about dreaming [...] modifies the subject by striking him with *stupor*. [...] But in modifying him, [it] does not disqualify him as meditating subject.[7]

The reader will undoubtedly have understood that it is not a question, here, for me, of taking sides for or against Foucault or Derrida. The Derridian analysis that concludes his reading of Foucault is *clearly* necessary. And the response given by Foucault, in reaffirming the singularity of the question of the dream in relation to that of madness, is *just as necessary*.

The history of philosophy is not a series of *matches* going 'back' and 'forth' between competing athletes, in which one is able to defeat the other, and where it would be possible for us to choose one over the other: it is a *process*, for which those recognized as philosophers are necessary moments – but *never sufficient*. Rather than choosing between adversaries who are only expressions of adversity, it is a matter of *striking, more or less belatedly, an iron that is more or less hot* (being the *pharmakon* and its ὕβρις at this or that stage of grammatization).

This process – for *this, too,* is a process, inscribed in what we call history, *as a différance of madness as much as of the dream*, that is, as the *realization* through which the dream *fades away*, allowing the emergence of a waking state and *arousing other dreams* that may always turn into nightmares – this process is that of the doubly epokhal redoubling. And this is what, in our absence of epoch, fails to occur, that is, no longer continues on, as if we no longer have the ability or the knowledge to pursue this process.

This is what makes Florian suffer. And it is what makes us all suffer alongside him. For we, *like him*, are orphans. We can no longer be content to be 'Foucauldians', 'Derridians' or 'Deleuzians'. We are living the ordeal of nothingness, of *being nothing* – which means that we must become the quasi-causal bearers of what remains to come – 'if there is any' [*s'il y en a*], as Derrida often said at the end of his work and of his life.

70. From Descartes' dream to the bifurcation towards the Neganthropocene (the ὕβρις of philosophy itself)

In the *contemporary* context – which is no longer that of the 'sixties' – in our time, in which the question of ordinary madness is posed, in which it imposes itself and does so in an extraordinary way, and in a way that would undoubtedly have been unimaginable for Foucault and Derrida (for if it were otherwise, they would have been led to think otherwise), in the context of disruption as a stage of the process of disinhibition and ὕβρις that is clearly crossing a threshold, with consequences as unpredictable as, and *undoubtedly* much more transformational than, those to which the 'discovery of America' gave rise – in this context, we must profoundly reconsider the question of ὕβρις and the question of its relationships to the dream and madness from the perspective of the dream of Descartes himself.

It is well known that to 'make ourselves [...] masters and possessors of nature'[8] is one way in which we might summarize the dream of Descartes: it is Descartes himself who formulates it thus. This 'dream' is a protention that will become collective, precisely in that Descartes will transindividuate the whole of philosophy, that is, the whole of science, taxonomic knowledge, and, progressively, morality, politics and economics. All this, and then art, starting from this foundation of the Cogito that *is* the *ego*, will become 'modern', that is, characteristic of the Modern Age – all this stemming, furthermore, from *grammatization*, which does not wait for Descartes but on the contrary precedes him.

The dream, which is thus the redoubling of this grammatization, begins neither with the *Method*, nor with the *Meditations on First Philosophy*: it begins in *Rules for the Direction of the Mind*.[9]

When we decide to put this dream into question, by considering it as a point of origin of what will then be *concretized* as the *end* of the classical age and the advent of the Anthropocene (through a process of transindividuation of remarkable complexity), we proclaim that *this dream will have been madness*. As for myself, what I will now maintain is that this madness was made possible by the new ὕβρις in which consists *writing, reading and rereading, conceived as the analytical conditions of thinking*, and in such a way that *calculation* becomes, as *ratio*, the *mathesis universalis* constituting the method of any rational philosophy.

This dream, which with Leibniz *will continue* beyond Descartes, *will be concretized* by those who will materialize it by *realizing* these ideas

– but *also* by *de-realizing* them, that is, by limiting them, and sometimes, and increasingly, by pharmacologically inverting them – with the analytical machine, the difference machine, tabulation machines, informatics and finally digital technology (as the set of reticulated computing machines) in the service of the data economy. Noam Chomsky is characteristic of this inversion.

I refer to *inversion* because, after Leibniz and with Kant – who undertakes his philosophy so as to free himself from the dogmatic thought of Christian Wolff, himself a disciple of Leibniz, who immeasurably extends the laws of speculation, this excessiveness being for Kant the ὕβρις *of philosophy itself*[10] – the analytic undergoes renewed reconsideration. This reconsideration of the analytic results in a *differentiation of the faculty of knowing* such that *the understanding and reason become the two inseparable but irreducible dimensions* of *analysis* and *synthesis* (where both are required for any true knowledge).

This complex process, where the relations between analysis (in Descartes' sense) and synthesis (in Kant's sense) continuously evolve in such a way as to *realize* the dream of Descartes – and, in so doing, to de-realize it by concretizing it through what, today, seems bound constantly to turn into a nightmare – is made possible by those statements by Descartes that establish the Modern Age of philosophy as the transindividuation of a new stage of grammatization induced by the proliferation of grammars, dictionaries, account books and forms of money.

It is Max Weber who will point to the banal phenomenon of merchant account books, showing that the emergence of accounting will penetrate every dimension of life, as accounting ratios come to be inscribed at the heart of the process of rationalization accompanying the 'spirit of capitalism'. Clarisse Herrenschmidt will then highlight the dissemination of the culture of money, which, through its circulation, comes to be placed into everyone's hands.

Sylvain Auroux has shown that the practice of these linguistic tools that are grammars and dictionaries conditions what, five years after the *History of Madness*, Foucault will describe in *The Order of Things* as the *epistēmē* of representation, of which the logic of Port-Royal (1662) will for him be the acme – which means that the 'classical age' is fully constituted on the basis of this 'general grammar', around the same time, then, as the 'great confinement' to which Foucault will then no longer refer.[11]

This logic of representation would not be possible without that which precedes it, namely, the ortho-graphic grammatization required by printed writing with movable type, which leads to the global expansion of what Auroux describes as 'extended Latin grammar' – just as *generalized tabulation*[12] required the appearance of *pagination*.

This stage of grammatization that is extended Latin grammar – informed by and taught via the linguistic technologies that emerge from ortho-graphic writing, and that stem from what, in 'Faith and Knowledge', Derrida called 'globalatinization' [*mondialatinisation*][13] – conditions the classical *epistēmē* of which Rules Fifteen and Sixteen for the direction of the mind are major expressions. Through these rules, Descartes paves the way for the analytical functions of the understanding to be organologically delegated to algorithmic automatons, and, in so doing, he frees the way for the unprecedented possibility of ὕβρις *typical of the Anthropocene.*[14]

It is this ὕβρις that constitutes the essence of the Modern Age in its relation to madness, and this is what Foucault and Derrida both ignore in concert – namely, the pharmacological and retentional condition of the extraordinary ordinary madness that now afflicts our disrupted world.

This *pharmakon*, as it is considered in Rules Fifteen and Sixteen, establishes the *factors and the factical (artificial, instrumental) conditions of the obvious and the simple* that constitute the first object of analysis, which, according to Descartes, resists doubt in the case of madness or dreaming – but not in the hypothesis of the evil genius. Foucault, like Derrida, ignores these organological conditions of the appearance of the simple, with respect to what it contains of ὕβρις *and* with respect to what it contains of the dream. They both ignore the question of the conditions of possibility by which the dream can pass into actuality, just as they ignore the presence of artifices in dreams that Freud so emphasizes (and yet, paradoxically, Freud himself pays no attention to this), and the transformation of instinct into drive by the mutability of the latter, responding and corresponding to the detachability of the artificial organs that themselves result from these dreams, and so on.

71. The Cartesian sources of disruption

In the Cartesian experience of doubt, which in some way prefigures that of the phenomenological *epokhē* as a bracketing of the world, what remain indubitable, at the stage of methodical doubt that is the 'hypothesis of the dream', are 'simple and intelligible' things: even in dreams, two plus three always makes five, and a square will always have four sides.

But the things in question are *in fact* 'simple' and '*intelligible*' only insofar as they are isolated *as such*, dis-engaged from the phenomena in which they present themselves *as* simple and intelligible, that is, precisely, *beyond* phenomenality alone – and as *constituting* the latter, precisely as its *elementarity*, that is, as its *irreducible granularity, itself un-decomposable*. For the square to *present itself* as 'un-decomposable' – as this *geometrical* figure that is a square (that is, essentially and in-dividually a square) such that it is *composed* of its 'four sides' – it is necessary that the notions of figure and, ultimately, geometry, be themselves evident and 'un-decomposable'.

This is undoubtedly what Descartes expresses when he distinguishes geometry and arithmetic from physics and the 'taxonomic' sciences. But this distinction *presupposes* this specific attentional mode – one *composed of the specific retentions and protentions* that are described by, precisely, Rules Fifteen and Sixteen. These are, however, the givens [*données*] of *noetic artefactuality* constituted by alphabetical writing, in relation to which, in addition, the ortho-graphy required by the printing press generalizes normed practices. And these practices will become the analytical basis of what, as universal characteristic, and more generally as algebra, will lead, through applied mathematics and its instrumentalization, and by constituting the universal element of the new industry that is the data economy, to the *algorithmic decomposition of the faculty of knowing*.

This algorithmic concretization of the Cartesian dream is what – as, on the one hand, instrumental and automated understanding, and, on the other hand, reason *lost*, because exceeded by this understanding that outstrips and overtakes it – *systemically* short-circuits reason, which is lost *by this very fact*. This short-circuiting of reason amounts to a kind of *metaphysical and speculative disinhibition* that proves literally *in-conceivable*, and, today, it is what allows calculation to be deployed without limits, and as the *fundamental principle of the age of disruption* (which is also to say, of transhumanism).

Disruption is an *incommensurable* stage of disinhibition. Its accomplishment occurs in a capitalism that has become *purely, simply and absolutely computational* – that is, it is accomplished as *absolute* ὕβρις: absolutely *freed of any and all limits*, bringing nihilism to fulfilment as the completion of the Anthropocene. This in-com-mensurable stage of disinhibition, paradoxically founded on a certain conception of measure (μέτρον), and now founded on calculation, results in an *immense process of demoralization* that is also a *massive denoetization*, an agent of *systemic madness*, and more than just systemic (or functional) stupidity.[15]

'Simple and intelligible' things are productions of the *passive discretization* that is grammatization. Nothing is more difficult to isolate than a simple thing, nothing is less immediate than the 'simple': it is through a tertiary attentional form opening the possibility for discreetly discretized traces to be 'examined at leisure'[16] that the simple can be dis-engaged from the composite. The tertiary retentional basis of Cartesian analysis and the *mathesis universalis*, which predates Descartes, conditions his analytical approach, which is founded on the 'indubitable' simple and intelligible elements produced by a 'passive synthesis' that is not psychological, but *techno-logical*.

It is obviously not a matter, here, of casting doubt on the indubitable dimension of these elements, but of showing that we can access their simplicity only on the condition of passing through technological passive syntheses such as those that Husserl described in *The Crisis of European Sciences* in terms of the *occluded and non-intuitive development of algebra*. The latter contained such syntheses from the earliest moments of analysis, pre-ceding the techno-logical grammatization that has today developed into the automated function of the analytical understanding cut off from all reason – that is, from all intuitive and speculative (in Whitehead's sense) synthetic possibilities.

This is something on which we must insist, because it evidently involves not just the possibility but the *necessity of a machinic* ὕβρις that is nothing but the concretion of the fact that ὕβρις in general stems from this *passive violence*, if you will. Consequently, the violence referred to by Derrida in terms of the madness of language – which consti-tutes the *possibility* of language, a theme that will turn up again in *Of Grammatology* with respect to writing and Lévi-Strauss – is equally that of an organological supplementarity requiring specific analysis. But such an analysis is what Derrida never undertakes, concerned that to do so would reintroduce the oppositions between nature and technics, animal and human, and so on.

Such a *neutralization* of the question, however, while being highly detrimental to the reactivation of the concept of the supplement, turns out also to be highly valuable in the context of disruption. Here, the conclusion that Thomas Berns and Antoinette Rouvroy draw concerning concepts derived from Deleuze, Guattari and Simondon is equally applicable to Derrida and Foucault: the implementation of these concepts somehow drains them of their content. In other words, it is a matter, here, of making a leap into the ordeal of this new kind of 'kenosis'.

At the heart of these questions lies that of the relations between the psychic individuation of Descartes and the technical individuation of hypomnesic tertiary retention, of which the Republic of Letters[17] is the transindividuated consequence as the collective individuation constitutive of an epoch referred to as the Modern Age. In other words, Derrida's objection to Foucault with respect to history is one question, but there is also an objection that must be made to Derrida himself – a necessity demanded so as to be faithful, in a way, to his own objection.

We must reconsider the Foucauldian question of structure in terms of the question of the doubly epokhal redoubling, which Derrida made thinkable with the supplement, but which he did not himself think. This amounts to the question of general organology. General organology is itself thinkable only in terms of a pharmacology that, too, goes back to Derrida, but which does not remain tied to the letter of his thinking. That this is the case derives firstly from the fact that this question is Socratic, not Platonic: we must distinguish these two figures, here, precisely on this point, and this is something that Derrida did not do. But it is something that we will develop in what follows.

72. Foucault, Asclepius and the death of Socrates

Disruption, inasmuch as it amounts to the epoch of the absence of epoch and the contemporary form of madness – a madness and disruption rooted in the history and archaeology of the Anthropocene – is what Foucault and Derrida could not have thought, but which they can aid us in thinking.

In his original preface to *History of Madness*, Foucault ignored the tragic Greek question of madness as ὕβρις, which is all the stranger given that what he claimed to be offering was precisely a tragic approach to madness. It is for this that Derrida reproaches him, and rightly so.

But the Derridian critique of the Foucauldian reading of Descartes does not do justice to the questions initiated by Foucault through an approach whose great originality derived from his exhuming from archives the conditions of subsistence and existence on the basis of which new consistences form. And this was carried out by Foucault as a kind of reconstruction of the material processes involved in the doubly epokhal redoubling – but this would not be truly developed until 1966.

From the *History of Madness* to the seminars held and published as *The Courage of Truth*, Foucault was led back to Graeco-Roman antiquity and eventually to the Greece of Socrates, passing through the questions of the *epistēmē*, disciplinary societies and biopower. When he came to the end of this journey, Foucault accorded major status to the question of writing – in this instance, epistolary writing and Seneca's discourse on reading texts, such that *from such practices, understood as techniques of the self*, powers of conversion arise.

It is on the basis of such processes of conversion that circuits of transindividuation are formed, through which the time of a doubly epokhal redoubling comes to be established, where the establishment of such epochs also amounts to the constitution of what Foucault will later call regimes of truth.

The processes of conversion establishing such regimes of truth are at work from the very beginning of philosophy insofar as it is firstly a *therapeia*, a care that the noetic soul takes of itself, a *tekhnē tou biou* of which Socratic dialogism is the first practical expression. And Foucault stresses, first in *The Government of Self and Others*, then in *The Courage of Truth*, that the *initiation* of such practice is a *parrhēsia*, such that, practised within a primordial friendship binding the speakers, it is bound to *wound* them – it is a *trauma*:

> *Parrhēsia* [...] involves a strong and constitutive bond between the person speaking and what he says, and opens, through the effect of the truth, through the effect of the wounds of truth, the possibility of rupturing the bond between the person speaking and the person to whom he has spoken.[18]

The truth that arises from out of the 'effects of the wounds' of *parrhēsia* is always *painful*, and, in practising it, the parrhesiast always courts *risk* – and in particular the risk that, as in the case of Solon, the city as a whole may consider him mad, a city that he wounds as this very whole.[19]

This pain comes from the *fatum* on the basis of which one must speak truthfully and frankly, so as to overcome (always temporarily and only ever intermittently) the fundamental ambiguity of the mortal condition. Is this condition a disease? Is it possible to be cured of such a malady? These questions constitute the framework of Foucault's final meditations, when he was himself ill and approaching his end,[20] and turned to the end of Socrates at the end of *Crito* – and to the moment when Socrates asks Crito to make a sacrifice to Asclepius on his behalf, as he drinks the hemlock.

During this seminar of 15 February 1984, Foucault attempts to interpret Socrates' end, and his sacrifice to Asclepius, by starting from the interpretation proposed by Georges Dumézil in *Le Moyne noir en gris dedans Varennes*.[21] I have maintained elsewhere and on several occasions[22] that, if Socrates sacrifices to Apollo at the moment when he drinks the poison, it is firstly because Asclepius – as a son of Apollo and a mortal,[23] and as himself a hero who becomes an immortal but who is struck down by Zeus for having wanted to give immortality to mortals – is the god of poison, which is *also* to say of medicine, of therapeutics.

> He was given the blood which had flowed in the Gorgon's veins by Athena, and while the blood from its left side spread a fatal poison, that from the right was beneficial, and Asclepius knew how to use it to restore the dead to life. [...] Zeus [...] feared that Asclepius might upset the natural order of things and struck him with a thunderbolt.[24]

Neither Dumézil nor Foucault give any thought whatsoever to the question raised by the fact that Socrates, in the moment of drinking the fatal poison, chose to offer a sacrifice to the god of poison. Following a long tradition that passes through Wilamowitz, the question posed by Dumézil and taken up by Foucault is instead to ask what illness it would be a matter of treating, for which Socrates, by his sacrifice, would be thanking Asclepius. They ask if life itself could be this illness, or whether it is some other illness that would be found within life, but that would not amount to life itself.

What is surprising here is that Foucault gives no consideration to Canguilhem's perspective, which Canguilhem himself takes up from Nietzsche, according to which noetic life is essentially a life of healing. But this life, *this healing* – through an irreducibly factical, artificial and therefore fictive and precarious normativity, because it is founded in technicity as the organological and pharmacological condition of this form of life – *turns these illnesses into new normativities that are also new regimes of truth. In doing so, however, it is bound to produce new forms of infidelity* – and, in this instance, infidelities of an organological and pharmacological milieu that is itself, therefore, always changing.

Foucault does quote a passage from *The Gay Science* in which, as he recalls without seeming to pay it much attention, Nietzsche himself raises the question of poison[25] and gives it fleeting consideration as an interpretation (a possibility he immediately dismisses). Despite this, however, Foucault never explores whether the filiation of Asclepius with Asclepias, milkweed, could have something to do with the *pharmakon*, even though he had shown that what, for Socrates and Plato, was a *pharmakon* (meaning by this something rather different from my own view), had become, for Seneca, a therapeutic practice.

Derrida, who contested the two pages of the *History of Madness* devoted to the *Meditations* of Descartes and the four lines on the Greeks and madness that appeared in Foucault's first preface, will himself exhume the question of the *pharmakon*, but he will never make it the object and the condition of a 'curativity' that lies at the basis of reason and does so through the ordeal of ὕβρις that the *pharmakon* can also engender in the form of madness.

The examination of such questions inevitably leads us back to the status of μηχανῆ (*mēkhanē*) in the two lines of Pindar quoted by Valéry in *Cimetière marin*, which are translated by Alain Frontier in a way that highlights the erasure of the question of *mēkhanē* in Aimé Puech's translation, which is utterly inscribed in this forgetting of the *pharmakon* that is a repression and a denial, both a *negation* [*dénegation*] and a *disavowal* [*déni*].

It is the pharmacological condition that governs the transductive relation that constitutes gods and mortals endowed with that divine fire which in the hands of mortals becomes the *pharmakon*, that is, *always ambiguous*, which is not the case for such *tekhnai* in Olympus. Unlike mortals,

> the gods [...] attain their goal *immediately*, [...] they are already in possession of that towards which men tend and strive. It is enough for Artemis to want to reach his target for it to be reached at once: there is no need to aim. There's no need even for a bow. The bow is there only for decoration. The prerogatives of the gods thus represent this ideal, virtual point (situated at infinity, so to speak) towards which a mortal (in the best case) can *orient* his action, by utilizing all the possibilities offered by the tools that lie at his disposal according to his field of excellence. To claim to have reached this point, situated at infinity, would be an absurd and vain dream (a 'sacrilege', says Pindar). [...] This impossibility [...] does not prevent mortals from acting, nor from having intelligence, ingenuity or the personal talent to develop themselves and grow.[26]

In this *transductive relation between mortality and immortality, constituted by the theft of fire*, lies the possibility of ὕβρις, at once as crime,
madness *and noetic fate.*

   This is what will have been denied since the beginning of the
'metaphysics' of the philosophers – Foucault and Derrida included:
therein lies the source of their many misunderstandings (which are
repeated by little Foucauldians and little Derridians in the quasi-noetic
menagerie of scholarly and literary monkeys and parrots).

73. Dream and anthropology in Foucault, reader of Binswanger

In the conclusion of 'Dream and Existence', which Foucault will
introduce to the French public through a translation by himself and
Jacqueline Verdeaux, Binswanger writes that, faced with the 'selfsame'
[*même*] that 'hits' [*arrive*] the dreamer in the dream 'he knows not how',
an

> individual turns from mere self-identity to becoming a self or 'the'
> individual, and the dreamer awakens in that unfathomable moment when
> he decides not only to seek to know 'what hit him', but seeks also to strike
> into and take hold of the dynamics in these events, 'himself' – the moment,
> that is, when he resolves to bring continuity or consequence into a life that
> rises and falls, falls and rises.[27]

The individual becomes starting from his dream. And he does so as a
movement in which the individual becomes, by taking hold, by intervening – in awakening – one who *sometimes rises, sometimes falls*: living
intermittently between the high and the low, like those birds that are for
Binswanger essential examples of dreams, and that should also be related
to the eagle or the vulture devouring the liver of Prometheus – that is,
the organ of black bile.

   In this *relation to the dream* that is also a *relation to the gods*, to that
which is *most high*, and which is a *transduction of sleeping and waking*,
the dreamer *makes* something:

> That which he makes, however, is not life – this the individual cannot
> make – but history. Dreaming, man [...] 'is' 'life-function'; waking, he
> creates 'life-history'.[28]

> [W]e do not know where life and the dream begin.[29]

These considerations describe the horizon of the scene set by Miyazaki
on the basis of *Cimetière marin*, *The Magic Mountain* (two texts that
he read also through Tatsuo Hori's *The Wind Rises*) and the life of Jiro

Hirokoshi. It is clear that they haunt Foucault, too, at the moment when he reads Descartes' *Meditations* and situates the dream not on the side of madness, but as coming from existence itself: 'Between the sleeping mind and the waking mind, the dreaming mind enjoys an experience which borrows from nowhere its light and its genius. [...] But the theme of original dimensions to dream experience [...] can easily be discerned as well in Cartesian and post-Cartesian texts.'[30] In other words, if, according to Foucault in 1962, Descartes has stopped listening to the madness that still reigns in the Renaissance, he nevertheless maintains a relationship with the dream that is the very issue taken up by Binswanger in 'Dream and Existence'. And, among the post-Cartesians who still participate in this *oneirology*, there is also Spinoza:

> in terms of dreams, premonitions, and warnings, he distinguished two sorts of imaginings: those that depend solely on the body [...] and those which give sensory body to ideas of the understanding [...]. The first form of imagination is encountered in delirium, and makes up the physiological fabric of the dream. But the second makes of the imagination a specific form of knowledge.[31]

Noetic existence, that is, existence capable of meditating, and not limiting itself to subsisting, is drawn by the attraction of those consistences of which the gods as read by Alain Frontier are markers for these *noetic patients* [*malades*] who are mortals worthy of the name: as thought by Canguilhem, noesis is in fact organological illness [*maladie*]. Canguilhem: who inspired Foucault, but where this is something that the latter seems constantly to forget (to the point that on occasion one wonders if he has actually read him).

The dream can be divine only because, day and night, it transduces the nocturnal and the diurnal. This transduction, which ties mortals to immortals, that is, existences to consistences, to eternities, is constituted in and by meditation.

The issue, however, is that the dream, insofar as it can be realized – and insofar as thinking is what thinks the conditions of the realization of the dream that it is, in which it *consists* – being pharmacological through and through, always brings with it ὕβρις and hence madness, and, in so doing, brings nightmarish inversions that are not just always possible, but always imminent.

This is why *knowledge* is perpetually required to contain the ὕβρις that it itself contains more or less poorly or well, and where knowledge amounts above all to a therapeutics – and firstly of techniques of the self.

This is why, if Foucault's gesture is not soluble into Derrida's objection (the dream is in continuity with madness, meditation always in any case harbours the imminent madness of the evil genius), this objection by Derrida must nevertheless be meditated upon, but so as to do justice to

Binswanger's question of the dream as the *ability to make history* beyond any 'life-function'. It seems that Derrida ignores the new questions this brings with it – organological and pharmacological questions that we must resituate in the still uncharted territory of noetic exosomatization.

74. Entropocentrism and neganthropology

Just as it is highly instructive to read the text written by Derrida in 1953 while still a *normalien*, on *The Problem of Genesis in Husserl's Philosophy*, so too it is striking to read the 'Introduction' that Foucault published in 1954 for the translation of Binswanger's 'Dream and Existence'. In both cases, the visions and the discourses are almost the opposite of what will later come to constitute these authors named Foucault and Derrida as we know them.

In the 'Introduction' to 'Dream and Existence', we see how the young Foucault's project is to construct a new anthropology, located within a kind of square whose corners are Freud, Binswanger, Heidegger and Husserl, and where the central question of the dream paves the way for Foucault's position with respect to the *Meditations* in *History of Madness*.

Foucault's works must be taken up in a totally new way, starting from a critique of the *entropology* evoked by Lévi-Strauss at the end of *Tristes Tropiques*[32] – a work written after *The Elementary Structures of Kinship*, the *Introduction to the Works of Marcel Mauss* and the texts gathered together in *Structural Anthropology 1*, and where it seems that the young Foucault allows himself to be diverted by this set of publications, as he becomes attracted by their structuralist vision. These works of Lévi-Strauss and of a triumphant anthropology built on the Saussurian method will mark the course of intellectual debate for the next fifteen years, dominated not just by Lévi-Strauss and his reference to Saussure, but by Althusser, Lacan and Barthes.

In 1954, the young Michel Foucault wanted to create a new anthropology founded on a therapeutics of the psyche, on a psychiatry and a psychology – where this anthropology and this psychology would constitute a therapeutics and a clinical practice, through a more or less 'existentialist' deviation from Freud with respect to the status of the dream, itself a primordial resource of meditation. At the other end of this immense journey, some thirty years later, Foucault will pose the questions of *parrhēsia*, the hermeneutics of the self and techniques of the self as conditions of meditation, to which he seems to trace back all primordial questions.

All this, for we who find ourselves plunged into the depths of disruption, which is also the *ordeal of denoetization that nihilism ultimately constitutes*, opens the path, which has not been traced but which is indicated (like a dream within a nightmare – as, in a way, in

Chris Marker's *La Jetée*, as seen by Jonathan Crary[33]), to a neganthropology – beyond the anthropology and humanism that in 1966 Foucault will attempt to overcome.

The positive anthropology of Lévi-Strauss and structuralism was constructed by systematically ignoring Leroi-Gourhan, who *is* mentioned by Derrida as well as by Deleuze and Guattari in the elaboration of their respective enterprises – hence referred to as *post*structuralist. A rigorous approach to exosomatization in the absence of epoch that is the disruption – as the Anthropocene reaching its extreme stage by generalizing the *entropology* in which this anthropocentrism, in fact, consists as *entropocentrism* – requires us to embark on the path of neganthropology. The question, for this neganthropology, is ὕβρις – where this whole question must be conceived in a tragic way, because it stems from a *problem of fact*.

This does not mean that we should return to a tragic way of thinking in the same manner as the Greeks. It means that, today, it is a matter of reopening – urgently – the hermeneutic question, as a *hermeneutics of the madness of the human that 'does not exist yet'*, as Jean Jaurès wrote on 18 April 1904 in the editorial for the first issue of *L'Humanité*. This is what we must set in opposition to the totalizing and pre-totalitarian ideology of the transhumanists and neo-barbarians.

In 'My Sunday "Humanities"', Derrida comments on this statement by Jaurès, which takes humanity as being first and foremost what, not existing, *has* the faculty and *is* the faculty of dreaming itself as that which 'does not exist at all yet or [which] barely exists',[34] and, in so doing, of *thinking itself as that which inscribes into becoming [devenir] the bifurcations of its future [avenir]*. As the rest of the editorial makes clear, what Jaurès believes in is the *realization* of this dream. He believes this dream will eventually be realized, but he ignores the fact that this realization is always also a de-realization: this is what forever separates us from him and from his epoch, whatever denials persist in this regard. This is the significance of the first paragraph of the editorial:

> The very title of this newspaper, in its breadth, marks out exactly what our party proposes. It is, indeed, the realization of humanity for which all socialists work. Humanity does not exist at all yet or it barely exists. Within each nation, it is compromised and fragmented by class antagonism, by the inevitable struggle of the capitalist oligarchy and the proletariat. Only socialism, by absorbing all classes into the common ownership of the means of work, can resolve this antagonism and make of each nation, finally reconciled with itself, a parcel of humanity.[35]

Derrida, unlike Jaurès, sees in this non-existence, which gives humanity its consistence, the processual *différance* of a promise remaining always to come, constituting what I try here (and in *Automatic Society*) to think, by reading *Derrida beyond Derrida* [*hors de lui*], as a neganthropic potential.

This is, indeed, a matter of reading Derrida *outside* the framework of 'deconstruction' as he defined it: for Derrida, différance would amount to negentropy in all its forms, that is, life in all its forms, without the need to specify what constitutes noetic différance. What is at stake in the words of Jaurès, on the other hand, is noetic différance, that is, exosomatized différance, distinguishing itself from mere bio-logical life, constituting knowledge that, necessarily exteriorized, is also exposed to the degradation of proletarianization (as the *Grundrisse* describes it).

We no longer believe (in this *we* I include Florian, but also, believing I may do so on their behalf, the members of Ars Industrialis) that socialism will *resolve* the tragic pharmacological situation, and in truth nobody truly believes in this anymore. We *do believe*, however, in the *possibility* of a renewal of noetic life that would make it more care-ful, more attentive and more worthy. And we assert that this is possible simply because it is *absolutely necessary*, and *therefore* rational. That a renewal of noetic life is possible does not mean that it is probable: the possible is often so improbable that it presents itself before anything else as the impossible.

This impossible can and *must* be realized, and as the dream of the improbable: to fail to achieve it would be to condemn non-inhumanity to disappear, either into an oligarchy who would dominate a nano-bio-technological anthill, in the sense envisaged by Leroi-Gourhan,[36] or, more probably, into an irreversible increase in the rate of entropy such that life itself will find itself threatened – at least in its more highly evolved forms.

Having posited that such are the terms of our problem, we must turn to the question of the dream in politics and the question of the politics of dreaming – where dreams would be the resource of any neganthro-pogenesis, that is, any exosomatic organogenesis *such that it would preserve its future by maintaining its noetic capacity*. The expression 'politics of dreaming' would no doubt sound pleasing to many fools who do not hesitate to refer to 'creativity' – for example, to the 'creative economy'.[37]

The deterioration of the place of dreaming in politics – which falls within that 'realism' denounced by Kant in *Critique of Pure Reason*[38] (see my commentary in *Automatic Society, Volume 1*[39]) – consists in claiming that promises (or dreams) are binding only on those who believe in them:[40] this pejorative deterioration amounts to a form of *denial*. What it denies is precisely the fact that dreams are the origin and the future of exosomatization as noetization, even if they can always turn into nightmares. And it is this fact that necessitates – as the thera-peutics of this pharmacology that is the organology resulting from an organogenesis of which politics is the organization – the call for a politics that would not be dissolvable into the law of the market. For the latter can only extrapolate calculabilities and computations, whereas politics must know how to manage bifurcations that are incalculable because they are improbable. It is from this perspective that we must reread and reconnect *The German Ideology* and the *Grundrisse*.

The betrayal of dreams insofar as they are the resource of all noesis and all neganthropogenesis is what, in the last three decades, particularly in France, has engendered an immense demoralization. Beneath the disruptive horizon, this demoralization seems to prevent any opportunity for an eventual peaceful bifurcation – as if what should be inscribed in becoming, and inscribed as the future of *Neganthropos*, can no longer be conceived except on the condition of passing through the experience of extreme violence.

Part Three

Demoralization

11

Generation Strauss-Kahn

75. The collapse of the 'American way of life'

In an article that reported and commented on the double murder committed by Vester Lee Flanagan and the statements he made just before his *passage à l'acte*, when he declared, 'I've been a human powder keg for a while … just waiting to go BOOM!!!!', the *Washington Post* pointed out that, from the beginning of 2015 until 26 August 2015, the day after the murders, there had been 247 mass shootings in the United States: more than one per day.[1]

Shortly thereafter, various articles reported that in Palo Alto, the small but famous city in the northern Silicon Valley, 'the 10-year suicide rate for the two high schools [in the area] is between four and five times the national average. [...] Twelve percent of Palo Alto high-school students surveyed in the 2013–14 school year reported having contemplated suicide in the past 12 months.'[2] Over the previous five years, eleven teenagers from Palo Alto threw themselves under the Caltrain that links San José and San Francisco.

On 9 November 2015, Paul Krugman published an article in the *New York Times* entitled 'Despair, American Style'. He pointed out that the life expectancy of white middle-class Americans has been in decline since 1998, while the suicide rate has increased considerably, as has self-destructive behaviour involving voluntary intoxication through the use of various substances, including heroin:

> Basically, white Americans are, in increasing numbers, killing themselves, directly or indirectly. Suicide is way up, and so are deaths from drug poisoning and the chronic liver disease that excessive drinking can cause. We've seen this kind of thing in other times and places – for example, in the plunging life expectancy that afflicted Russia after the fall of Communism. But it's a shock to see it, even in an attenuated form, in America. [...] There have been a number of studies showing that life expectancy for less-educated whites is falling across much of the nation. Rising suicides and overuse of opioids are known problems. [...] But what's causing this epidemic of self-destructive behavior?[3]

The 'American way of life' is no longer what it was – and, in truth, its days are over: while the children of the wealthy inhabitants of Silicon Valley are haunted by suicidal fantasies, and are four or fives times more likely to act them out than those in the rest of the country, the white middle class, whose suicide rate is steadily increasing, consumes heroin, and in particular in rural areas, where this *pharmakon* can be obtained for ten dollars a hit.

Addiction to hard drugs no longer just afflicts those on the margins, or impoverished African-American populations, and this was a subject of debate during the run up to the 2016 American presidential election: 'The explosion of [heroin] use – referred to in the United States as an epidemic – has become a public health issue and is being widely discussed in the American media. [...] In September, Hillary Clinton [...] announced a $10 billion plan to stem the rise of heroin use.'[4] By 2011, there were already 250,000 medical emergencies 'related to heroin abuse'. By 2013, 'more than 500,000 Americans had reported using heroin in the last few years, an increase of 150% in six years. [...] In Montgomery County, [...] 127 overdoses were recorded in 2014 alone, and heroin-related deaths have increased by 225% since 2011.'[5] After the ravages of more than a decade of the 'conservative revolution', and after the birth in 1993 of the World Wide Web, a new model of development began to spread across North America, and was then exported to the whole world with digital technology and its disruptive effects, knowingly calculated and implemented by the new entrepreneurs of Silicon Valley.

While across the Atlantic inequality gaps not seen since 1929 were bringing the 'American dream' to an end, the European Commission, totally devoid of any understanding of the situation, tried to imitate this 'model'. They failed to see that the United States was heading down a path to violence, notably racial violence, 2015 having borne witness to a succession of racist crimes, including by the police, but also of random killings, seemingly without 'reason'. Nor did they see that this path would lead Europe itself into ruin, insofar as it did not possess the means to conceive and realize its own development policy for a *digital industrial economy that would be non-destructive of social structures and psychic apparatuses.*

In France, the few studies related to industry policy commissioned by President Hollande after his election were, on the whole, totally blind to this state of affairs – and in particular the Gallois report[6] and the Pisani-Ferry report.[7] The Gallois report had nothing whatsoever to say about the fundamental transformation of the economy by digital networks, and hence drew no consequences, and the Pisani-Ferry report was an incredible compendium of waffle of the kind that only France can produce, and for which it pays top dollar.

As for the *Mission d'expertise sur la fiscalité de l'économie numérique*, it does indeed identify the novelty of the situation.[8]

But as we have already seen, Nicolas Colin, finance inspector (and co-author of this report along with Pierre Collin, state councillor), has come to adopt, without distancing himself from it in any way, the perspective of the Californian disruptor – in the end defining himself as a 'barbarian'.

In order to resituate these questions in the contemporary historical context, let us recall some figures that describe what has unfolded on our planet over the past two decades:

- In 1997, as mentioned at the beginning of this book, there were a billion televisions being watched by a large majority of human beings, who at that time numbered six billion.
- In 2010, there were two billion computers and just under five billion mobile phones – which over the past few years had mostly become smartphones, that is, handheld computers, mainly produced by Apple and Samsung.
- According to a 2014 IPCC report, summarized here by Jean Jouzel,

  > we have already emitted about two trillion tonnes of CO_2, and, if we want warming to be limited to 2°C, we must not exceed an accumulation of 3000 billion tonnes of CO_2 in the atmosphere: we do not have the 'right' to emit, in the future, more than 1000 billion tonnes of CO_2. At the current rate of emissions, this quantity will be reached in less than thirty years. [...] This means that we must not use more than 20 percent of our fossil fuel reserves (conventional and unconventional gas, petrol and coal) if we want to limit the warming to 2°C. We can see that this task will be difficult. It implies a profound change of our mode of development.[9]

- In September 2015, less than ten years after the appearance of Facebook, this social network had acquired approximately 1.55 billion active users, and the planet had approximately 7.35 billion human inhabitants.

76. The catastrophic start to the twenty-first century

11 September 2001 was the date that tragically marked the transition from the twentieth to the twenty-first century, after a series of catastrophic initiatives undertaken in the Middle East by the United States in the last quarter of the preceding century. These initiatives were undertaken after the secret CIA policy in the 1980s that consisted in the attempt to support and manipulate Osama bin Laden, carried out in connection with Saudi Arabia and those who constitute what Fethi Benslama calls the petro-families in that country[10] – the Bush family being itself such a petro-family in the United States.

On 25 July 1990, April Glaspie, the American Ambassador to Iraq, let Saddam Hussein understand that the United States was not concerned about his intention to invade Kuwait, which Iraq had long claimed was its territory.[11] Nevertheless, following this invasion, George H.W. Bush entered the First Gulf War – thus giving the whole thing the appearance of being a trap.

After 9/11 – that is, after bin Laden had turned against the United States, which had constantly manipulated Islamist movements during the Soviet occupation of Afghanistan,[12] and after he had turned against his native Saudi Arabia – George W. Bush, allied notably by the United Kingdom and Spain, made this a pretext to annihilate Iraq during the Second Gulf War, on the basis of the claim that Saddam Hussein had produced weapons of mass destruction.

It was later established that this claim, which Dominique de Villepin rejected during a remarkable speech before the United Nations General Assembly, was a lie – for which Tony Blair received much approbation in the United Kingdom. The military coalition involving twenty-two nations and headed by the United States thus destroyed a developed country that would then become a base of operations for Daesh. The latter coalesced via the regrouping of former Iraqi military and agents, the Naqshbandi Army, the Islamic Army in Iraq, and Al-Qaeda, ultimately leading to the formation of Islamic State in 2006. And it was during this terrible war that the prisons of Guantanamo and Abu Ghraib appeared, where American mercenary armies committed their atrocities.

After Iraq was totally disintegrated by this military coalition – which, thanks to Dominique de Villepin, France was wise enough to avoid, as was Germany, unlike the Spain of José María Aznar and the Britain of Tony Blair – the operations of Daesh shifted across to Syria, following the uprising against Bashar al-Assad and his subsequent retaliatory war against his people, in relation to which the French media had, for a long time, been denouncing the atrocities of the Ba'athist regime. In June 2014, Islamic State, henceforth named Daesh, proclaimed a caliphate in the Sham region,[13] established both in Iraq and in Syria.

Daesh then established itself in Libya, a few hundred kilometres from Tunis. On 19 March 2011, encouraged by Bernard-Henri Lévy, who was clearly engaged in electioneering, Nicolas Sarkozy attacked the regime of Muammar Gaddafi, as if trying to make himself appear strong in the face of massive public disapproval. Gaddafi was himself rattled by the 'Arab Spring', which, between the borders of Tunisia and Egypt, unleashed a war through which AQIM (Al-Qaeda in the Islamic Maghreb) was established. This war would soon migrate to sub-Saharan Africa, in particular to Mali, where, in January 2013, François Hollande in his turn decided to intervene, so that he, too, might boost his ratings in the opinion polls.

It was on the basis of these Western military interventions in the Middle

East, Near East and Africa that Daesh, through a discourse forged by former Iraqi agents, managed to capture and channel the suicidal despair of young French minds.[14] As for the hundreds of thousands of refugees coming from Syria seeking help throughout Europe, they are fleeing the immense chaos that has afflicted this region since 1990 – looked upon from afar by the United States, while observing the rise of the European far right.

As they have evolved over the course of the last ten years, all these catastrophes surrounding the Middle East, with their combined effects on a portion of French youth, and, more generally, on the moral state of France, must be seen in terms of a global context marked by an incredible combination of calamities:

- The so-called 'financial crisis' of 2008, for which no solution was found other than the reinstallation of the same system.
- In 2011, the disaster at the Fukushima nuclear plant, which is still leaking, and the 'securing' of which, if it ever happens, will require many years of research and many years to achieve decontamination.[15]
- The revelations of Edward Snowden on the activities of the NSA and the new Silicon Valley giants conducted through the PRISM system, leading to his being charged with espionage and theft.
- The Greek crisis, which humiliated this country and its fragile democracy, where, ultimately, the European Commission judged it guilty for having debts that, in fact, it had itself largely encouraged.
- The extreme fragility of the European Union itself and the doubts about the future of the euro.
- The conflicts and anxieties tied to the 'uberization' of the economy, that is, its disruption, everything that accompanies it, and the carelessness and negligence of public powers in the face of 'techno-logical unemployment', the stakes and coming consequences of which are systematically denied.
- The political traumas provoked in France by the rise of the National Front, while the progression of the far right is visible throughout almost all of Europe, including, in Germany, the neo-Nazis.
- The refugee crisis in the Middle East and its share of postponements, reversals, evasions, infamies and cowardice, but also destabilizations, notably in the Balkans.
- The attacks in France and the irresponsible reactions of the Valls and Hollande governments, which legitimate in advance the future installation of explicitly authoritarian regimes, through a political disinhibition that is utterly in line with the 'uninhibited right' [*droite décomplexée*], brought up to date by Nicolas Sarkozy, which is what led to talk of his 'madness'.[16]
- The manifest non-response of COP21, presented as a success, but where this was understood by everyone as a deception perpetrated

for purely electoral reasons – despite the warnings of the fifth IPCC report, released in 2014:

> The SYR [*Synthesis Report*] confirms that human influence on the climate system is clear and growing, with impacts observed across all continents and oceans. Many of the observed changes since the 1950s are unprecedented over decades to millennia. The IPCC is now 95 percent certain that humans are the main cause of current global warming. In addition, the SYR finds that the more human activities disrupt the climate, the greater the risks of [...] long-lasting changes in all components of the climate system. [...] [S]tabilizing temperature increase [...] will require an urgent and fundamental departure from business as usual. Moreover, the longer we wait to take action, the more it will cost and the greater the technological, economic, social and institutional challenges we will face. [...] As such, the SYR calls for the urgent attention of both policymakers and citizens of the world to tackle this challenge.[17]

> Warming of the climate system is unequivocal [...]. The atmosphere and ocean have warmed, the amounts of snow and ice have diminished, and sea level has risen. [...] Each of the last three decades has been successively warmer at the Earth's surface than any preceding decade since 1850.[18]

- A thousand other symptomatic facts and gestures that should be mentioned here – including the dreadful 'Strauss-Kahn affair'.

77. Becoming without future: when the world is without meaning

The long series of catastrophes that have struck the whole world since the beginning of the twenty-first century, and in particular France, have plunged every generation into a state of demoralization unique in the history of humanity. Never before has there been such planetary anxiety, turning little by little into dangerous despair.

Never before have scientists from all over the world made pronouncements – with rational arguments exposed to peer criticism and founded on logic and quantification, and without any tinge of 'millenarianism' or 'apocalyptic prophecy' – concerning the likelihood of the disappearance of noetic life on earth and a significant part of the planet's biodiversity, that is, the disappearance of life in general – all this within a short period: on the order of a few human generations.

It is this about which Florian speaks. It is also in this global context that, over the years since the 2002 Nanterre massacre, we have seen the attacks in Madrid on 11 March 2004 and in London on 7 July 2005 that claimed hundreds of victims, and in Norway in 2011 the attack by Anders Behring Breivik, who killed seventy-seven people including sixty-nine teenagers, while in 2014 Maxime Hauchard, a young man from Normandy, who according to his neighbours was 'kind, courageous and well-bred', was 'recognized on a video as a Daesh executioner'.[19] On 7 January 2015, Chérif and Saïd Kouachi murdered eleven people including eight at the offices of *Charlie Hebdo*, among them Bernard Maris. On 24 March the same year, Andreas Lubitz steered the airliner he was flying, with 144 passengers and six crew on board, into a mountainside. Many other examples of mad behaviour occurred in France during this period, some of which were discussed above in §39. On 13 November 2015, nine terrorists attacked in Paris and Saint-Denis, killing, either by execution or by detonating explosive belts, 130 people – the first time that France had experienced suicidal attacks.

These are the *worst* symptoms of a *world going mad* that manifests itself in a thousand other ways, and that is possible only because *we all* increasingly live, like Florian, as bearers of a negative protention of a becoming without future. Unlike Florian, however, we prefer not to say so: we *do not want* to know about it.

In this absence of epoch, cynics practise the reflective madness that takes the maxim 'after me, the flood' as a *principle* – for speculators as for most electoral candidates, ruining any remaining democratic credit just as it does fiscal or public solvency. This is what I began to describe in the *Disbelief and Discredit* series.

As for the better-adapted deniers, they turn this misery into a spectacle by wallowing in television programmes, social networks and other systems designed to capture attention and retention. The desperate consume alcohol, anxiolytics, sedatives or other drugs, and sometimes commit suicide. As Carlos Parada has pointed out, 'jihadism', which often occurs after a period of using 'hard drugs', is also a kind of substitute for drugs.[20]

As long as there seems to be no hope of producing a future in a becoming that has, for everyone, everywhere, become overwhelming, from Palo Alto to Saint-Denis by way of the Carlton hotel, at least for those present who were merely 'material' in the eyes of 'Dodo the Pimp' as in those of Strauss-Kahn himself,[21] there will be more and more suicides – even among children between five and twelve years old, including by blowing themselves up in the midst of a crowd.[22]

This is the case because there are more and more young people and parents who themselves think what Florian says, most of the time without admitting it, while at the same time the parents are less and less able to truly *raise* their children, being totally delivered over to a marketing that knows no bounds. This impossibility of educating and raising[23] is clearly *the* key factor contributing to impulsive outbursts [*déchaînement pulsionnel*] of all kinds, and of the death drive in particular.

The growth of despair [*désespoir*], that is, of the loss of *reasons for hope* [*espérer*], and hence the *loss of reason*, amounting in the final reckoning to the growth of madness, is the inevitable result of factors that, when they combine, compound their potentials in a way that goes beyond every limit. Among these limits are the following:

- the *scientific anticipations* that demonstrate that the metasystemic crisis established as and by the Anthropocene could prove fatal to the biosphere, as foreshadowed in 1972 in the Meadows report;[24]
- the effects of the *outright liquidation of public power and of the defence of public interest* due to disruption;
- *general economic insolvency*, of which 2008 was only the first shockwave, and the risk of which will be intensified by the predicted collapse of employment;
- the *destruction of the affective spheres*, and, through that, of the intergenerational and transgenerational relations that are the various forms of education (familial, national, religious, cultural, artistic, political and so on);[25]
- the extreme increase in *violent behaviour* based on resentment and the designation of expiatory victims, which stems from all these causes, and the 'new form of barbarism' anticipated by Adorno and Horkheimer.

Such a combination, where the *pharmakon* that *creates suffering* leads to the designation of a *pharmakos* that must be *made to suffer*,[26] sets up a vicious circle and an infernal spiral of violence responding to violence that inevitably leads to a widespread feeling of a world going mad.

By creating a context in which the future can no longer be antici-pated other than as inconceivable chaos, this turn to madness makes even those who still escape it crazy. It *maddens* [*affole*] everyone – by the most *ordinary* pathways as well as the most *extraordinary*, and also through that '*reflective* madness' of which President Hollande is a kind of repulsive incarnation.

In fact, the measures to change the French constitution in response to the terrorism of Daesh are themselves examples of incredible madness, which in addition reflects very badly in terms of electoral calculation, and which can only *institutionalize* the spiral of violence, and sacrifice the future in order to 'save' a power that is totally discredited and widely held in contempt: Hollande beats Sarkozy hands down on this score. Both of them have *alarmed* [*affolé*] French society – which means that they have fundamentally weakened it, in a context that was already extremely difficult, and which more than ever requires, on the contrary, great clarity of thought.

How can one not go mad when becoming seems to bear no future, so that *the world no longer seems to have any meaning*? Going crazy, this mad becoming stems from an immense *demoralization*, itself aggravated by processes of denial of all kinds. Demoralization is what creates a loss of morale. And morale [*le moral*], which is also trust [*confiance*], is the condition of any rational action, given that reason is always borne by reasons for *hope*.

78. Morale, 'morals', moral being: diseconomy and demoralization

Global demoralization is a planetary loss of reason. It amounts to a new age in the history of madness – and precisely not to a new epoch, since it is the *absence* of epoch that provokes despair and the loss of all reasons for hope. Like Florian, *we are all* possessed by visions that are increasingly dark, and sometimes completely black: we have the feeling of 'descending into darkness'.

This is so because – despite countless processes of denial, based on psychic repression mechanisms[27] and systematic practices of subjugation and of the destruction of lucidity and will, through marketing as much as propaganda, with maddening ideologies and practices instituted as government methods – we suffer the effects of a negative causal series that we absolutely cannot see how it would be possible to reverse.

We do not 'have' morale. What is 'morale' anyway? And what are 'morals'? What is morality? What is immorality? What relationships are there between a state of morale and a state of morality, if not morals or the absence of morals?

Bernard Maris, as I have already mentioned, suggested that the way the death drive functions in capitalism is its most daunting limit.[28] The counterpart of the death drive, which Freud revealed in 1920, is the life drive. The life drive is what *contains* the death drive in the sense of *limiting* it. This is possible, however, only because the life drive is itself inscribed in a libidinal economy outside of which it itself becomes destructive, and where, in some way, it itself returns to the death drive, which it contains also in this other sense: it thus returns to this '*selfsame*' that is death.

This is why Freud shows that the life drive also harbours the death drive, as, for example, in the case of sadism:[29] the life drive is not the opposite of the death drive, but rather that with which the latter must compose so as not to destroy the being that contains it.

In this respect, the Strauss-Kahn case is *perfectly typical* of the terrifyingly generalized misery to which our absence of epoch abandons, in particular, the younger generations, in the face of the *systemic unbinding of the drives*. The latter is less a matter of the deviance, for example, of the former managing director of the IMF, than it is the ordinary, normal and perfectly demoralizing functioning of drive-based capitalism, which

Strauss-Kahn unreflectively embodies, just as Hollande embodies its reflective madness.

Strauss-Kahn has shown himself to be a participant in the libidinal diseconomy, embodying its power[30] and setting up the French stock-option system. Among the younger generations, the libidinal diseconomy induces disgust, disrespect and an incommensurable state of demoralization, leading to desolation, which renders this misery incomprehensible, that is, implacable.

This state of fact, which is one of immense suffering, increasingly manifests itself in ways of acting out [*passages à l'acte*] that involve more or less suicidal tendencies. To this we must oppose an *increase* of reason, that is, a *new* understanding of what, in the world, seems to lead to a befouled worldlessness [*immonde*].

> I cannot deny all respect to even a vicious man as a man; I cannot withdraw at least the respect that belongs to him in his quality as a man, even though by his deeds he makes himself unworthy of it. [...] On this is based a duty to respect a man even in the logical use of his reason, a duty not to censure his errors by calling them absurdities, poor judgment and so forth, but rather to suppose that his judgment must yet contain some truth and to seek this out, uncovering, at the same time, the deceptive illusion (the subjective ground that determined his judgment that, by an oversight, he took for objective), and so, by explaining to him the possibility of his having erred, to preserve his respect for his own understanding.[31]

This is exactly what Manuel Valls – who is like the orchestra conductor of the destruction of all reason, and, along with it, of all reasons for hope – cannot understand. After the crimes of Mohamed Merah, Gilles Bernheim and Henri Guaino claimed that monstrosity has no explanation.[32] On the contrary, we must always seek to understand those who would try to deprive us of our respect and our reason because they themselves absolutely lack respect and reason. Nothing is worse than giving in to the 'sleep [*sueño*, which also means dream] of reason'.[33]

> For if, by using such expressions, one denies any understanding to a man who opposes one in a certain judgment, how does one want to bring him to understand that he has erred? The same thing applies to the censure of vice, which must never break out into complete contempt and denial of any moral worth to a vicious man; for on this supposition he could never be improved, and this is not consistent with the Idea of a *man*, who as such (as a moral being) can never lose entirely his predisposition to the good.[34]

Here, a very difficult question arises: what meaning should be accorded to the expressions 'moral being' and 'be improved'?[35] And from what, in political societies, and in their 'morale', stems, not 'morals' – which designates a specific, bourgeois epoch of 'morality' – but the mores and virtues constituting *moral philosophy*, which proposes *rules for retentions, contentions and protentions* that allow these fundamental dynamic factors of noesis that are relations to life and death to be saved and economized? Such an economy is indispensable because, when these relations de-compose, that is, in Freud's sense, *diseconomize*, these dynamic factors become outright powers of destruction.

Such a diseconomy, leading to such a demoralization, is systemically and systematically produced with the exasperation of the drives in which drive-based consumerist capitalism consists. The latter is characterized by *neoconservative disinhibition*, which replaces public power with marketing and thus realizes the functional and systemic arrangement of the drives of consumers and the drives of speculators – of which Strauss-Kahn is a kind of logical and pathological extension.

It is remarkable that it was Dominique Strauss-Kahn, Lionel Jospin's minister of economics and finance, who, in France, installed the stock-option system, which structurally binds the interests of top corporate management to those of shareholder capital, itself completely subject to the speculative capital born of 'financialization' in the wake of the 'conservative revolution' – in the name of a pseudo-'modernity' of the left that Valls now less elegantly reclaims.

79. 'Morals', education and credit

Vincent Peillon wanted to reintroduce moral teaching into schools.[36] This would doubtless also have necessitated introducing a new critique of political economy – at least if we understand Dostaler and Maris correctly: a new critique of political economy ought to inscribe moral philosophy deeply into its economico-political horizon, and vice versa.

Any new critique of political economy should constitute a new economy, not of the 'morality' of capitalism, as Sarkozy claimed he was doing (through one of those lures whose secret he knew – but this kind of secret quickly fades), but of morale and of the *moral being* to which, in fact, every *noetic being* amounts, that is, every non-inhuman being. For this, we should read Adam Smith's *The Theory of Moral Sentiments*, along with Spinoza, Marx, Nietzsche and Freud, in the twenty-first-century context, that is, faced with what *none* of these authors could ever have known or even imagined – no more than could Foucault or Derrida.

Whatever prospects this may hold, to which we will return in the conclusion, how would Vincent Peillon – who is a philosophy graduate – have recommended that the teaching staff of his Ministry of Education respond, during the lessons in morality that he advocated, to some or other 'digital native' who inquired about the morality of the former head of the IMF, who was also a putative candidate for the 2012 French presidential election?

How would Vincent Peillon have recommended responding to questions that the younger generation would not fail to ask, faced with the flagrant lies of the French president throughout his five-year term, from beginning to end, and which constituted his very method of governing?

According to Condorcet, ignorance always leads to servitude. That this statement is a traditional teaching of academic morality is what inscribes every morality in the question of a will and a freedom that stems from this knowledge – will and freedom reputedly being unconditional.[37] It is the in principle positing of this affirmation that distinguishes any moral philosophy from 'morality' as the bourgeois, hypocritical epoch of the submission of minds to dogma,[38] which also becomes the justification of the confinements described by Foucault.

Moral philosophy posits that freedom, together with will and knowledge – outside of which a non-inhuman and noetic life is impossible – *contain* the drives. The contemporary language (*pulsion, Triebe,* drive) adopted here to describe an epoch of thought that did not yet possess the *concept* to which these words refer implies that reason, which stems from desire, and therefore from hope, inasmuch as it is the primordial condition of the noetic motives for living, is always an *epoch* of a libidinal economy – this epoch being composed of those collective secondary retentions that transmitted knowledge enables the younger generations to acquire.

The notion that there must be reasons for hope in order to project motives for living is not the prerogative of religious thought. The discourse of emancipation conceives itself as emancipation from religious dogmas and their degradation into the constraints of 'bourgeois' morality that, too, cultivates 'reasons for hope'. *These reasons, however, collapsed, one after the other, when the irreducibly pharmacological character of modernity and its epokhal collective protention – progress – was revealed.*

The accomplishment of nihilism as the madness of purely, simply and absolutely computational capitalism is what the *intrinsic toxicity of the Anthropocene* reveals, which is itself an epoch of the organological condition of every form of noetic and non-inhuman life. This organology is itself the reality of exosomatization that characterizes hominization as the realization of its wildest dreams [*rêves les plus fous*]. These questions, which require a redescription of factuality in terms of entropy, negentropy and neganthropology, will be further elaborated in *La Société automatique 2. L'avenir du savoir* – where virtue, ἀρετή (*aretē*), strength, will seem to constitute *neganthropological potential as such*.

The will to knowledge and the freedom of knowledge are the conditions of this potential, and they must be cultivated as such. Hence they themselves require, not just the possibility of reason, but the *imperative exigency of truth as motive of all motives to live reasonably* – and as a motive that is, in its structure, in *excess* over the *epoch* that gives it its shape. This is something that necessarily ties the experience of *alētheia* to that of *parrhēsia*.

It is *this exigency* that constitutes reason, and it is on this basis that knowledge is the condition of freedom. It is only in its name that one can solicit the reason of one's fellows, whoever they may be – including, more than anyone else, in France, the President of the Republic. It is this exigency that the most recent representative of this supreme function, François Hollande, will have betrayed, which is very bad news for the future of France and the French people.

80. Politics and moral philosophy

The knowledge that stems from such an exigency is a knowledge *in law*: it is not, it has never been and it will never be a knowledge *in fact*. This is both because it is unfinished, always provisional and precarious, that is, 'metastable', and because, in fact, *most of us ordinarily behave in a manner contrary to the imperative that it incarnates*. In other words, it is *intermittent*, and this is the case not just for so-called 'rational' knowledge in the sense of reason understood as logical: it is also true of life-knowledge [*savoir-vivre*], that is, mores and morals in general, as well as of work-knowledge [*savoir-faire*, knowledge of how to make and do].

It is, however, at the level of so-called rational knowledge that the *juridical* character of knowledge *in general* is explicitly, thematically and interrogatively formulated (as an imperative to *respond to* what *will always remain* not only as a question, but as a challenge, a putting *into* question). *Juris*, here, means:

- aiming at a rule that is never actually realized, hence '*questio quid juris*';
- leading (as *regulatory* idea) this aim towards what Kant called an *idea of reason* that is promised by the *difference* between fact and law, which is also a *différance* that *founds* the state of law as infinite promise in *this* sense, and promised as free and without possible limits.

This difference is a process, and this is why Derrida describes it as a *différance*: it founds the state of law by installing a state of fact that is the constant collapse of this state of law, with no other horizon, ultimately, than its constant relaunching, reconstruction, rebirth or reactivation (*Reactivierung*) as *anamnesis*, which constitutes history in the strong sense, and the notion of which emerged some 2,700 years ago, on the coast of Ionia, on the Aegean coast of present-day Anatolian Turkey, 400 kilometres south of the Homeric city of Troy.

As a new era of *noetic différance*, becoming that noetic faculty called 'logical', the *logos* that is its condition is made possible by the tertiarization in letters of mental events.³⁹ On the basis of this fact, such as it establishes a public, positive and 'logical' law, that is, a law governed by conflictual argument, a new way of life arises, founded on the constant search and need for *alētheia*. This search, insofar as the Greeks saw in it the convergence of the just and the beautiful, is also called ἀρετή, translated into Latin as *virtus*, and into French as *vertu* [virtue]. All this is concretized as the foundation of the *polis*. And it is all this that defines *politics*.

It is also all this that is being ruined by politicians who submit to the fulfilment of nihilism insofar as it reduces all values by dissolving them into calculability, even though they have value only insofar as they express and affirm an incommensurability that, having no price, can *legitimate* all commensurable values. These issues, which are generally related to the theory of money, are not reducible to it – and this is why political economy must be an economy of 'moral sentiments' that includes Bataille's questions of general economy and Freud's questions of libidinal economy.

Through digital tertiary retention, fully dedicated to the reduction of every trace of singularity to calculability, and, through this fact, to the disintegration of retentions and protentions, the fulfilment of nihilism makes the difference between fact and law *totally vain, totally devoid of consistence*. Such is the absence of epoch. Disruption multiplies legal vacuums, which amount to so many existential deserts. In so doing, it relentlessly systematizes and instrumentalizes this state of fact without law, through which it invades the world like a sandstorm raging across a desert within which centuries of civilization lie buried, ruining all credit and thrusting youth in want of idealization towards the mirages that this desert does not fail to elicit.

Hence there is a repetition, on a planetary scale, of what, at the scale of Athens, sophistry accomplished in the Socratic epoch – exposing the toxicity of literal tertiary retention and dooming the city to στάσις (*stasis*), that is, to civil war. It was in this context that Socrates continually reaffirmed that the character of virtue as ἀρετή lies in the need to make oneself compatible with what is true, what is just and what is beautiful.

In other words, the question of ἀρετή is the principal question through which philosophers will oppose sophistry. Having seized hold of the noetic *pharmakon* that is literal tertiary retention, itself conditioning logical différance, sophistry, according to Plato, reverses its virtuous potentialities, that is, veritative potentialities, by reducing them to efficiency, making them serve what Aristotle will call efficient causality.

Aretē is on the contrary fundamentally constituted by final causality – whereby the protentions of noetic souls converge within what Kant will call the transcendental affinity between noesis and the world.[40]

In 1916, after Baudelaire, Flaubert, Manet and so many others in all the fields of what will become art's modernity, Dada eventually establishes the vanity of bourgeois morality by teaching the emptiness of the 'beautiful' submitted to the lies of the 'true' and the 'just', which had become alibis of the crimes committed at that time, during the First World War, which, as I have attempted to show elsewhere, led to the revelation, for Valéry, Freud and Husserl, of the pharmacological dimension of knowledge itself in all its forms (not just in science but including, precisely, as Valéry states in 1919, the 'moral virtues').[41]

As capitalism becomes mafiaesque, it factually liquidates these bourgeois 'values', while the speculative art market 'recuperates' them by buying up all these inclinations to escape the 'devaluation of all values'. The result is negligent carelessness, dictated by a financialization that shamelessly unbinds the mechanisms of the drives and amounts to the concretization of nihilism, of which Dadaism was an active and affirmative form (in the senses of both Nietzsche and Deleuze).

In the final stage of nihilism, which *is* the absence of epoch, everything that is 'active' is captured and channelled in order to *divert* noetic possibilities, that is, in view of their devaluation. This is what happened in France after 1968, and throughout the world.[42] Lyotard conceived the history of modern and contemporary art starting from the category of the sublime rather than the beautiful, and after having long lingered over the question of the drives: that he did so is no doubt one indication that what is coming is not merely a new epoch, but another era – where, just as the beautiful is exceeded by the sublime, good and evil will be projected *beyond* their opposition, as that which, being good, is nevertheless *composed* of and with that which is bad, and which, when they are decomposed, that is, unbound, like the life and death drives, become in fact bad.

What Lyotard observes in the movement towards the sublime is that the experience of the beautiful as completeness and well-roundedness of form shifts towards that of the figure and sublimation as confrontation with excess and in default. This experience of excess, as what surpasses not just the understanding but the imagination itself, is also the experience of an irreducible madness, and it is this that Slavoj Žižek points to in the following:

> in the *Critique of Judgement*, [...] Kant conceives of the Sublime precisely as an attempt to *schematize* the Ideas of Reason themselves: the Sublime confronts us with the failure of imagination, with that which remains forever and a priori un-imaginable – and it is here that we encounter the subject *qua* the void of negativity. In short, it is precisely because of

the limitation of Heidegger's analysis of schematism to transcendental analytics that he is unable to address the excessive dimension of subjectivity, its inherent madness.[43]

But this excess is equally the experience of a default that itself constitutes the condition of the schematism of the imagination in the transcendental analytic:[44] what makes the schema possible, inasmuch as it allows the data of intuition to be connected to the concepts of the understanding, is tertiary retention. Furthermore, the unimaginable is not a failure of imagination, but precisely its interminable condition (in Blanchot's sense) insofar as it is always constituted by default – in that default of origin epitomized by exosomatization.

Art or the work are, therefore, this experience of the support, of its surface, its depth, its virtuality, its 'interactivity' or its 'interpassivity', as Žižek also says, and so on.[45] Obviously, this does not erase beauty itself inasmuch as it is always well-rounded and delimited: it is what, in the absence of epoch, projects *beyond* this plenitude, a beyond where it gives itself in the infinite sublime of excess and the default (which is also Bataille's question).

As long as the Western eras (tragic, Homeric, then pagan in Graeco-Roman antiquity, followed by monotheistic) had not been exhausted – completely devalued – by nihilism, it was on the basis of the rational convergence of the true, the just and the beautiful, which tie the *epistēmē* to individual and collective ways of life, that Graeco-Roman philosophy could be considered, well before Christianity, *as essentially a moral philosophy*, as this is conceived by Pierre Hadot and Michel Foucault.

Now that nihilism has reached its own limit, this ancient conception of morality – as this *praxis* guided by techniques of the self that lie at the origin of the *legitimate* government of self and others – has become totally incomprehensible, because it is in fact reversed by a disruptive tendency to turn everything that has resulted from the realization of the noetic dreams of the West, as what Kant called the Ideas of Reason, into a desert. Today, the state of fact engendered by this new era of technological shock no longer seems capable of establishing the state of law in which the second stage of the doubly epokhal redoubling always consists.

81. For example

During the so-called 'Carlton trial', what *morale* was possible for the Strauss-Kahn generation, who were subjected on a daily basis to details of the escapades, indiscretions and impulses [*frasques pulsionnelles*] of this former minister, former head of the IMF and former favoured candidate of the social-democratic left? And, furthermore, what morale was possible when the Cahuzac affair remained fresh in everyone's minds,[46] all of this unfolding as the symbolic credit of the French presidency is ruined by having an occupant of the Élysée Palace elected by default precisely because of Strauss-Kahn's ineligibility, even if the latter continued to advise the prime minister?

On the basis of the moral and mental state resulting from all these infamies, of what 'morality' would this generation be in need – if we wish to preserve the kind of outdated terms that Vincent Peillon wanted to dust off? What did he himself think of Strauss-Kahn, this Vincent Peillon who had declared his support for Strauss-Kahn's candidacy – Strauss-Kahn, whom Michel Rocard claimed was suffering from a 'mental illness' that prevented him from 'controlling his drives [*pulsions*]'?[47]

There is every reason to believe that the Strauss-Kahn generation, which is also the generation of disruption, known as Generation Y or the generation of 'digital natives', can only *with great difficulty* accede to reason in such conditions – before ever having had a chance to lose it.

There is every reason to fear that this generation *will no longer be able to accede* to reason, insofar as the latter is always necessarily and *firstly* the *reason to live* and *to hope* in an *almost miraculous way*, given the *counter-exemplary* conditions provided by putative or actual representatives of the Republic, who have now totally succumbed to the drives in the most deplorable way.

To accede in a 'miraculous' way to reason, however, is to accede to it through divine revelation. We must go through despair, writes Kierkegaard, in order to reach God – but this despair affects the man who, having been able to access reason before having experienced the infinity of God, has passed through what Kierkegaard calls the aesthetic stage, then through what he calls the ethical stage.[48] This is what the Strauss-Kahn generation has never had the fortune to know, which means, perhaps, that they are precisely *not* a generation. Hence they struggle to project themselves as *genitors*, thereby giving rise to a harsh awareness of the fact that God is dead.[49]

It is distressing to observe that Alain Juppé, in whom many Frenchmen still hoped to find a buffer against the worst, found it necessary to declare: 'I do not believe in exemplarity. No one is beyond reproach.'[50] To admit this mild strain of cynicism, however benign it may be, is already to testify to the fact that we no longer understand in what education, transmission and ultimately 'virtue' consist. Nor do we understand in what politics consists, especially in the epoch of disruption, the stakes of which clearly escape the mayor of Bordeaux: exemplarity *never* meant purity, innocence or 'irreproachability' – *quite the contrary*.

The one who is exemplary is not beyond reproach, that is, without fault [*défaut*] and therefore perfect, but the one who struggles *with* his or her faults, and, *as much as possible*, turns them into *that which is necessary* – whether these faults are moral or organological, that is, tied to the condition of mortals, who are neither animals, whose own

organogenesis completely escapes them, nor gods, whose power, represented by divine fire, is not pharmacological, a power that gives them,
along with heroes, that is, demigods, the ability to mark out the stakes
of the ἦθος (ethos) of mortals.

Perfection is the transhumanist fantasy par excellence, and the
symptom of an immense moral poverty [*misère*]. In stark contrast to
this moral vacuum, the reversal of the fault [*défaut*] into that which
is necessary is both exceptional and banal: it stems from what Freud
described as the libidinal *economy* that transforms the drives necessarily
contained in each of us into social energy invested in a thousand ways,
through the most ordinary dreams as well as the wildest ones – which,
as we shall see in conclusion, are the condition of the effective exercise
of freedom for the young Foucault reading Binswanger.

To defend exemplarity, to be exemplary, is to be 'worthy of what
happens to us': it is not to *oppose good* to *evil*, but to *extract good
from evil* by *composing* with it – and doing so, firstly, within oneself.[51]
This is what is meant by techniques of the self, and, more generally,
the knowledge that stems from them, that is, that stems from a μελέτη
(*meletē*), from a discipline, from a rule, from a μέτρον (*metron*), and
which is conceived as all the therapies and therapeutics derived from
these techniques insofar as they are also the condition of any moral
philosophy – *all* philosophy being such a morality.

As for the *example*, not just as a pedagogical virtue but as what
shows what cannot simply be demonstrated[52] (and this is the case for
any therapeutic prescription, which is why medicine is not a science but
a technics, as Canguilhem emphasizes), it is not a substitute for rational
exposition or a facilitator of understanding as regards the teaching of all
that is moral: it is the index of the fact that *only singularities can fully
embody moral necessities*.

'Virtue', in the sense of ἀρετή, always manifests itself in a singular
way. And this is a fact that always constitutes *an aporia for 'morality',
in that it always tends, as a poorly understood rule, to prescribe an
average norm that dissolves singularities*[53] – singularity being precisely
what deviates from the average. Morality, as a body of rules, is in this
sense always on the verge of betraying itself, that is, preventing the
expression of the singularity that, alone, is virtuous. Ethics is what, in
not compromising with this finitude of morality conceived as a set of
mores followed firstly out of habit, always brings out the singularity of
each ethical situation: of each ἦθος.

82. Economy and function of reason at the turn of the twenty-first century

What since ancient Greece we have called λόγος, which we have translated
as *ratio*, then as reason, is not some faculty located somewhere in the

cerebral apparatus: *tying together*, as a *logical necessity*, a *common series of elements that are not just discrete* (as in mathematics) but *singular* (we have just seen why), it is through reason that the *noetic potentials* that are the psychic apparatuses of all non-inhuman beings can pass to the *logical noetic act*, which, since the Pre-Socratics, was something made necessary by the literal (lettered) stage of grammatization.

With the spontaneous analytic in which literal tertiary retention consists, logical reason sets up the duality of analysis and synthesis – where synthesis is the interpretative act (*hermeneia*) through which singularity is *individuated*, and *reason along with it*, that is, *motive*, whose function is to provoke a *bifurcation* in entropic becoming so as to affirm its différance.

From *Phaedrus* to the *Critique of Pure Reason*, the duality of analysis and synthesis organizes the *economy* of reason. This economy is the pharmacology of an exosomatic organology. This is what we can and must learn from disruption, but we can do so only by allowing ourselves to be carried *beyond all our masters*.

The series of noetic singularities that are formed through these masters as the 'history of reason' – whose *work*, that is, the *process of collective individuation*, will be theorized as *logical* and *ontological* analysis and synthesis firstly by Aristotle in the *Analytics*, then by Kant in his Transcendental Analytic, then by Hegel in his *Phenomenology of Spirit* that, in effect, serializes reason – this series begins with Thales setting up the ἀλήθεια that we call truth, doing so as the regime of thinking in its becoming λόγος and imposing itself as criterion in all the ways of life of citizens worthy of the name.

Through what Husserl called the 'we of geometers', this series links together those who, if they do not 'tell the truth', as least dedicate their lives, from one generation to the next, to conforming to this *criterion of any virtuous life* that is the search for truth inasmuch as it has been conceived since the origin of geometry.

Truth, conceived and experienced in this way, cannot be confined to this or that specialty – physical truth, biological truth, social truth, juridical truth and so on. This division into isolated disciplines stems from the industrial division of intellectual labour and coincides with the Anthropocene as a *separation of analysis and synthesis* – which, arriving always too late, reflects the ill-considered risk-taking and legal vacuums generated by the process of disinhibition.

The experience of truth that begins in the epoch of Thales is transgenerational, and those who devote themselves to it – as citizens, jurists, artisans, parents, soldiers, priests, workers, athletes, builders, artists, activists and so on, as well as philosophers or mathematicians – remain in memory (more or less) as those 'parrhesiasts' whose memory marks education: as *Bildung* cultivating the love of truth insofar as it is *above all a strength* [*force*] – and, more precisely, *neganthropic* force.

The form of education that Western civilization will until recently have constituted – until the collapse of the 'School' in all senses of this word, including that understood by Vincent Peillon in his project of 'refoundation' – ensured the transmission of those therapeutic experiences that are the disciplines in all their forms insofar as collective secondary retentions are the carriers of protentions that are themselves collective, thereby constituting epochs, that is, *common reasons for living and hoping*.

Reason, therefore, is *clearly and firstly* a moral state, a motivation: it is this motive for living that, for noetic beings, in each epoch, in a singular way articulates subsistence, existence and consistence. Consistences bind the generations together, and it is by passing them on that one generation asserts its influence [*conquièrent leur ascendant*] over its descendants, amounting to what Blanchot described as a *curvature*, and thereby gives them the feeling that life is worth living, and is so in and as the right and the duty to seek the truth, in all things, and in all freedom.

The combination that destroys all reasons for hope is characteristic of the end of a geological era that will have been exceptionally short, and that we call the Anthropocene. The Anthropocene – these being the terms in which it is today publicly and centrally problematized, not as 'resistance to innovation' or denunciation of ill-considered risk-taking, but as a geological era without epoch, and through that the end of an historical era, if not of History – is what seems, in the very course of this thematization, to amount to an *unbearable, unliveable and unviable* episode, from which we must find an exit by all means possible – even though the careless and the cynics of all stripes have already given up on this in advance.

Today, we are *experiencing* this extreme urgency because we have discovered that the Anthropocene is an Entropocene – as Lévi-Strauss saw, when he proposed understanding anthropology as an entropology, but without drawing any neganthropological consequences. Faced with this reality, which has barely begun to be thought, the cosmic dimensions of entropy having hardly if at all affected philosophical thinking (with the exceptions of Bergson and Whitehead), reason is what proves to constitute a negentropic function.

This is what Lévi-Strauss could not think. Reason as negentropic function begins with the noetic faculty of dreaming that Binswanger observed and studied, and that is the faculty of realizing dreams. This realization is a neganthropology, that is, an organology bearing pharmacological alternatives. This seems to me – more than sixty years later – to be the hidden stakes of the reading of 'Dream and Existence', an issue that Foucault himself did not understand, because it was not comprehensible in his epoch or by his contemporaries, who had yet to live through the absence of epoch.

With respect to this lack of comprehension, insofar as it also bears within it the inability to anticipate the radicality of the leap involved in disruption, Heidegger amounts more or less to an exception: the neganthropological alternative, which can only consist in a *bifurcation* inaugurating a new era *in the geological as well as the historical sense*, constitutes the horizon of what the author of the 'Rectorate Address' will call, twenty-nine years after this text in which he also discusses Prometheus, the 'turn', or again the 'leap' of an 'event' that establishes an advent.

This is particularly clear in *Identity and Difference* – published five years before 'The Turn'.[54] Heidegger's exceptionality in anticipating the absence of epoch, to which these texts bear witness, is not full and complete. It seems, therefore, to stop short before the turn that it proclaims: having never integrated the question of entropy and negentropy, Heidegger could only surround the theses characteristic of his late work with the obscure discourse of the 'fourfold' – which Graham Harman claims to elucidate in his book *The Quadruple Object*.[55]

83. Indiscretions, deceits, falling prey

In the hands of *disintegrated* psychic apparatuses (in the sense I have given to the word 'disintegrated' in *Automatic Society*[56]), technological but domesticated objects such as smartphones, computers and automobiles, or for that matter airliners, can also become as destructive as automatic weapons or explosive devices, as well as being able to greatly facilitate the latter's transport and even manufacture. In such circumstances, *pharmaka* become massively toxic. Now, psychic apparatuses *as a whole* are tending towards this *disintegration*, which is not merely or even firstly psychic, but organological and social.

Without a *more than epokhal* turn, psychic and social disintegration in a disruptive and entropological situation will inevitably thrust the Anthropocene towards an enormous conflagration: the disruptive industrial model, as it is currently developing today, is *based* on the exploitation of this madness without precautions, without thought, without scientific discourse, without the elaboration of a new critique of organological and pharmacological reason, which alone could prescribe therapies and therapeutics capable of cultivating new reasons for hope.[57]

The epoch of the absence of epoch is above all the complete destruction of all symbolic power and every *positive* process of identification – identification having become a colossal economic sector, firstly with the culture industries and now, on another basis, with the 'data economy', and which, in both cases, results in *filial and intergenerational dis-identification*. In this way, the new form of barbarism glimpsed by Adorno and Horkheimer comes to assume gigantic proportions.

This disaffiliation engenders a *negative identification* that in its most accomplished forms constitutes the absence of epoch's homicidal madness. Positive identification can be built only on the basis of a candour that always singularly embodies the harsh and wounding need to tell the truth of an epoch, and as the example of courage and of what is admirable in human existence in general. Only on the basis of such an admiration can ascendancy be conquered: positive identification is, in the West, what the historical power of παρρησία procures as the 'courage of truth'.

The 'courage of truth', which Foucault made the subject of his final meditation, is not generally the strong point of the politicians of our time. Most of the time it is, *more than anything, what they lack* – just as do ordinary men and women in other fields:[58] denial, malaise, submission and crudeness make up our quotidian existence. This is something that no one escapes. Clearly, it is something we mostly notice in others, and this is how it is bound to be. It is for this reason that we have need of others, when, as friends, they tell us our παρρησία.

As for deliberate lies, knowingly weighed, calculated, 'reflected upon', this has become the very method of governance in the absence of epoch. This attenuates Florian's faculty of dreaming, and François Hollande will go down in history as the one who went furthest along this path that is infinitely perilous for all: he deliberately lied when, claiming in his speech in Bourget on 22 January 2012 that he would tackle the financial world, he gave hope that a *courageous, which is to say normal*, president would finally tackle the real issue. By arguing that the possibility of being deprived of citizenship [*déchéance de nationalité*] should be inscribed into the Constitution, he has taken this duplicity to the level of true historical betrayal – and he has, in a way, himself fallen prey [*déchu*].[59]

Lying is one of the worst moral faults, if it is true that, *in principle*, citizenship and the city within which it can flourish depend on the experience of that truth for which we must have the courage – which is the moral aspect of any political philosophy.

Foucault shows that modes of veridiction, and the regimes of truth that arrange them, characterize this or that epoch,[60] binding together the government of self and the government of others. These arrangements generate the collective retentions and protentions that form the truth-based knowledge [*savoirs véritatifs*] of which political noetic life is composed.

Since the appearance of analogue then digital retentions, these arrangements have been unravelled and short-circuited in an industrial way.

It is in this still largely unconsidered context that, after Sarkozy, and between the deceits of Hollande and the indiscretions of Strauss-Kahn, politics is degraded, falls prey [*déchu*], in this way losing its soul, that is, its movement, its dynamic, its strength, its virtue – the great beneficiary of which is Marine Le Pen, who exploits every moment of this in a meticulous way.

84. The epidemic of which Strauss-Kahn and his disease are merely symptoms: on moral philosophy

The disindividuation that strikes the whole world – including and firstly Alan Greenspan,[61] and 'leaders' in general – is obviously also shown by the sexual, moral and affective poverty [*misère*] of Dominique Strauss-Kahn, this man of power (in Pascal's sense) who could be exceptionally seductive.

I recall admiring his smile and his twinkling eyes on the evening of the 1997 legislative elections – a period when I still watched television. I have known many people who admired him – and this must be borne in mind when we refer to admiration, which is also part of what seduction seeks, and is that through which it functions.

Any admiration – which is always a kind of dream – can turn out to be such a mirage. The immense disappointments these mirages can generate, such as the dreams dishonestly incited by corrupt politicians (by lies, by money, by both, by a thousand other evils, all this falling within a diseconomy of the drives), are what slowly but surely kill off all reason for hope. Incarnated by personages of high rank, the deceptive mirages of our admiration seem to *demonstrate* the *vanity* of reason without which no real hope is possible.[62]

The vanity of reason, which we experience in the ordeal of extreme desymbolization that is inevitably also a disidentification, is engendered by the immeasurable emptiness of that desert of nihilism that elicits the disease that afflicts Strauss-Kahn, bringing this oh-so-refined gentleman to the point of 'possessing' 'material', that is, sexual livestock furnished by a procurer known as Dodo the Pimp. But this disease is itself the symptom of another malady, one that is immensely contagious: that of a time with neither future nor past, the time of the absence of epoch, radicalized by a form of disruption that takes the process of disinhibition to its extremes, and founded on the industrial exploitation of the drives.

The Strauss-Kahn case is typical of this absence of epoch in which the unbinding of the drives is the state of fact that establishes the basis of ordinary madness evoked in 'Aux bords de la folie'.[63] Before Freud, ancient moral philosophy and modern 'metaphysics of morals' [*métaphysique des moeurs*] were already attempting to conceive the conditions that, as the *binding* of the drives, constitute the social bond. After Freud, the moral apparatus of psychic apparatuses turned out to contain a ὕβρις constitutive of the psyche, for which an economic conception of morality and mores was therefore required – 'morality' taking the name of super-ego.

From that point onwards, to consign moral questions to the archaism of 'Christian morality'[64] or 'Judaeo-Christian morality' becomes dangerously *fatuous* – opening the door for the negative sublimation through which the delusional suffer from a more or less criminal form of the extra-ordinary, that is, they are possessed by ὕβρις in its most destructive forms. To reopen the file on moral philosophy, as Foucault strove to do at the end of his life and work, nevertheless requires – for us, that is, in the ordeal of the absence of epoch – that we reconsider the question of the drives beyond so-called poststructuralist philosophy, a task whose outlines I attempted to sketch in *Pharmacologie du Front national*.[65]

Through the immense process of *unlearning* [*désapprentissage*] of which automatic society is the most advanced stage, we have unlearned the beautiful and the ugly as well as the good, the bad and the sublime. And only the sublime can *reverse* these terms and cause them to shift from one to another. As *negative* sublimation, which is the consequence of negative identification, this reversal can take place in two directions.[66]

We have unlearned the differentiation that is *différance* by confusing differences with oppositions. A society is founded on learning [*apprentissages*] that supports the capacity to make differences. These differences are the basis of organic solidarity as Durkheim understands and studies it, and they present themselves horizontally and vertically as *affects*.[67]

The unbinding of the drives is precisely a *disaffection* [*désaffection*] as well as a *withdrawal* [*désaffectation*] that generates disbelief, miscreance and discredit, that is, an *uncontrollable* becoming of societies of control and hyper-control[68] – a becoming panicked that is inevitably a becoming mad, and that aggravates all attempts to deny the gravity of the situation via new extensions of control.

Western society has globalized itself as a process of disinhibition of which nihilism and the Anthropocene are the geopolitical and biospherical concretizations. By doing so, and by theorizing itself, after Nietzsche and Freud, via what is referred to on the other side of the Atlantic as poststructuralism, Western society has deconstructed the constructions that constituted the pharmacological basis of organic solidarity through which oppression and domination are legitimated. It did so by showing precisely that these *are* constructions, and not divine or transcendental givens.

85. Pathogenesis and moral philosophy

All this was problematized in the twentieth century without ever really *conceiving* the possibility of the absence of epoch, despite the fact that the 'Exergue' of *Of Grammatology* says that 'absolute danger [...] can only [...] *present itself* as a kind of monstrosity',[69] despite the questions raised in Heidegger's late work and despite Deleuze's late reflections on control societies.[70]

In 1998, after the Bosnia-Herzegovina wars and before 9/11, Jean-Luc Godard opened his *Histoire(s) du cinéma*:

> For nearly fifty years, in the dark, people in darkened theatres have been burning the imaginary to warm up reality.
>
> But from Vienna to Madrid, from Siodmak to Capra, from Paris to Los Angeles and Moscow, from Renoir to Malraux and Dovzhenko, the great directors of fiction were incapable of controlling the vengeance they had put on stage twenty times.[71]

We now discover that, *unless we project ourselves* onto another plane – which could only be the *bifurcation of something genuinely new*, as a motive formed by the future in becoming, beyond innovation that changes everything in order that nothing changes – the deconstruction of the constructions of différance leaves us abandoned in the desert of computational barbarism that ransacks the last vestiges of civilization, on whose outskirts the barbarians roam, those who no longer calculate because they are enraged – mad with rage – and thereby become *themselves incalculable*.

The critique of late *Christian* morality, which fabricates images of hell and paradise, and, still more, the critique of this morality become *bourgeois*, becoming the reality of nihilism and the object of deconstruction, is the critique of the *opposition* of good and evil. Good is constituted *as* good by being *opposed* to evil, while evil nourishes the good that contains it: the good is a trans-formation of evil, which cannot but be substantialized or hypostatized (as 'good'), because it is pharmacological – such is the tragic condition of exosomatization. To be ill [*malade*], that is, to suffer from an evil, if not *from evil*: such is the *constitutive* condition of non-inhuman being and of its specific forms of malady that all stem from ὕβρις.

This is why, after quoting the psychiatrist Henry Ey, who posits that '[m]ental health contains disease – in both senses of the word "contain"',[72] Canguilhem summons up Thomas Mann:

> Thomas Mann writes that 'it is not so easy to decide when madness and disease begin. [...] Never have I heard anything more stupid than that only sick can come from sick. Life is not squeamish, and cares not a fig for morality. It grasps the bold product of disease, devours, digests it, and no sooner takes it to itself than it is health. Before the fact of life's efficacy ... all distinction of disease and health is undone.'[73]

It is with Nietzsche that the question of evil becomes that of disease, which Canguilhem will turn into his thought of the noetic living thing properly speaking as the question of normativity.[74] The opposition of good and evil *denies* this constitutive character by turning it into the fallenness [*déchéance*] of sin. Canguilhem's reappropriation of Nietzsche, however, goes beyond 'great health' by introducing organological technicity into the condition of noetic health – where fallenness instead becomes the always threatening cost of the pharmacogenesis that is exosomatization.

What results from the organological dimension of noetic life realizing its dreams is that test and trial [*épreuve*] of the infidelity of those milieus that noetic life produces by realizing its dreams – a noetic life that constantly disadjusts from itself, constituting the chronic character of its disease and its pharmacology. It is this of which disruption is an unprecedented ordeal [*épreuve*]: the infidelity of the noetic milieu is brought to its highest culmination when it becomes automated understanding that overtakes reason as analysis without synthesis and as absence of epoch.[75]

As the trial and ordeal of the infidelity of the milieu and source of normativity, noesis is itself a kind of illness, and as such it requires a nosology and a nosography. Noesis, which always passes through synthetic and intermittent moments of 'madness' – which is *also* to say, here, of the dream, of dreaming that under the impact of analysis is thereby transindividuated – constitutes the circuits of transindividuation that tie the first stage of the doubly epokhal redoubling to the second, through which a new normativity is established.

The nosology and nosography of noesis as disease can also be practised as *clinical questions of epokhality* such that, sometimes, under the effects of the infidelity of the milieu, the epochs tied together by one form of reason or another find themselves exhausted, engendering the ab-surdity of life and thrusting them into the drive of destruction in all its forms.

As for us, the infidelity of the milieu, carried to its extremes as disruption in the Anthropocene, requires a leap beyond the illness of an age that still does not know how to locate its normativity, because it does not know how to think that new and immense question which comes to us on doves' feet: the question of entropy, which in the twentieth century becomes that of entropology.

12

Thirty-Eight Years Later

86. The political function of dreaming

After 21 April 2002, and over the course of the various national and European elections during which the rate of abstention continued to increase just as did the successes of the far right, it was said that the discourse of campaigning politicians failed to evoke any dream, especially on the left – as if the left should know how to dream better, and be able to dream better, than the right. This kind of statement can only provoke sneers from those who, after Henri Queuille and Edgar Faure, keep repeating that when politicians make promises to their constituents, they are commitments only if they believe in them.

To sneer is not to laugh – it is even the very opposite: a sneering wisecrack of this kind no longer makes anyone laugh, except fools. Because what this wisecrack describes has become the norm, generating an immense feeling of revulsion at the deception and producing a profound exasperation that exacerbates antagonisms and phantasms. The political reality to which this kind of joke refers plunges the population into dangerous disarray: instead of making people laugh, it enrages them.

What was previously a 'wisecrack' [*mot d'esprit*], however, does raise important questions: what is the status of the promise in politics, what does it incite insofar as it is a kind of waking dream, and what are the conditions of its eventual realization?

A noetic dream is a dream that can be realized – as exosomatization, the condition of which is thus the dream. This occurs firstly as the first stage of the doubly epokhal redoubling, and through the production of a new *pharmakon*. In Western civilization, politics and the dreams from which it stems essentially organize the process of collective individuation via circuits of transindividuation through which a social group *fully and positively realizes* this exosomatization – and does so by creating a new social reality, as a new social *body*, a new way of coming together [*faire corps*] by collectively individuating exosomatization: by epokhalizing it.

Such collective individuation is a therapeutics that organizes the arrangement between *pharmaka* – and that can clearly also limit, condition or even prohibit *pharmaka*, as is the case with many drugs. This therapeutics comes out of an organological situation in which psychosomatic organs, technical organs and social organizations constitute an epoch as a stage of an accomplished exosomatization – and where the arrangements between these are the objects of general organology.

The exosomatization of *pharmaka* itself evolves (as the first stage of a new epokhal redoubling, bearing a new epoch) under the close or distant effects of an infidelity of the milieu. It always tries to resolve the new and pathological tension (pathological in Canguilhem's sense) induced in the technical milieu as it becomes unfaithful [*infidèle*], that is, as it disadjusts itself or its initial exogenous conditions (for example, in terms of natural resources) through a new organological realization that is always itself pharmacological – that is, pathogenic. This is what *Civilization and Its Discontents* describes.[1]

The social realization of a new stage, or some new consequence, of exosomatization is political insofar as it conforms to positive rules of law, where law [*droit*] is itself prescribed (as a 'therapeutics' characteristic of an epoch – of which mores are the implicit rules) by legislators legitimately appointed according to rules that are called laws [*lois*], constituting a regime, and establishing or continuing an epokhality concretized by new rules and meta-rules of common life.

The noetic dream, therefore, is *realized in layers*: the first stage of the epokhal redoubling stems from a work-knowledge [*savoir-faire*] that overturns the existing organological horizon, while the second stage of the epokhal redoubling stems from conceptual knowledge [*savoir concevoir*] (knowledge of critical noesis), which is itself the source of new forms of life-knowledge [*savoir-vivre*] (mores – which themselves vary more or less according to the variations in the knowledge of how to do and conceive: according to the tensions that are always maintained within these different spheres, and that constitute their dynamic potentials).

Hence it is that the dream is the *condition of politics*. But it can constitute a political *promise* only on the condition of being noetic, where noesis finds its source in this faculty of dreaming that is always also a technesis.[2] Let us call such a promise a political noesis. It is then possible to interpret Henri Queuille's disillusioned wisecrack in a slightly different way from that which was imposed on France on the basis of the so-called 'radical-socialist' tradition.

If the commitment to promises is on the side of those who *believe* in them, the latter, who try to *constitute a new law*, who *promise themselves* new laws or rights, must not simply delegate to their representatives the legislative and executive functions of the realization of promises, but actively *contribute* to the *conceptualization* and the *concretization* of the noetic dreams from which they derive: to their *effective noetization*, and as the formation of circuits of transindividuation. Such effectiveness of the concept and the promise from which it stems is what Hegel called *Wirklichkeit* as the fulfilment of 'actually effective' history.

This last consideration, which thereby advocates a *contributory politics*, presupposes that 'capabilization' [*capacitation*],[3] which lies at the heart of the *contributory economy* such as it is conceived by Ars Industrialis, also becomes the condition of *transindividuation as the realization of a political noesis* – establishing a new epoch of political debate, giving rise to the emergence of new psychosocial individuations of citizenship and defining new democratic and republican rules and laws.

The evolution of today's 'dissocieties'[4] (which are disintegrated and 'disrupted' hyper-industrial societies) towards such a social reinvention, which obviously seems *absolutely improbable* and *unexpected*, is nevertheless the *condition* on which any bifurcation beyond the Anthropocene depends, as a redefinition and reinvention of knowledge in all its forms, and as a way out of the process of self-destruction to which the current disruption is leading. And this gives meaning as never before to Fragment 18 of Heraclitus: 'One who does not hope for the unhoped-for will not find it: it is undiscoverable so long as it is inaccessible.'[5] The realization of noesis insofar as it stems from noetic dreaming is what all institutions have the responsibility of producing (by instituting it) and maintaining (by reproducing themselves). But in the age of disruption, the *reproduction* of institutions must give way to their *neganthropic evolution*. Among political institutions, establishments for research, teaching and training are the *sine qua non* for the reproduction, *transmission* and *trans-formation* of circuits of transindividuation.

In the context of disruption – where, short-circuiting the whole process of adjustment between the technical system and the social systems (which is the concrete and effective reality of transindividuation as *social cohesion* [*faire corps social*]), institutions are as such and in advance invalidated (since they always arrive too late, enabling the growth of the desert of legal vacuums to which these short-circuits give rise) – *research, teaching and training establishments must be reinvented from top to bottom*.

This requires that *knowledge* itself be *redefined from top to bottom* on the basis of an organological and pharmacological perspective, according to new models of the goals, organization and functioning of these establishments of capabilization.[6] I will return to this subject in *La Société automatique 2* from the perspective of 'digital studies'. The latter consists above all in reconsidering the social nature and role of *forms of knowledge as forms of life containing (in both senses of the word 'contain') the* ὕβρις *that exosomatic organogenesis constitutes in a structural way*.

The faculty of dreaming lies at the origin of noesis because the latter is *thoroughly* and *originally* neganthropological. Noesis, as the source of *all* forms of knowledge (work-knowledge [*savoir-faire*], that is, knowing how to exosomatize so as to subsist; life knowledge [*savoir-vivre*], that is, knowing how to live noetically in exosomatization so as to ek-sist; and conceptual- and spiritualizing-knowledge [*savoir concevoir et spiritualiser*], knowing how to transindividuate in coming together [*faisant corps*] in exosomatization in the direction of consistences), is what opens opportunities for bifurcations that are inscribed in the real, but that are accessible only as and through the epokhal *derealization* of this real. This derealizable real is complex, because it is constitutively out of phase [*déphasé*] – being *at once social, vital and physical*. Always *in default*, this phase-shifting [*déphasage*] is ὕβρις as such (*delinquere*).[7]

The real is processual. Real processes are overwhelmingly entropic when they are purely physical, and negentropic when they stem from processes of vital individuation or psychosocial individuation. The ability of living things to produce *différance from entropy* is what diversifies itself in all living species arising from evolution, whose noetic form, as the *creation of a social body* [*faire corps social*], is not just negentropic, but neganthropological, because it is not just organic, but organological: exo-somatic.

What the current disruption renders impossible is the noetic faculty of dreaming precisely insofar as it is neganthropological – which it replaces with a constant incitement to *fantasize in vain*, for nothing: *nihil*. In this way, it amounts to a fulfilment of nihilism that we can call *dis-integrating*, as a submission to drive-based and mimetic protentional models that short-circuit the faculty of dreaming, while giving rise to immense and extremely dangerous frustrations.

These frustrations are themselves systemically exploited and intensified – which installs both the *denoetization* that stems from *systemic stupidity and madness* (as the 'propensity to madness' and 'functional stupidity' inherent to the process of disinhibition) and the *economic insolvency* that leads rapidly to planetary ruin – so long as no bifurcation has occurred that would amount to a 'turning point' in this disruptive period that is at present pushing the Entropocene to its extremes.

87. Dreaming together

As Canguilhem shows with respect to biology and as Whitehead shows with respect to reason in general, forms of knowledge are functions (in Canguilhem's case, these functions are organological, but this is not so for Whitehead) of the psychosocial forms through which noetic life takes shape [*fait corps*].

What Valéry and Binswanger (as read by Foucault) tell us, in very different registers, as does Pindar interpreted by Miyazaki, is that the *epokhē* of the real in which noesis consists, as the power to provoke a neganthropic bifurcation, presupposes the realizable dream. In other words, it presupposes the capacity for *waking dreaming in meditation*, that meditation whose oneiric tenor Foucault explored in Descartes. Hence it presupposes the capacity to conceive – that is, to *engender* – the theoretical and practical conditions of the noetic realization of this *epokhē*.

From the side of collective existence in all its forms, the faculty of *dreaming together* – for example, the *geometric point*, the dream of which is collectively maintained by learning geometry, and where this point, like all consistences, does not exist, no more than does justice, which is a promise in that, while it may be inaccessible, it nevertheless guides all non-inhuman behaviour – this faculty of *dreaming so as to cohere* [*faire corps*] is the condition and the resource of noesis *that requires tertiary retention* (where what follows from this is that we must revisit the status of imagination in Kant, but in a different way than that attempted by Žižek), which is thus both the result of exosomatization and its condition.

The consistence of justice, and, beyond that, of all motives – which, ever since the beginning of exosomatization, have more or less *metastabilized* themselves through all the forms of noetic life – is not the inscription of an *eternal* necessity constituting being *beyond* the world-as-becoming. It is what, *in* the entropic process that is the *cosmos*, constitutes conditions of (im)possibilities (pharmacological conditions) of *localizations of negentropic différances*, of which we ourselves, as non-inhuman beings, are the neganthropological form.

As such, these forms are protentions. This is why *Being and Time* is concerned, not with a new formulation of the question of being, or with a new conception of being as the 'questioning of being', but with the specification of the question of time qua futural capacity. The latter, however, is conditioned by the noetic faculty of dreaming: it is upon this possibility of the possibility that Binswanger meditates [*songe*], which he does on the basis of the dreams [*songe*] of his patients – including Aby Warburg.[8]

In general terms, consistences are motives of oneiric projections *onto screens* (operating through an atranscendental archi-cinema), and, as such, they are the *sine qua non* conditions of processes of neganthropological différance – where empathy for the living and repugnance at destroying it or making it suffer are principles of this *economy* that is différance.

After Freud, the dream is also that which constitutes the scene of the unconscious. This means, on the one hand, that reason must be conceived as the motive of the *desire* that is noesis, which in any case philosophy takes as its starting point (from Diotima to Aristotle), and, on the other hand, that the noetic dream is above all the fruit of a libidinal economy that must be *cultivated and maintained*, being perpetually exposed to the possibility of its diseconomy, whether local or general, a possibility that amounts to that of ὕβρις – but where, nevertheless, the latter is itself always the potential bearer of a fruitful bifurcation.

It is starting from this primordial noetic oneirism and its archi-cinema[9] that we must pose the question of the status of the promise in social relations in general, and in political and institutional relations in particular, as well as in the economy today, both in terms of the market in general and the financial market in particular.

Starting from the question of the will inasmuch as this question was completely reconfigured by Schopenhauer, Nietzsche made the promise a specific trait of the noetic soul, and Derrida made it his constant concern at the end of his own life and work. Dreaming and promising are, here, modalities and conditions of protention and also of retention (of the retained [*retenue*]), and of their realization (of holding out [*tenue*]) as expectation and attention. This is why, in *On the Genealogy of Morality*, the animal who promises (the *Neganthropos*) is also the one who cultivates mnemotechnics.

The *fulfilment* of promises and the *realization* of dreams, as well as the *différance* of promises (as the consistence of what, not existing, is therefore all the more necessary – as 'regulatory idea'), stem from knowledge in all its forms. The industrial economy made conceptual knowledge its principal production function, and did so by short-circuiting life-knowledge, work-knowledge and theoretical knowledge – via machinism in the nineteenth century, via marketing and the mass media in the twentieth century, and via algorithmic governmentality in the twenty-first century. This meant, however, that the political function, in its failure to take the measure of this situation, found itself dispossessed of any critical capacity founded on the noetic faculty of dreaming.

The faculty of dreaming is then turned into a process of denoetization – and, in France, a history that extends from Mitterrand to Sarkozy and Strauss-Kahn bears witness to the summoning of marketing in the service of politicians against politics, that is, against citizens. So far as I understand the situation, those who claimed that we must make it possible for voters to dream in order to struggle against voter abstention raised the question of political dreams after the nightmare orchestrated by Jacques Séguéla.[10] But in the very way this question was raised, and in such circumstances, it was already drained of meaning. Were this not the case, they would not even have needed to ask it: they would themselves have dreamed it. But they do not dream.

In France, this was especially true after François Mitterrand, who in this way became the very model for both François Hollande *and* Nicolas Sarkozy. Mitterrand, however, still cared enough to ensure that his manipulations of the faculty of dreaming served his own historical dreams: he still dreamed of strutting the historical stage – which is, indeed, a succession of dreams.

The effect of the professionalization of politicians has been that they couldn't care less about history, in which they no longer believe any more than does Florian. Politics has thus turned into just one more job market among others, where politicians are subject to the evolution of the economy of promises, as is the case in all other markets, and, where, in the wake of the process of disinhibition (miserably exemplified by the fall of Strauss-Kahn), the improbable and neganthropic promise turns into the calculated risk of the nihilist in search of opportunities.

In other words, calculated promises are not genuine promises (this is the meaning of Derrida's work on this score). And this is why they are virtually condemned to become lies – such is the downfall [*déchéance*] of Hollande calling for the downfall of those who thereby become scapegoats (here I am obviously referring not to terrorists themselves, but to those who are designated as being so in potential).

Promises that arise from noetic dreams cannot be fulfilled or realized by mere calculations, however complex and well-equipped [*outillés*] they may be, but only by literally improbable bifurcations.[11] Pseudo-promises, promises that do not arise from noetic dreams, decompose almost immediately into lies, just as, in Charles Perrault's 'The Fairies', the words of the evil daughter turn into vipers and toads that fall from her mouth.

88. Flowers, pearls, diamonds and the King's son

If dreaming is the condition of exosomatization,[12] and if, as noetic hallucination, it allows not only the generation of new technical organs, but also, as the transindividuation of the epokhal redoubling, the generation of works, knowledge and the organizations that these require (institutional systems, economic and monetary apparatuses and instruments, collective enterprises of all kinds, and so on), if all this is true, then the *question* of politics is, indeed, one of fostering dreams in those to whom it is addressed through significant historical figures, who become so only to the extent that they succeed in *realizing* such dreams.

It is in this way that Martin Luther King leaves his mark in the global memory of the twentieth century. As a sublime figure of non-inhuman being, he embodies the dreamer whose dreams, in being realized, become global. The noetic dreamer realizes his dreams firstly by playing: play is a kind of waking dream.[13] The noetic dream constitutes an epoch through a generation, and as its *awakening*:

> Awakening is a gradual process that imposes itself both in the life of the individual and in that of the generation. Sleep is the first phase of this process. For a generation, the experience of youth has many things in

common with that of the dream. Its historical figure is an oneiric figure. Each epoch has such a side, turned towards dreams, which is its childish side.[14]

The play and games through which the child individuates itself and constructs its psychic apparatus become, in the course of adult development, *work* as the *power to make other adults dream (the power to open up)* in these transitional spaces that, in some way, forms of knowledge always constitute.[15] If this were not the case, then the arbitrary rules in which life-knowledge (mores) consists would have no ascendancy over those who share it, those who, in so doing, trans-form it by trans-forming themselves.

The destruction of the faculty of dreaming is necessarily also that of this ascendancy, and it thus leads to the generalization of incivility – from the managing director of the International Monetary Fund to the 'savages' that Jean-Pierre Chevènement believes he can and must refer to with this term, via the shameless lies of President Hollande – while Florian, who is not uncivil, practises this wounding truth that arises from every παρρησία, which, afterwards, and as the bifurcation that it foreshadows, transforms into flowers and diamonds.

> As soon as she had arrived at the spring, she saw a lady, magnificently dressed, approaching from the wood, who came up and asked for a drink. She was the fairy who had appeared to her sister, but she had made herself look and dress like a princess, so as to see how far this daughter's rudeness would go. 'Do you think I've come here just to give you a drink?' said this proud, rude girl. 'I'm supposed to have brought a silver jug on purpose, am I, for Madam to drink from? As far as I'm concerned you can drink straight out of the stream, if you want.'
>
> 'That is not very polite', said the fairy, without getting angry. 'Very well, then; since you are so disobliging, the gift that I give you is this: at every word you say, a toad or a viper will come out of your mouth.'
>
> As soon as her mother saw her, she cried out: 'Well, daughter?'
>
> 'Well, mother?' replied the rude girl, and spat out two vipers and two toads.
>
> 'Oh Heavens!' exclaimed the mother, 'what's happened? This is all because of her sister; I'll see she pays for it.' And she rushed off at once to give her a beating. The poor child ran away and escaped into the forest nearby.
>
> The King's son, who was on his way back from hunting, met her there, and seeing how beautiful she was, he asked her what she was doing all alone, and what had made her cry. 'Alas, sir! it was my mother, who chased me out of the house.' The King's son, seeing five or six pearls and as many diamonds coming from her, asked her to explain how this could be. She told him the whole story. The King's son fell in love with her, and,

considering that the gift she had was worth more than any dowry that another girl could have, he took her back to his father's palace, where he married her.

As for her sister, she made herself so hateful that her own mother chased her out of the house, and the wretched girl, after a long time going from place to place without finding anyone to take her in, went off to die at the edge of a wood.[16]

89. Politics and interpretation of ὕβρις

Today's disruption destroys the faculty of dreaming that, when it dreams its own conditions of realization, whether in a nocturnal way, as in the case of Valéry's mathematical dream, or in a diurnal way, as the Cartesian meditation presents itself, opens up the possibility of effecting [*opérer*] bifurcations through which anthropic *Anthropos* becomes neganthropic, thus preparing the way for its own future by making it irreducible to becoming, to which, as an agent of entropy, it nevertheless also contributes in a pharmacological manner.

Ever since the French Revolution, the parties of left and right have fought over this faculty of dreaming inasmuch as it revealed itself to indeed be the faculty of 'changing the world', and, along with it, of changing nature, so as to become its 'masters and possessors'. This in turn was realized as the modern dream, and, with the industrial revolution, as the entry into the Anthropocene qua intensification of the process of disinhibition that began with the Renaissance.

Like many of my peers (those of my generation), I affiliated myself with so-called left-wing thought because I saw in it a promising collectivity, one that was alive and struggling for the realization of the noetic dream that emerged from the Enlightenment. To defend a politics of the left was firstly to defend a theoretical perspective, that is, a right – a *legitimate right* [*un droit en droit*] intrinsically in advance of the real, itself intrinsically correctable and perfectible, but which, *in fact*, and, as we said at the time, 'dialectically', itself precedes the law.

This seemingly circular causality is in reality the hermeneutic phase-shifting contained by the threefold transduction induced by exosomatization, and as its ὕβρις: such is the dynamic of the three strands of individuation (psychic, collective and technical[17]) that organology tries to conceive in terms of a threefold transductive relation.

With the concept of transduction, it is no longer a matter of conceiving the terms of a 'dialectic', but, on the basis of the *phaseshifting* [*déphasage*] of this circular causality as it contains ὕβρις, of *interpreting* it as a pharmacological *spirit* from which it is always necessary to *extract the quasi-causality of a rule and a law* (which is a difference in the différance of a repetition the conditions of which are altered by tertiary retention in each new epoch).

A 'left-wing' thought is what considers in facts that which exceeds them as the laws that they conceal, that they require, and which fall within a function of reason that sets them up as the *condition of possibility, après coup*, of such facts. It is necessary to redress facts with rules of law, so that, indeed, in law and not just in fact, they can *last and intensify the durability of forms of life that emerge therefrom* – in the sense of 'forms of life' referred to by Canguilhem. Of course, there is 'right-wing' thought that thinks this way – and it often goes much further than 'left-wing' thought.[18]

To admit this does not mean that right and left will thereby be dissolved into one another. It is, again, a matter of doing justice to the quasi-causal logic of the *pharmakon*. In this pharmacology, what continues to distinguish right and left today is the status of *calculation*, and this is what keeps me firmly anchored to the side of the latter – which, by its very name (left, *sinistral*), says something about what, in the *accident*, is not reducible to the probable, and requires the improbable of which the dream is the reserve (the μέτρον).

Nevertheless, calculation here is not what must be rejected or treated pejoratively: it is what, through critique, must be limited by reason. It is precisely the rejection of calculation (like the rejection of the determination of the 'indeterminate' that is being-for-death, which I have here called archi-protention) that, in *Being and Time*, keeps Heidegger locked within the metaphysics he is trying to deconstruct – as well as within, not just the 'right-wing' of his age, but the National Socialist far right.

Conversely, Marxist thought, which sees in the historical effectiveness of the relations of production the 'infrastructural' causal model of 'superstructural' relations, believes that it can reduce the latter to such calculations.

Under the influence of marketing, which is essentially the technology by which control is taken of the faculty of dreaming (something that Foucault failed to address in his analyses of neoliberalism), it is noesis as the faculty of dreaming that is individually and collectively short-circuited and proletarianized by the specialists of the dream industry, where the latter also produces the cinema discussed by Godard in 1998.

Politicians have failed to critique or to think the limits of law emerging from the faculty of dreaming in the epoch of this dream industry. As representatives of political parties that have themselves become temporary employment agencies, they have learned to ignore the historico-political function of the dream. Hence is installed the absence of epoch and its epidemic of the drives, whereby the social body decomposes and starts to fester.

In so doing, politicians have destroyed their parties. A political party is an organism of *partial transindividuation*, that is, it represents a part of society that distinguishes itself from another part, itself represented by some other party. The 'marketization' of politics that began in France in the 1980s with Mitterrand and Séguéla, manufacturing the 'Mitterrand generation' like industrialists of the dream, is what inaugurated the movement 'à la française' that systemically destroyed the faculty of dreaming.[19] Hence it was that minds were *functionally locked into dreams that can never be realized.*[20]

90. Worstward Ho

In 1994 or 1995, I suggested to Sylviane Agacinski that we discuss the political situation at the time, and the prospects for which I felt it was possible to hope. I had known her at the Collège international de philosophie in the 1980s. She had since become the companion of Lionel Jospin. She was a lively, beautiful, friendly and warm person, and she was someone of whom I was extremely fond.

We did not at all, however, share the same political analysis: her ideas on the subject were typical, in my view, of the great critical weakness of French social democracy. But she was intelligent and sincere, and we had over a long period of time built up a relationship of trust. Convinced that Lionel Jospin would play an important role in the coming years, I wanted to enter with her, and with him through her, into a dialogue on the future consequences of the development of what was then still referred to as 'information technology'.

I had met Jospin with Sylviane during a post-thesis party [*pot de thèse*], and I was struck by his cordiality. Furthermore, like everyone else, I listened to his discourse attentively, especially with respect to his desire to take stock of Mitterrandism and to exercise a 'right to make an inventory'. Richard Beardsworth and George Collins, with whom I spoke a great deal during that period – they were translating *Technics and Time, 1* into English – had urged me to take the initiative at this meeting.

They and I suspected that the left would come back to power, and we believed in the absolute necessity of Jospin and his entourage understanding and integrating that of which the three of us were convinced: the great economic, social and political questions of the years and decades to come would be *overwhelmed (disrupted)* by the completely new conditions in which technology was going to develop with the digitalization of the technical system – of which the World Wide Web had been the initiating factor in April 1993.

I met Sylviane in a bar opposite the École des hautes études en sciences sociales, where she taught. After catching up on her news, I began by explaining to her – by trying to explain – what it was I wanted to talk about with Jospin. But she quickly interrupted me, saying in an impatient tone: 'Listen, if you want to talk about these things, all you have to do is join the Socialist Party. Nowadays, the party has a quite brilliant member, Pierre Moscovici, and he's the guy you really need to speak to.'

As politely as possible, I brought the conversation to an end. I knew Moscovici well. Ten years earlier, when I was at the firm TEN, to which I had been recruited by Claude Neuschwander,[21] he had been there for a time himself, as an ENA trainee.[22] He was the living caricature of the pretentious and cynical careerist, totally devoid of ideas about the possibility of changing the French or European situation, and with absolutely no interest in reflections of this type – which were for him useful only as election campaign trickery.

I left Sylviane disappointed and concerned. Eighteen years later, Moscovici was a minister under Hollande. Today, he is a European Commissioner. Since that time, I ask myself what lessons Jospin and Sylviane, who became his wife, have drawn from these miserable developments, from the escapades and indiscretions of their Dominique and the lies of little François.

In April 2005, George Collins, Marc Crépon, Catherine Perret, Caroline Stiegler and myself founded Ars Industrialis, an international association for an industrial politics of technologies of the spirit. The manifesto[23] of this organization referred to the political economy of 'spirit value' developed by Paul Valéry in *Regards sur le monde actuel*.[24] In *The Re-Enchantment of the World*, published in French in 2006, we again took up and developed these ideas.

From the beginning, we at Ars Industrialis argued that consumerist capitalism and its culture industries had systematically exploited, exhausted and ultimately destroyed the libidinal economy, while installing a financialization guided by the drives that liquidates investment and replaces it with speculation. We argued that this was bound to lead to a generalized diseconomy – and a general violence.

In 2010, after our expectations were confirmed by the 2008 crash, we published a new manifesto. It emphasized the psychic, social and economic divide resulting from 2008 – which was for us not merely a financial crisis, but a demonstration of the insolvent state of consumerism, implying the need to shift to what we began calling a contributory economy.

Over the course of those years, we focused not just on analysing and denouncing the limits of carelessness and neglect to which the conservative revolution and the generalization of systemic stupidity had given rise, but on proposing a pharmacological politics capable of *reversing* the effects of those technologies that had accompanied all these evolutions, by giving them the economic and geopolitical means to become 'quasi-causes', which in this case means to be capable of trans-forming facts into new laws and knowledge.

We affirmed, and we continue to affirm, that these economic and geopolitical means are clearly within reach for the European continent, and that the latter will survive as an actor involved in the global future only provided that it implements an alternative industrial politics, one capable of preserving its interests and those of the entire planet, instead of following the prescription of American ideologues as the European Commission currently does in suicidal fashion.

On 14 May 2011, I presided over a meeting of the Ars Industrialis board of directors. At the time, France was in pre-election campaign mode, and the two frontrunners were Nicolas Sarkozy and Dominique Strauss-Kahn. Before getting to the formal agenda, I opened a discussion on the economic and political situation, and, in relation to the latest news, I expressed my discomfort at seeing Strauss-Kahn riding around in a Porsche belonging to 'Ramzi Khiroun, one of DSK's communications gurus and closest followers'.[25] My problem was not that it was a luxury car, but that it was a grotesque symbol of the most vulgar parvenus of 'communication', and in that way quite in keeping with Nicolas Sarkozy's Rolex – and with Séguéla's industrial destruction of dreams.

My comments produced a strong reaction from one of us, who objected that these issues were of no interest. Another invited me to temper my point of view, pointing out that this was really a private matter. The next morning, news broke of what would come to be called the Sofitel affair, which was also a private matter but quickly became public, because it was soon to become a legal matter.

Strauss-Kahn withdrew his candidacy. A primary was held through which it was determined that Hollande would be the candidate. In February 2012, we published a dossier on the *Télérama* magazine website, arguing that 'the true issue of 2012 is 2017'.[26] The theses developed there were taken up in 2013 in *Pharmacologie du Front national*, which Manuel Valls did not read.[27] A week before Hollande's election, 38 per cent of those polled stated that they were in agreement with the ideas of Marine Le Pen. With the publication of her book, this figure rose above 40 per cent.

91. Laroxyl and writing

In the summer of 2014, five months before the *Charlie Hebdo* massacre and in a social and political context that was becoming overwhelming and deleterious, I found myself retracing the steps that, long before my incarceration, had once led me to a psychiatric clinic in Loir-et-Cher: seriously depressed, I made the decision to consult the psychiatrist who had treated me in 1971, when, at the age of nineteen, I had a 'psychotic episode' – after which a psychoanalyst had sent me to this particular clinic for an apomorphine treatment for alcohol detoxification.

At the beginning of August, three weeks before the beginning of the fourth summer academy at Épineuil-le-Fleuriel, entitled 'For a new critique of anthropology: dream, cinema, brain', finding myself increasingly obsessed by death, that is, by what I *projected* as being my death, and by the latter as my deliverance, waking up every night haunted by this suicidal urge, I called, somewhat at random, this clinic where I had received treatment. I asked for urgent help, seeming, so I thought, to be suffering from some kind of early dementia, and succeeded in getting hold of a psychiatrist on duty who agreed to see me immediately.

I got into my car and drove the 200 kilometres that separated Épineuil from the La Chesnaie clinic, which I had not seen again after what had turned into sleep treatment and then into a course of psychotropic drugs. Early one morning in late autumn 1971, I had walked out of La Chesnaie and hitchhiked my way out of there, after a conversation with a patient made me realize that, if I didn't leave, then, as had happened to him, I would still be there ten years later.

The doctor who saw me at La Chesnaie during the summer of 2014 went back to my file, listened to me, recommended I see the psychiatrist who had treated me in 1971, and prescribed an antidepressant called Laroxyl.[28] Two months later, I made an appointment with my former doctor, who in 1971 had been quite young. He was now retired, but he would still give the occasional consultation.

He prescribed a fresh bottle of Laroxyl. I told him that, of course, I would take this antidepressant, which I clearly needed and which had allowed me to get some sleep, and that I had indeed come for that, too. But I added that I was worried that this *pharmakon* might prevent me from writing, that is, from working – that is, from treating *myself*. And I began to tell him how I had tried to treat myself in prison through this 'technique of the self' that we call writing, but also through reading – two practices of another *pharmakon*.

I began writing in prison, and I did so almost without being aware of it – feeling my way into the invention of a therapy that, fortunately, I could continue to practise after I was '*levée d'écrou*' [taken off the prison register, that is, released], as they say in prison administration. What I am here calling 'writing' is not at all what, having arrived in prison, I believed in doing or wanted to do at the end of two or three weeks of incarceration, which consisted in a pathetic attempt to 'do literature'.

During my adolescent years, in addition to playing the saxophone, writing poems and novels had been one of my unrealized dreams – or what we call fantasies. After arriving at Saint-Michel, I spent the first months trying in vain, and on countless pages now lost, to tell a story that never took any form other than the same fruitless effort to write.

This did, however, have the advantage of leading me to see that, ultimately, I had *nothing whatsoever to say*, nothing to tell or write to anyone else – not even to myself. Hence I began to learn the difficulty, the futility [*vanité*] and the necessity of writing, three moments that must be encountered in order to begin to write, by forgetting, in this confrontation with the difficult, the vain and the necessary, what we had started writing and would like to write.

My relationship to literature and to reading was thrown into complete disarray by this failure, which had nevertheless allowed me, while awaiting something better, to not feel totally stuck in the incredible void that opened up as the yawning gulf of my imprisonment. This eventually led me to *read* – to read with an intensity greater than I had ever imagined myself capable of, to the point that I would almost entirely *become* the texts that I was reading, eventually discovering that, despite so many readings I had believed I had previously undertaken, perhaps I had never before begun to read *in actuality*.[29]

Every evening, after my 'bowl', which happened around six o'clock, I read novels. Having come to understand the fact that I would not be a writer, parallel to the novels I read in the evening, in the morning I began to study, with an immense and increasing appetite, the works of poetry I have already mentioned,[30] after closely rereading Saussure's *Course in General Linguistics*.

I wanted to know everything about linguistics. And finally, after having passed the special entrance exam for those who had not finished secondary school, I enrolled in this discipline at the Université de Toulouse-Le Mirail, through the distance education programme – my teacher's name was Nespoulous. By then, I had already abandoned my project to make linguistics, semiotics, poetics, narratology, textics,[31] intertextuality or ancient or modern letters the main subjects of my studies: I had decided to study philosophy. That would be the path through which my adolescent dreams would be realized.

92. The ordeal of the *pharmakon* as the fall into insignificance

For I had understood, or believed I understood, that the question of language, in the golden silence in which I lived, had finally become my principal interest. I had begun to think that, rather than the signified/ signifier pair, it was necessary to conceive and especially to *practise* what was then called the signifier [*signifiant*] (after Saussure, Lacan, Lévi-Strauss and Barthes) as the *tension between the insignificant and the non-insignificant*. This later led me to translate ἀ-λήθεια (*a-lētheia*) by signi-ficance [*signi-fiance*], a significance *making* signs as *non-insignificance*, as *coming out of* [*sortant*] the insignificant, and as the extra-ordinary *step out of* [*sort*] the ordinary.

I had understood, or believed I understood, that all this, as a *question* of language, began with philosophy, which was, moreover, as this theory and this practice of significance after the insignificant, a primordial and constant issue for me, who, no longer speaking, found myself incessantly practising the soliloquy.

And from this fact, the phenomenological experience of the world in the absence of the world was also, first and foremost, that of language in the absence of speech. It seemed that all that remained for me, in the absence of the world, was the memory of the world and the mnesic, visual, sonorous, olfactory, tactile and verbal traces through which I kept it within me.

The question of language became that of these ghosts of which language in the absence of speech is the most common experience, and yet the most extra-ordinary, the most oneiric, the most noetic and the maddest. This is how I read Mallarmé, and how it would be necessary to interrogate Hölderlin, Nerval and all those madmen who would make philosophy think so much about the death of God – beginning with the solitary Nietzsche, obviously.[32]

The question of language – as the ordeal of insignificance and its continuation throughout the long history of the West as the accomplishment of nihilism, that is, as the destruction of all significance and the expansion of insignificance to the point of encompassing everything, up until the absence of epoch into which we were entering at the moment I was entering prison – was imposed in Greece, in Socrates, and in the trial of sophistry, itself stemming from a *falling prey* to literal grammatization that, nevertheless, lay at the origin of the new relation to language that appeared with the *polis* under the authority of what the Ionians had already named λόγος.[33]

Investigated as such, considered by philosophy as such, thematized by Plato in *Cratylus*, the question of language is decisively constituted *beyond language* and as the question of *consistence* in λόγος in Aristotle's *Analytics*, on the way to becoming onto-theology, before it becomes the business of the grammarians (there is also an Indian grammar, which I was not aware of at that time), then of linguistics as 'science of language'.

In the course of my experiments with reading, writing and meditating with language, with or without traces as the elementary condition of noesis, I discovered that the material and organological trace, spatializing the mnesic and cerebral temporal trace, is the condition of logical noesis, 'for meditating without leaving any traces becomes evanescent'.[34] Through these readings, *temporalizing* the space and the volume of letters read, and through the writings that I began to *draw* from these readings, as one draws a line in industrial design, which *spatialized* the time of my readings as retentions themselves literal, engaging me, without my perceiving it clearly, in a dialogue with myself – practising, in the manner of Bill Evans revisiting Thelonious Monk,[35] a kind of *re-recording* of what thus becomes myself-an-other [*moi-l'autre*] – I parted ways with Saussure, the only theorist whom I had read prior to my stay at Saint-Michel, and threw myself into what Granel said was to be my philosophy of language (I was at the time studying Wittgenstein with Élisabeth Rigal and Granel, and he told me constantly to read Humboldt).

Michael [*Michel*] fights the *diabolical dragon* in the name of the *crucified symbol*. This is what onto-theology would soon become – the dramatization of this combat in which the forces of good and evil are *opposed* to one another. This is what I would reinterpret as the play of those tendencies referred to by Saussure as the diachronic and the synchronic, but which he does not approach *as* tendencies: he treats them as dimensions of a dichotomous methodology that will then become the structuralist method. In becoming a principle of method, the dynamic of what I here call transindividuation will be erased.

The signified/signifier pair, based on Saussurian thought, which had at first seemed so enlightening, gave way, after reading Derrida, to the question of the *compositional relation* between diachrony (or diabols) and synchrony (or symbols), constituting for psychic and collective individuals the dynamic of two transductively-linked tendencies: significance (*sign-making*) and in-significance (différance as indifference and the vanity of self-reference).

It soon became clear that these tendencies also lie behind Dionysos and Apollo, just as they organize the Freudian pairs, pleasure and reality and Eros and Thanatos, as well as Bergsonian dynamics. Studying the physics of order and disorder with Roger Cavaillès, it was not long before I conceived these tendencies in primordial relation to entropy and negentropy. And, upon reading *On the Soul*, the play of tendencies composing with and against one another finally became the question of *noetic intermittence*, that is, of what I have since called the allegory of the *flying fish*[36] – of which the experience of language is one of the possible versions, and which made this pair of tendencies a question of *rhythm*.

Coming before linguistic questions, *therefore*, are the questions of *noesis* (of the noetic soul), of being, of its 'logic', that is, its categories, of rules of predication and of onto-theology, the latter being presented in Aristotle initially as the question of the 'manifold ways that being expresses itself' in the course of a noesis that is itself intermittent – all of this being reappropriated by Thomism and thereby rendered illegible: this is what Heidegger made it possible for me to understand.

I then began to devour the work of Pierre Aubenque,[37] while following a course given by Annick Jaulin on Aristotle's *Analytics*, *Metaphysics* and *On the Soul* – with the help of Hegel's *History of Philosophy*.

93. Taking notes, consistences and prophets of doom

I have already explained how the question of language began to take shape in the silence of this phenomenological laboratory that was my cell, and in the absence of all communicative practice, where, without respite, I experimented with the extremes of soliloquy while in the morning I read, after a poem by Mallarmé, Husserl's *Logical Investigations*, and, in the evening, Proust's *In Search of Lost Time*.

All this took on immense noetic proportions without my really being aware of it. The uninterrupted character of my tacit experience of language was like an alternately nocturnal and diurnal dream, which, *projected through annotations, summaries and notebooks* that I began to fill, traced its 'guiding thread' while following it out – trying to *realize* it, that is, to spatialize it, however *locally*. My cell was filled with traces: the books brought to me by Granel and that I received by mail, plus the notes and cards I produced and that accumulated on the floor of the cell.

This cell, within which I barely existed, allowed me to discover that to exist is not simply to subsist: it is to project oneself, through intermittences, towards consistences, inasmuch as *any* existence accedes to them by *experiencing them* [*éprouvant*] while being *tested by them* [*éprouvée*]. Ordinarily, the initial experience of consistences is through *relations* that may be filial, affectionate, familial or friendly, but that, after that, they are also experienced through all kinds of events, and *as* these very events.

Events form nodes in a process of transindividuation within which affects are produced (or what I call traumatypes and stereotypes).[38] Affects of all kinds, on the occasion of which consistences are encountered, but also on the occasion of which and in the wake of which they are forgotten, repressed, denied – these affects fall within what Aristotle called φιλία (*philia*). But φιλία is not just, as is said all too often, friendship: as I learned from Jean Lauxerois in his translation of Book 9 of the *Nicomachean Ethics*,[39] in Greek, friendship is φιλότης (*philotēs*), not φιλία.

I was deprived of such relations – but, through the annotations, commentaries and syntheses of my readings, I recounted the great texts that made me dream and that in this way amounted to another type of φιλία, and more precisely the φιλότης discussed by Peter Sloterdijk in 'Rules for the Human Park'.[40] This relation occurred inside an apparently sealed box, and this was possible, indeed, only because, exteriorizing what I was reading, I made it ek-sist, and hence made myself ek-sist, and as an other. This relation did possess, however, the power to go through walls, or to remove them – so much so that my cell became enormous, if not unlimited. Such was the madness that protected me from madness.

The noetic beings that we try to be all accede, precisely insofar as we are noetic, to such consistences, and always do so, more or less, in one way or another, through the innumerable occasions for it offered by knowledge – knowledge of living, doing and conceiving – *so long as we do not prevent it, which is barbarism.* The *generalized* proletarianization that is *characteristic* of the end of the Anthropocene when it becomes the age of disruption and the Entropocene, however, consists precisely in *disintegrating and annihilating knowledge.* Barbarism then becomes *inevitable* – and, along with it, so too do cowardice and submission of all kinds, including that of literary degeneration [*déchéance*], so well embodied by Saint Michel Houellebecq.

Cultures, and those who cultivate and frequent such consistences, follow various paths along which spirits arise, magical forces of the supernatural, the gods of antiquity, the God of the 'people of the Book', or of Buddhism, along with many other forms of spirituality. In ancient Greece, from the seventh century BCE, the relation to the consistences engendered by λόγος is tragic – and it is Prometheus who establishes the lot of mortals.

The Latin words *intellectus* and *spiritus* translate the Greek νοῦς (*nous*), from which the term 'noetic' derives, which thus refers to *both* of these (at once *intellectual* and *spiritual*). This νοῦς, *esprit*, comprises these two dimensions, which with Kant become understanding and reason, as fundamental tendencies and conditions of the 'faculty of knowing', and it does so up until the Enlightenment, realizing itself as revolution – 'French' then 'industrial' – deifies Reason at the very moment when the latter becomes, with the birth of machinic capitalism, a 'production function', installing the tertiary retentions of what Guattari tried to describe as the asignificance of the 'machinic unconscious'.[41]

Hence reason becomes *ratio*, conceived ever more exclusively as computation, and *through that* as this *disinhibition that is then referred to as Progress*, and which, today, at the end of the Anthropocene, in the age of disruption that turns it into an Entropocene, seems *to exhaust all consistences* (that is, all reasons for living, acting and hoping) in the growth of a desert where legal vacuums accumulate and prophets of doom proliferate.

From out of this nightmare comes the drought and the thirst from which emerges a new form of ὕβρις – terrible, suicidal, homicidal – which the political decay [*déchéance*] of the dreams of the left, turning into the nightmare of the deprivation of citizenship [*déchéance de nationalité*], can only aggravate *to even greater extremes*, leading to *ever greater extremes* on the side of 'jihadists', on the one hand, and, on the other hand, of those, increasingly numerous, who take any opportunity to put their persecution of scapegoats into practice.

Disruption is also and above all what is today concretized in the life of everyone via the 'data economy', through which the five billion terrestrial inhabitants who have a mobile telephony subscription[42] *more or less unconsciously produce the traces and metadata through which they annotate themselves*, enabling *automated glosses* to take over and short-circuit their protentional potentials – and, along with that, any interpretation, any individuation, any hermeneutics and any consistence, that is, any *value*.

94. Taking notes, from the birth of θεωρία as the hypomnesic spatialization of ἀνάμνησις to the 'tags' of the data economy – via Saint-Michel

When ancient Greece, which is pious, gives birth to philosophy by passing through the law-givers and founders of cities – poets, geometers and 'physiologists' discoursing on what remains and becomes what we today call nature, which they called φύσις (*phusis*) – consistences come to be *considered as such, for themselves*, and no longer in terms of spirits or gods.

This does not mean that this consideration, which is also called θεωρία (*theoria*), overthrows idols: the gods of Greece are precisely *not* idols. As we have seen with Alain Frontier, they are *markers*: those who mark the *limits* in and through tragic culture, *cultivating* the knowledge of its proper limits, and as the culture belonging to what Ruth Benedict and Eric Dodds will call 'shame cultures', cultures of what the Greeks called αἰδώς (*aidōs*)[43] – which with monotheism will become 'guilt cultures'.[44]

I began, in that cell, to trace back over the traces of the path that, starting from the *hypokeimenon prōton* (ὑποκείμενον πρῶτον) and going through the primordial element first conceived by the Pre-Socratics as fire, air, water or earth, leads from ancient atomism to contemporary physics, passing through the *Rules for the Direction of the Mind* and the *characteristica universalis* before reaching those formalisms that are now increasingly algorithmic, and that automate the treatment of those 'tags' of every kind that are today's 'traces'. I methodically explored the landscapes of the life of the mind and spirit, into which I penetrated with the help of the academic institution and by taking notes, just as Malcolm X did thirty years earlier, and even D., almost forty years later.[45]

I annotated the books I read and I transferred these notes onto loose sheets that I ordered into folders, from which I made cards, which I synthesized. In other words, I reformulated them in the form of arguments that enabled me to take stock of what I had actually read, *effectively* read, to the extent that I was able to *spatialize* these arguments. I transformed everything that I read into writings that served me both as a way of preparing for my exams – since I had become what one calls a student – and as a way of *annotating myself*, in my turn, in order to seek out the other lying there, in front of me, or, rather, around me, literally roaming around, like a spirit, as a promise to come, as a protention of a possible future.

What I am here calling 'the other lying there' did not consist just in *my* notes taken from my readings. It was also the whole library itself (which my cell became), from which I discovered that, even before having read it, it had already framed and woven who I was, as my 'already there', as that 'past that has always already preceded me',[46] the bifurcations produced by all these works having been *incorporated and socialized*, that is, *transindividuated* as *realizations of those noetic dreams* that became the Greek *polis*, the *civitas* of the Roman Empire, the Christianity of Saint Michael, modernity – until the epoch of the absence of epoch, as the nihilism that was the *disembodiment and decomposition of all that this had been.*

In the end, what I understood is that what I had *effectively* read was what I would write and would know how to write on the basis of what I had read (texts that also stemmed, already, from reading notes, from *hypomnēmata*, as I would later discover with Fouault[47]), and that I experienced firstly in a kind of dream. Hence, without being aware of it, without worrying about it, I began to write, in order to realize, in the space of dreams that I created during my readings, the dreams that those who wrote these books had countlessly realized, and which, for centuries, had responded amicably to each other, in the noetic crypt from which it is possible to escape only intermittently.

> Words, of their own accord, are exalted in many facets, recognized as the rarest and most worthy for the mind, the centre of vibratory suspense; which perceives them independently from the ordinary sequence, projected, as on the walls of a cavern, as long as their mobility or principle lasts, being that part of discourse which is not spoken: all of them, before their extinction, being quick to take part in a reciprocity of fires, either at a distance or presented obliquely as a contingency.[48]

What I discovered, walking through the lettered traces of the world and from the language already there before me, is that, in these highly specific circumstances of being deprived of the presence of others, and in *echoing* what had been kept in my memory, existence could come to intensify itself to the extreme, and that it could do so in the direction of consistences.

95. My circuits are screwed up

Of course, this happened only at the limits of madness: as *limits*, and as the absence of all bouquets, these consistences of the (de)fault [*défaut*] that in 'My Books Closed' Mallarmé called a lack (a 'learned lack'[49]) became *alterities in the absence of any other*, non-existent motives of the default of origin, that is, of primordial ὕβρις, forming stars in this perpetual night, constantly 'transfigured'.

Hence I elaborated a method for living, existing and consisting in an infra-existence that still tended to appear, between these 'four walls',[50] like a supra-existence. In addition to reading and synthesizing techniques that I began to punctuate with physical exercises on the floor of my cell, I adopted a daily programme to which I strictly adhered.

It was based on close *rereading* and *rewriting* – on a *systematic practice of repetition.*

After reading a text for the first time in a cursory way between 7:00 a.m. and 9:00 a.m. – which I did after reading a poem at sunrise – I reread, at 9:00 a.m., the text I had read the previous day, while going back over the notes I had taken the day before, where I would generally notice interpretative gaps that were, in truth, often the beginning of what I now call experiences of 'surprehension'.[51] The next day, at around 2:00 p.m., I would prepare a synthesis – after a cup of Ricoré chicory coffee and a Gauloises cigarette.

Through these experiments with reading, writing, rereading and rewriting, I came to redefine the question of signi-ficance (of non-insignificance) as participating in what, in *Meno*, Socrates called ἀνάμνησις (*anamnēsis*). Proust, too, accompanied me along this path – in the evenings, just as Mallarmé did in the mornings. ἀνάμνησις is the fruit of a *disturbance* [*trouble*], which Socrates called ἀπορία (*aporia*), through which, as affect, the set of circuits of transindividuation emerging from an epistemic collective individuation (for example, geometry) re-presents itself in its very presence – in its *Anwesen*, as Heidegger will say.

The *Anwesen* of ἀνάμνησις, and, more generally, of any signi-*ficance*, of any moment through which what *signifies* [*fait signe*] arises from the insignificant, constitutes the *salience* (I borrow this term from René Thom[52]) provoked by a catastrophe in transindividuation inasmuch as it reaches and affects the one in whom the sign is made. A catastrophe, in Thom's theory of catastrophes, is a phase in a morphogenetic process.

Hence arise *semantic storms* – that is, conflicts of interpretation – through which the process of the collective individuation of a μελέτη (*meletē*, commonly translated by meditation, exercise or discipline) is reinvigorated through the one who undergoes the experience of ἀνάμνησις, hermeneutically reactivating circuits that are already constituted, circuits that had hitherto been thought to be well-known, well 'understood', and which, in this reactivation, and on the occasion of this hermeneutic conflict, once again become constitut*ing*, and *reconstituting*, sur-prising, and no longer simply understood: reopening a *future*.

This, however, is what has now mostly become an unfamiliar experience, one that is missing [*fait défaut*], failing to incite the work of opening towards that which is necessary.

My circuits are screwed up,
There's some kind of build up there,
The current can no longer pass through.[53]

96. Detention, retention and protention

I assigned myself a μελέτη that, as a 'technique of the self', defined for each day the hours at which, over the course of weeks and months, I invariably obliged myself to read, annotate, comment and finally write, and then to read again, which always sent me too quickly back to bed for a short night's sleep. I got used to getting up early – with the return of the electric light in the early morning. The prison practice was indeed to turn the lights out pretty early and turn them back on at 5:00 a.m. – but all that must since have changed, unfortunately for the inmates, after television sets were placed in cells.

Apart from a few letters in which I ventured to interpret what I had read, and that I sent to Granel and some of the distance education professors, the little texts that I was producing had no readers other than myself. I wrote them only to reread them, to reduce them, to turn them into commentaries, then commentaries on commentaries, increasingly compacted, which were ultimately organized in large folders that no one would ever read – but which still drive everything I write thirty-five years later, and which are sitting right now, covered with dust, above my desk.

During this *epokhē* of my life, I still did not know that all these amounted to tertiary retentions that I was generating on the basis of the secondary retentions from which my memory was woven: I had not yet read Husserl's *On the Phenomenology of the Consciousness of Internal Time*.[54] And, in reading 'The Origin of Geometry' for the first time, and Derrida's introduction to it,[55] I did not yet understand all the issues that linked this text to what would become the latter's central problematic in *Speech and Phenomena*.[56]

It was not until about 1987, when I began to write the thesis that would form the origin of *Technics and Time*, that I discovered the key difference that Husserl posits between primary retention and secondary retention – whose stakes I have expounded upon many times, and will do once again here.

After Kant, Husserl posits that perception is temporal, and that, given that what is perceived is what is present, the temporality of perception requires that the sensible data that I *retain* in what I perceive (during what he calls impressions) be distinguished from the memories I have of previous perceptions: the latter are from the past, whereas what is perceived, and therefore present, and retained as such by perception, is not, unlike that which is past, the fact of a more or less imaginary reconstitution on the basis of memory traces.

Primary retention amounts to the material of perception, and therefore of the present inasmuch as it presents itself, which is to say that the present is a dynamic process of *presentation*. *Secondary* retention is that of memory, that is, of the past, of what is absent and represented by a dynamic process of *imagination*.

In my own analyses – which are very different from Derrida's – I have tried to show that primary retention is always contaminated by secondary retention, because it is a *selection*, the criteria for which are, precisely, secondary retentions, but where these criteria are susceptible to being controlled through what I call tertiary retention. That primary retention is a selection is made clear by the fact that, if I reread a text a second time, I do not 'primarily' retain exactly the same thing: what I retain depends on what I am *capable* of retaining, and this capacity is constituted by secondary retentions, that is, by my past.

As for tertiary retention, this is what makes it possible to control the play of primary and secondary retentions, and to control this play in terms of selection, either to critique this play by spatializing it – precisely in the form of tertiary retentions (this is what I did while I was taking notes) – or in order to certify a particular path of reasoning, such as, for example, in the case of the protogeometer evoked in 'The Origin of Geometry', or, thirdly, to manipulate the processes by which secondary retentions lead to the selection of primary retentions. This is the accusation that Socrates makes against the Sophists (in a completely different language, of course), and it is how I argued, in 'To Love, to Love Me, to Love Us', that Le Lay has turned television into an industrial activity generating a 'new form of barbarism' – which is what I was teaching my students at the Université de Compiègne on 11 September 2001.

In this cell of Building A at Saint-Michel – from out of which, at the beginning of 1981, I had to shift my books, notebooks and cards, moving them to a cell in Building F of the Muret detention centre – the following process was underway:

- an inmate was locked up with his secondary retentions, more or less reconstituted and imaginary remains of a past that was moving further away with each passing day, a past that, when the detention began, had lasted twenty-six years, and that amounted to the images of a world that was absent because totally gone;
- books, notebooks and cards were, however, added, some written centuries ago, such as the works of Plato, others written while under lock and key, but in either case amounting to various types of literal tertiary retention;
- by reading these literal tertiary retentions, the inmate mobilized his psychic secondary retentions, on the basis of which he tried to interpret the texts that he read or wrote, by selecting what Husserl called primary retentions;
- rereading the next day what he had read the day before, he found that the previous day's selections were not quite identical to those that formed during the rereading;
- trying to interpret these variations from one day to the next, he came

to see, not just that what he had read the previous day modified the organization of his memories, but that, by reading these books, he was also reading and interpreting the more or less textual fabric of his own memory.

Although when I was an inmate I did not yet have the Husserlian concepts of retention and protention at my disposal, I was already trying to describe reading as an interpretation by the reader of his or her own memory through the interpretation of the text that he or she had read. In so doing, I took inspiration from the following analysis by Proust:

> In reality every reader is, while he is reading, the reader of his own self. The writer's work is merely a kind of optical instrument which he offers to the reader to enable him to discern what, without this book, he would perhaps never have perceived in himself. And the recognition by the reader in his own self of what the book says is the proof of its veracity, the contrary also being true, at least to a certain extent, for the difference between the two texts may sometimes be imputed less to the author than to the reader. Besides, the book may be too learned, too obscure for a simple reader, and may therefore present to him a clouded glass through which he cannot read.[57]

I did not know then how to theorize the production of the protentions that resulted from this discipline so that *motives of reading and writing* are projected that in the end generate the book currently being read, generated by someone who does not just try to understand [*comprendre*] what it means [*signifie*], its *significance* [*significations*], but to 'surprehend' what it is that creates signs [*fait signes*], which is its *meaning* [*sens*], this reader trying, therefore, to surprehend this meaning in himself or herself, which Socrates called 'thinking for oneself' and which 'you, reader',[58] as Calvino writes, can find only as *he-the-other* or *she-the-other*.

The creation of the meaning [*faire sens*] of what is significant in the sense of being not insignificant is what – in the *politeia*, then in the Republic of Letters, and eventually in the French Republic dreamed of by Condorcet and concretized with Jules Ferry as the Third Republic – as *protentional projection*, can create an epoch for a community of readers. And this is what is no longer accessible to Florian.

What follows after this is the rise of cinema, which Jean-Michel Frodon called 'national projection' (I am not sure that he really understood the full implications of what he was saying[59]), which then turns into control by electromagnetic broadcast, as Adorno and Horkheimer saw. And this is what, without an epokhal redoubling worthy of the name, that is, without a thought worthy of the name and of the epoch, bears within it the new form of barbarism, where, today, we have a clearer understanding of what this entails.

97. Suffering

When, in the early autumn of 2014, the doctor prescribed for me a new bottle of Laroxyl, I told him of my concern that the effects of this *pharmakon* may diminish my writing activity. I explained to him that my writing was a way of treating myself, and that if there was an aspect of this care that probably maintained my psychic suffering, nevertheless I cultivated the latter in such a way that it could be transformed into a *noetic culture* capable of sublimating it – which must be *regularly practised*, just like physical culture, and practised as a rule, in the broadest sense of this word, which therefore also implies a *meletē*.

I needed this suffering in order to transform it into 'hypomnesic tertiary retentions', which, in the form of what we refer to as books, participate in the writing of new circuits of transindividuation, through which I ek-sist by creating signs [*faisant signes*] pointing towards, as potential collective protentions, that which consists, and consists on the basis of those countless pharmacological shocks retained in the accumulation of tertiary retentions.

It is, *therefore*, through these tertiary retentions, producing our own primary, secondary and tertiary retentions and protentions, that we individuate ourselves psychically and collectively. And we do so because concealed within them are pharmacological shocks, which are retained, just as in the jar of Pandora. These shocks may be more or less ancient, and, because they remain active, they may *return* to affect us – so that we ourselves, being disturbed [*troublés*], become troubling. In being affected, we become 'affecting', and this *affection* is precisely that in which the profound dynamic of transindividuation consists.

Since the end of September 2014, I have had regular discussions about these questions with this doctor, who has played his part by continually reminding me of what is clinically known about depression, delirium and madness, while listening attentively to what, in the contemporary context of negative protention, of disruption and of the Anthropocene, I have stubbornly insisted on reminding him with respect to thinking and what fundamentally ties it to ὕβρις, to μωρία (*moria*), to μανία (*mania*) and to everything we refer to as madness, and equally to the *pharmakon* that treats, cares and heals, but which also destroys, and which, as an agent of the *delinquere* that constitutes the lot of the exosomatized beings that we are, is at stake in all moral suffering as in every joy.

The essence of these 'therapeutic conversations', as the doctor referred to them, dealt with (from my perspective as the patient) what there is that is common, ordinary, banal and sometimes extraordinary in madness, inasmuch as it is precisely not what psychiatry calls, for example, depression [*mélancolie*]. Hence this bore a resemblance to what the psychotherapist would refer to as 'literary melancholy', but hence also, when I discussed the great bifurcation to come, when I spoke of a *shift*, and of its cost as generating so many negative protentions and consequently leading to violence of all kinds, the psychotherapist would respond by describing it as my 'pessimism about the future, typical of depression' – which, if I understand correctly, would then no longer be in any way literary.

I replied by saying that the issue is the containment of ὕβρις, which makes us suffer insofar as we are or try to be – whether we know it or not. We all cultivate a way of containing this ἐλπίς to which ὕβρις always also amounts, and which, when it becomes a *negative* ἐλπίς engendered by knowledge of the negativity of the *pharmakon*, requires a therapeutic culture of the positivity of this *pharmakon*, a *therapeia* that takes care, and that, treating and healing *while inscribing circuits of transindividuation*, generates positive collective secondary protentions.

Today, however, melancholy and more generally all forms of ὕβρις are shaped by the retentional and protentional specificities of the *pharmaka* of the absence of epoch, and this absence of epoch produces disruption inasmuch as these automated retentions and protentions are treated by algorithms operating more quickly than any form of care. Because of all this, madness, which becomes ordinary and general, is today a question that is inextricably medical, economic, juridical, political and industrial, that is, technological.

To confront this crucial fact, a fact that is more than just epokhal because it paves the way either for a new era or for the end (this is the issue of 'the turn'), Freudian theory and its Lacanian extensions are no longer sufficient. This is what, to treat myself, to take care of myself, I argued during these therapeutic conversations, and this is what I will come back to at the end – in order to conclude by opening up a dialogue, unfortunately posthumous, with Bernard Maris.

13

Death Drive, Moral Philosophy and Denial

98. The absence of epoch as demoralization

The immense demoralization that strikes the absence of epoch manifests itself in various ways. There are those who transgress without even being aware of doing so. There are those who lie shamelessly, without shame[1] [*sans vergogne*] or modesty or honour: without αἰδώς. There are those who have lost their morale, which is to say, their reasons for living, and therefore their reasons for acting – and, along with them, the feeling of existing – and those who do not see any future for themselves or for their generation.

Disruption can only intensify this state of fact and bring it to its highest point – and it is only just getting underway: the programme of transhumanism, starting with 'medicine 3.0', is to extend it to all dimensions of life, and to accelerate it still further by taking it to the stage of the control and intensification of *exosomatization that is exclusively subject to the selection criteria of the market*, accompanied by an *endosomatization of artificial organs*, in particular through the use of neurotechnologies.[2]

In *Automatic Society, Volume 1*, I argued that this state of fact requires a state of law, and that this implies rethinking law itself from an organological perspective – and, along with it, a law of law, that is, *the positively constituted knowledge on which law is founded*: positive law and theoretical knowledge share the same *experience of truth* through which are produced those 'regimes of veridiction' that characterize the epochs, as well as the eras within which they are successively linked together.

The fulfilment of nihilism in which disruption consists is the radicalization of the process of disinhibition described by Peter Sloterdijk and Jean-Baptiste Fressoz. It is the 'disruptively' (radically) disinhibited fulfilment of nihilism that leads to the extreme (and extremely dangerous) demoralization described here as that which leads 'to the edges of madness' (as the loss of reason that causes despair, the absence of any reason for hope). Various forms of political 'unburdening' ('*décomplexion*', as Sarkozy put it in France) accompany this extremization and its various extremisms.

What disruption renders impossible is any transindividuation capable of leading, through a doubly epokhal redoubling, to the constitution of a new epoch. It radicalizes the process of disinhibition by *liquidating all kinds of social systems insofar as the latter are always moral systems that metastabilize behavioural rules by forming and transmitting forms of knowledge* (of how to live, do, conceive and so on). The outcome is an *organological disintegration* that ruins the organic solidarities[3] without which there can be no social cohesion [*faire-corps social*], that is, no society.

Unless we open up the prospect not just of a new epoch but of a *new era*, unless we make possible an era of new epochs through an *integrated* organological basis amounting to *a new stage in the process of exosomatization*,[4] in turn capable of forming *new processes of organic and organological solidarity*, that is, *new social forms* amounting to *new forms of life*, unless we do all that, the disruptive fulfilment of nihilism is bound to unleash an immense murderous violence, which will be protean and unpredictable because devoid of all rationality – that is, of all *attachment to life*.[5]

The noetic soul is de-moralized because, in the ordeal of the absence of epoch, it creates the ordeal of the absence of moral convergence between noetic souls, giving rise to a denoetization that is an ordeal not just of systemic stupidity (functional stupidity) but of asystemic madness, that is, a madness that destroys systems, 'at the edges' of these systems, and as the 'edges of madness', whether ordinary, extraordinary or reflective. All these are consequences of the fundamental risk that the noetic soul itself *constitutes*, as its default of origin, which Heidegger called *Abgrund* (in *An Introduction to Metaphysics*, for example), and which since ancient Greece has been called ὕβρις: excess, madness and crime.

Faced with this demoralization, which cannot last – if not by its submission and containment then by paths that can only aggravate it still further – it is imperative to *rethink the meaning, history and fate of moral philosophy*. The morale of the moral being that we are insofar as we are noetic – in the absence of which we 'lose morale', sometimes to the point of losing reason (which in Aristotle is called melancholy) – is *that which is cultivated*, that which a culture forms: morale is *that of which we must take care*, and this care stems from what the Greeks called αἰδώς (*aidōs*) and δίκη (*dikē*).

99. Exosomatization as interpretation: on the meaning of ἦθος

Αἰδώς is a *fundamental feeling* – an existential, as Heidegger would have said – that founds all *Stimmung*, all feelings [*sentiments*], all 'sentimance' [*sentance*] and all sentences inasmuch as they are always inhabited by a mood. Αἰδώς and δίκη are, in *Protagoras*, what Hermes brings to mortals (*oi thanatoi*), who were at that time in the course of destroying themselves, and he does so at the request of Zeus, who is disturbed by it.

Αἰδώς – shame, modesty, honour – goes nowhere without δίκη, justice, which itself goes nowhere without αἰδώς. Justice requires law as a difference from fact, that is, as the *différance of facts*: as différance from the facts of the drives [*faits pulsionnels*]. And justice requires this law inasmuch as the interpretation of the injunction to be just *makes* [*fait*] this différance counter to the facts, thereby elevating law right (up) against [*tout contre*] the facts, and where the law itself always returns to being a fact insofar as it is an artefact. Différance interprets, as justice beyond the law (*nomos*), and as the consistence of existing law, as its 'spirit', that injunction in which consists the primordial feeling of δίκη brought by Hermes, and of which αἰδώς is the obligation: that which *binds* [*lie*] it by *ob-ligating* it.

The interpretation (*hermeneia*) of the organological and pharmacological situation – of which the αἰδώς and δίκη brought by Hermes are the 'feelings for the situation', a situation itself provoked in *fact* by the fault of Epimetheus (forgetting) and compensated for by the fact of the fault of Prometheus (theft), the consequence of which is the default of being of mortals, who, thanks to this fact, 'have to be' (*Being and Time*) because they are unfinished – is what, here, by reading tragic mythology in the light of scientific knowledge of life and human evolution, we describe as the *state of fact of exosomatization*, of which each epoch proclaims a state of law, and of which each era upsets the moral foundations.

Faced with a state of fact in which it cannot fail to take part, since it is a situation of which it is both the agent and the patient, the noetic soul, insofar as it 'has to be', is endowed with the function, plainly both logical (constituted by *logos*) and moral (both adopting and prescribing rules), that we call *reason*. Reason as a function of the noetic soul *must* confront the irreducibly pharmacological character of the exosomatized organology *forming* the non-inhuman being. It must do so by constituting therapies and therapeutics that we commonly call forms of knowledge – of living, doing and conceiving, that is, *transindividuating and spiritualizing by projecting the collective protentions that are consistences*, without which existence is not worth the pain and effort of subsisting.

Zeus proclaims through Hermes that all mortals must be hermeneuts, that is, capable of interpreting, each time singularly, what is just and what is not, and equally, what is shameful and what is honourable. The singularity of each hermeneut's interpretations cannot be eliminated, because exosomatization does not cease redistributing the effective conditions of the exosomatization of this noetic form of life, and this is what jurisprudence takes heed of, keeping in reserve the possibility of inscribing, on the basis of the heritage of adopted rules and in those rules, the *neganthropic bifurcations* required by αἰδώς and δίκη faced with the fact of the ever-changing *pharmakon*.

Hermeneutics itself stems from the *pharmakon* that, as *hypomnesic* tertiary retention, opens the possibility and the obligation of *hermeneia* as such, just as it opens the possibility of the positive formalization and transmission of rules and laws – through literature, which with these hypomnesic retentions increases the accumulation of tertiary retentions of all kinds that are exosomatic organs in general. With the hypomnesic retentions that arise starting in the Upper Palaeolithic there is formed that noetic *we* in which Bataille was able to recognize himself at Lascaux as *Homo ludens*,[6] constituting the new stage of the *realization of dreams*[7] in which noesis consists.

From the first flint tools to what Heidegger calls *Gestell* (which became for him, in 1962, technics as 'being itself'[8]), the organology emerging from exosomatization has trans-formed the milieus of life. Through those 'objects invested with spirit'[9] that are books – and always in relation to the hypomneses that are works in general – through its eras and its epochs, which are the eras and epochs of the Book,[10] the West accumulates examples, experiments and theorems over the course of the realization of its dreams. These condition the transindividual constitution of all forms of knowledge, while subjecting them to the new critique that reason requires as it is reformed by the apodictic experience of *alētheia*, and which constantly puts the West organologically and pharmacologically into question.

This putting into question affects the theoretical, practical and axiological[11] dimensions of the noetic form of life. It is in *this* way that moral philosophy inherits moral imperatives that, under the names of αἰδώς and δίκη, establish the place of mortals by setting its limits and marking out its place within the *kosmos* – this place being *transitory, temporary*, and in that constituting the abode [*séjour*] of *oi thanatoi*: the abode is that which is called ἦθος (ēthos).

100. Ἀρετή, *Sittlichkeit* and neganthropological courage

That this abode is only a sojourn [*séjour*], that is, a finite stay, whereas the stones marking out the pantheon of the immortals last beyond the generations, is what, in the mythological turn typical of tragic Greece, expresses the *fact* that *all negentropy is transitory*: it is bound to dissolve into entropy.

Insofar as it conforms as much as possible to the feelings of αἰδώς and δίκη, the moral being in which the noetic soul *consists* (through being impregnated with the consistences towards which it aims) is virtuous. Virtue, ἀρετή (*aretē*), is what must be *cultivated* in these diverse ways[12] covered by the knowledge of how to live, do and conceptualize – in order to consider, *to observe, to contemplate, to theorize and to spiritualize*, that is, *to transindividuate beyond generational differences*, by maintaining and supporting[13] the *transgenerational coherence* that is *reason as motive and for which the synthesis of recognition is the keystone, as the synthesis of the future [avenir] in becoming [devenir]*.[14] As the *kingdom of ends*, reason is itself the function that cultivates these *motives for living, acting and hoping*.

It is, however, only *with* the *pharmakon* that knowledge takes *care* of the *pharmakon*. This is why such care – which is moral in the sense that it installs a νόμος (*nomos*), mores, a *Sittlichkeit*, but which always turns into its opposite – requires 'justice beyond law'.

Nomos and *Sittlichkeit* gather together the various kinds of noetic knowledge: not just the sciences and the *epistēmē* founded on the experience of *alētheia* as *apodeixis*, but everything that, as mores, forms culture conceived as *the whole manifold of life-knowledge* – which passes through rituals as well as *habitus*, trades and forms of worship [*cultes*] that are found in any society insofar as it consists in a social cohesion that is a collective individuation, cultivating, through these always diverse pathways, the organic and organological solidarity[15] in which it consists.

In the age of disruption, these questions present themselves negatively, by default, and as the ordeal of the radical unbinding of the drives installing the demoralization that destroys all knowledge [*savoir*] by draining all the flavour [*saveurs*] from existence: as absolute computational totalization, this ultimate stage of the Anthropocene makes it apparent that *nihilism is fulfilled as the entropic decomposition of differences* – and, at the same time, of *noetic différance* as such.

The moral question that then arises, therefore, presents itself above all as that of *neganthropological courage*. Neither denying nor repressing the disaster, practising *parrhēsia*, neganthropological courage does not sink into melancholy (even if it does not cease to challenge it and be tested by it, which it therefore does not deny): it *confronts* the pharmacological situation *as such*, and it *conceives* morality as what *cultivates*, both theoretically and practically (that is, organologically), an *economic différance* that presents itself at once as a *libidinal* economy and as a *political* economy – as a new *oikonomia*.[16]

101. Guilt and transvaluation

Transindividuating mores and manners, what in the West is called 'morality' opens the dimension in which the morale of the moral being is formed as the 'feeling of existing'. If the word 'morality' has become either obsolete or reactionary, this is because it is understood above all as that which enunciates the good by *opposing* it to evil. When we use the word '*morality*', in other words, we are really referring, by metonymy, to 'bourgeois morality' – thereby taking the part for the whole.

Morality becomes bourgeois when the bourgeoisie become the dominant class through a reappropriation of aristocratic morality that at the same time adapts it to the imperatives of its own class. In short, bourgeois morality is subjected to the valorization of calculation – for which it nevertheless tries to maintain a zone of incalculability, both by rooting it in religion and through practices of *otium*, at the risk of philistinism.[17]

The bourgeoisie thereby founded an immense new culture that comes to be referred to as *modernity*, whose *metaphysical possibility* was opened up by Descartes, while the epoch of the 'classical age' assigns a new place to madness – in direct relation with a new morality of risk and of calculation, which is also to say of disinhibition and of its denial. In the nineteenth century, this modernity will be transformed, precisely through the advent of the bourgeoisie becoming the dominant class, because the latter also gives rise to the figure of the artist who breaks with morality. It is for this reason that Flaubert and Baudelaire will be accused and indicted by the Second Empire.

The culture of risk and disinhibition will, however, ineluctably dissolve the protected zone within which calculation had hitherto been kept, and within which bourgeois morality cultivated its religious references to good and evil, that is, to guilt before the incalculable. In this secularization, religion will give way to the vitality of the modern spirit called the 'life of the mind' – until spirit-value itself melts away into the market, as Valéry foresaw in 1939. The bourgeoisie then cease to exist, leaving behind only the 'unburdened' [*décomplexés*] of all stripes – the disinhibited.

Bourgeois religious culture inherits what is typical of Christian morality, and more generally of monotheism – namely, guilt. The latter is a transformation of the moral question that *delinquere* constitutes for any society as ὕβρις, which the Greeks thought in terms of αἰδώς and νόμος and in original relation with δίκη: Christian monotheism transforms Greek shame into guilt – by subjecting its meaning to the revelation of original sin, where the default becomes the fault, and where the snake [*serpent*] becomes the devil (the dragon). The diabolic is then *opposed* to the symbolic.

The *problem* posed by bourgeois morality is not its *moral* dimension. Who would dare say, today, that we have no need for morality? In the name of what could we condemn those murderers who eliminate their fellow men and women as they would a surrogate animal [*animal de substitution*] – a scapegoat [*bouc émissaire*] – if we have declared the moral dimension of morale, within existence, to be nothing more than vain and futile?

The philosophical *question* opened by the vanity of bourgeois morality is not just a matter of interpreting the meaning of the disappearance of all values: it is the question of the *transvaluation* of all values.

What poses a *problem* for, and creates an obstacle to, the transvaluation of values is not moral discourse: it is the discourse of guilt. And it is this that forms the subject of *On the Genealogy of Morality*. The great *weakness* of monotheism, whether it is Jewish, Christian or Muslim, lies in being *founded* on guilt. And this is also the great weakness of Freud, who cannot envisage desire and its economy outside of this regressive disposition [*complexion*] that is guilt – which is in his eyes unavoidable.[18] Guilt always seeks the guilty on whom to discharge its 'faults'. Hence it is that scapegoats are designated – one after the other: Jews, Muslims, Christians, the secular, Roma and those who vote for the National Front.

As it gives rise to what will become the bourgeois morality of guilt, monotheism always seeks the guilty, which it designates and symbolizes by demonizing them – an entire section of Christian art is devoted to the representation of the diabolical. In so doing, it conceals the fact that *original sin is not a sin, but, precisely, a process of exosomatization*. By transforming the default of origin into original sin, monotheism denies exosomatization – blaming [*culpabilisant*] those whom it turns into scapegoats, and rendering unthinkable the pharmacological problem posed by each new innovation emerging from organogenesis.

Christianity and more generally monotheism (which is always *first of all the affirmation of the unity* of scattered tribes or of empires threatened by their very expansion) thus express in terms of the *opposition* of good and evil – an opposition that is substantialized by the denial of the pharmacological situation that is this discourse of guilt founded on a 'pharmacosophy'[19] – what is in fact an inevitable test and ordeal [*épreuve*] thanks only to the facticity that conditions technical life. Much more profoundly than sin, the default of origin is the ordeal of the duplicity of the *pharmakon*.

Monotheistic care tries to overcome the possibility of going mad by turning a default into a fault, and by making the blame fall on the φαρμακός (*pharmakos*) that is the scapegoat. It thereby denies that the *necessary possibility* of becoming mad, which is *always* contained in technical life (in Canguilhem's sense), lies precisely in the organological and pharmacological condition of noetic souls, which *are* only intermittently – because the curative *pharmakon* always ends up becoming toxic.

It is not just the noetic soul in potential that is noetic in actuality only intermittently: the epochs and eras that gather these souls together in the unity of their collective protentions, too, can always evaporate, because their basis, which is always organological – for example, as the era of books that is monotheism, in which the 'people of the Book' recognize themselves – is the 'abyss' [*sans-fond*] that contains ὕβρις.

In the age of disruption, the 'people of the Book' no longer recognize themselves, and for good reason: as the accomplishment of the disinhibition that nihilism would always have been *by denial*, the 'values' arising from monotheism *no longer have any value*. This is what makes fundamentalists of all stripes go crazy – including those of secular fundamentalism.

102. The apprenticeship of life 1: the cosmological dimension of noesis

Like knowledge in general, and because it is above all the *flavour* [*saveur*] of this knowledge, the moral dimension of the noetic being is not inscribed by God in the soul of this being: it must be *learned and cultivated*. This dimension relates to the technicity of life, which, fundamentally accidental, artificial and contingent, stems from a primordial diversality (of which babelism is one version) on the basis of which consistences are projected. Such consistences, as protentions to the infinite, affirm a *coherence to come*, in the encounter with and counter to the *incoherence of becoming*.

In the first version of the 'transcendental deduction of the categories' in *Critique of Pure Reason*, the *ability to unify that which remains to come* lies in the third synthesis of the imagination, called the 'synthesis of recognition'. The function of the latter is to inscribe the flow of the phenomena that arrive to consciousness (a flow in the course of which consciousness constitutes itself) in the *unity* of this flow – that is, as the affirmed unity of the past, present and future, both as 'unity of apperception' and 'transcendental affinity'.[20]

The third synthesis, however, presupposes the schematism, and I have previously tried to show why this in turn presupposes tertiary retention, which means that it always remains caught up in the artefactuality required by noesis.

This is why noesis in general, and the morality of the moral being in which it consists, are universal only in terms of the *universal fact that exosomatization is the condition of any noesis*, where the noetic soul, because it *is* only exosomatically, because only in this way does it *ek*-sist, must *begin by acquiring knowledge*. It must acquire that knowledge through which it will know how to ensure that the *pharmaka* that together form the organological apparatus which emerges from exosomatization, and which it inherits, are:

1. *beneficial* to its *existence* rather than detrimental; and
2. *beneficial* through its existence *to the universe of living things in totality*, such that, caught in entropic becoming, it nevertheless preserves a negentropic future, one that runs counter to entropic becoming, and that is neganthropological.

The second constraint – which is induced by the transcendental affinity – has the consequence that the *noetic soul is responsible for living things, within the universal becoming* in which it *organologically* and therefore *pharmacologically* inscribes its negentropy.

This responsibility – which, in the still tragic language of *Protagoras*[21] has, then, the names of αἰδώς and δίκη – must be cultivated and learned. Knowledge (of how to live, do and spiritualize – how to transindividuate) is transmitted along diverse pathways through which we must *learn how to live*. And the object of moral philosophy is the *formation* of this moral being who must learn how to live.[22] This does not mean that virtue can be taught: we could teach it only if it was not constitutively diversal. Virtue is improbable and unteachable because it is neganthropic. But it is exemplifiable – and we will see why.[23]

Teaching ensures the transmission of conceptual knowledge. Life-knowledge [*savoir-vivre*], however, is not conceptual. It is 'ethical' in the sense that it cultivates the responsibility that the technical form of life has towards its milieu and its place in the cosmos by irreducibly 'diversal' and 'historial' pathways, that is, pathways conditioned by the specificities of the process of transindividuation induced by a temporal and spatial stage in exosomatization, and the transmission of which stems from a culture that always involves a form of worship [*culte*], that is, of artifice (in the sense that Hermes is both the god of writing and the god of communication between mortals and immortals).

The concept of *life-knowledge* [*savoir-vivre*] should therefore refer to everything that constitutes, transmits and shares the mores of which the morality of the moral being is, in the modern West, the epokhal prescription. That is, it is a prescription organized by the convergence of positive collective protentions[24] attached to a stage of exosomatization and projecting to infinity within entropic becoming, and as that which can reverse its course – even if always temporarily and locally, that is, intermittently, in time as well as in space, and as neganthropy.

103. The apprenticeship of life 2: the transformation of the cosmological dimension into universal knowledge

'Bourgeois morality' is the degraded version of that morality required by ὕβρις as αἰδώς and δίκη. Witnessing the sudden expansion of

European capitalism in Germany, Britain and France, Nietzsche, who *saw* this degradation and its sudden acceleration, described it in terms of nihilism.

We ourselves see this nihilism fulfilled and accomplished in the twenty-first century as disruption, via machinic calculability that *today effects disinhibition in a thoroughly automatic way* (as the automation of protentions), installing societies of the hyper-control of individuals, whose traces are followed on a one-to-one basis, and *who no longer interiorize any law*: they are in this sense *dis-affected*. Such is the price of generalized proletarianization.

These societies of hyper-control, however, are dissocieties: they are *more uncontrollable than ever*, because they impose generalized anomie (another name for the barbarism preached by the 'new barbarians' as well as by the adepts of 'jihadist' chaos), that is, the unlimited liberation of ὕβρις – as much madness as crime.

Since Freud, the conditions of the elaboration and transmission of moral being, as the knowledge of the desiring being and the symptoms of its demoralization, have been studied on the basis of clinical psycho-analysis within that *process of epokhal transformation that is the rise of the bourgeoisie* – and, beyond the bourgeoisie, of capitalism, that is, *of generalized calculation as the operator of disinhibition, provoking a thousand pathologies that the twentieth century will call 'neurotic'*.

Freud made it clear that identification, idealization and sublimation are the conditions of both *moral cohesion [faire-corps moral]* (the *subjectivation of the body*) and *social cohesion*, through the process of super-egoization that would become 'morality'. Transcendental affinity in Kant's sense passes, here, through the *psychosocial affinity* of what is no longer just a soul, but an apparatus: the psychic apparatus. Within this apparatus, functions form, which, before being those of the logical individual (that is, the 'rational' individual in the Greek sense), constitute the conditions of possibility of a libidinal economy that is also an economy and ecology of the spirit.

Psychosocial affinity is possible only on the condition that, in each individual, it enables idealization, and does so by way of a social moral being that is itself possible only according to a set of therapeutic life-knowledge prescriptions with and within exosomatization at the temporal and spatial stage of its development. In other words, the constitution and sharing of an epokhal transindividuation, enabling the gathering together of the social body in its cohesion, requires an idealiz-ation that inscribes the imperative necessity of this cohesion within each of its members, in each body, and as the sharing of a neganthropic protentional horizon.

This is why Fethi Benslama – after having recalled that 'two thirds of the radicals identified in France [...] are between 15 and 25 years old, and a quarter are minors: the vast majority are in that zone of suspension that, in the transition to adulthood, borders on persistent adolescence' – emphasized (in an article published at the very moment the 13 November 2015 attacks in Paris and Saint-Denis were being carried out) the degree to which idealization is necessary for the constitution of the adult individual being formed through this process we call adolescence: 'This period of life is driven by a hunger for ideals on the background of a difficult reworking of identity. What is today called "radicalization" is a configuration of our epoch's disturbance of ideals.'[25] It is idealization that binds the generations together, and unites peers belonging to a generation [*congénères*]. Idealization is the intergenerational, congenerational and transgenerational cement. This transgenerationality is called the transcendental in Kant, God in monotheism, the return of spirits in magical societies, and so on.

It is through idealization that psychic individuation occurs *as* collective individuation – that is, both *in* collective individuation and as a *contribution to* this individuation insofar as it is a becoming oriented towards and by a future the idealization of which is its protention – which constitutes the noetic economy of the dream in which *each noetic soul is responsible for the whole*, that is, in charge *of* and charged *with realizing* its dreams.

This is why Benslama adds that it is through 'ideals [that] the individual and the collective is tied together in the formation of the human subject'.[26] When disinhibition reaches the final stage of nihilism, as the Anthropocene turns into the age of disruption,

> the jihadist suggestion [*offre*] captures young people who are in distress due to significant identity gaps. It offers them a total ideal that fills these gaps, enabling a kind of self-repair, even the creation of a new self, or in other words a prosthesis of belief that suffers no doubt.[27]

Through processes of *negative identification, idealization and sublimation*, this 'prosthesis of belief' comes to fill in for what is missing [*fait défaut*] in those who have been called 'adulescents', who are *both the descendants who fail to become adults and the ascendants who have failed to leave childhood behind, and with whom it is impossible to identify*. It is clear that, in recent years, such adulescents have come to occupy the head of the French state.

Adulescents are all those who – having failed to become adults, having failed to construct their dreams by idealization, through the concretization of which they surpass the stage of idealization and enter into a process of social cohesion (which generally reveals the duplicitous character of any idealization) – have either been deprived of the possibility of identifying with adults (through people or through works), or have regressed into infantilization (that is, into irresponsibility) by placing themselves in the service of this unleashing of the process of disinhibition to which consumer capitalism has amounted, especially since the 1980s.

104. Disgust, contempt and despair

Idealization can, clearly, concretize itself by default. It is not, however, lost: it thus constitutes an experience through which the psychic individual discovers, as a noetic soul, that the dream that conditions exosomatization only ever produces *pharmaka*, and that the latter can always (and usually do) end up turning their beneficial effects into toxic 'side effects'. In so doing, they reverse their sign: from positive, they become negative.

This becoming-toxic of what was dreamed does not change the fact that the oneiric and noetic origin of *pharmaka* lies in its participation in the bifurcations in which neganthropic différance consists. On the contrary, this amounts to saying that neganthropic différance is not divine, and that it is bound, sooner or later, to find itself disintegrated by the entropic becoming of the universe – for which it is commonly said that religion is a 'consolation', which shows a very poor understanding of what religion contains of knowledge, and of the meaning of the denial in which, most of the time, it does indeed consist.

When an idealization is maintained despite the duplicity revealed by its realization, it *constitutes*, beyond this *disappointment*, a *knowledge*, which, beyond facts, affirms *a right and a duty of the moral being* – of which αἰδώς and δίκη are the ancient names – *not to sink* into demoralization, that is, into de-idealization, a right and a duty that it asserts *by constituting itself as a critical power*.

The strength of adolescence and of what I have called the *Antigone complex*[28] is to constantly reaffirm this critical necessity again and again, thereby reminding adults of their obligations stemming from their own becoming-adult – where adolescence is what struggles against obsolescence, just as neganthropy fights against anthropy.

Αἰδώς is above all the knowledge of the precariousness of the noetic (oneiric and ideal) scaffolding that supports δίκη as the *art of life. It is the art of life within the constant transformation in which noetic life as exosomatization consists*, realizing itself in the most varied forms, and necessitating *the making of epochs, that is, social affinity in the cosmos*, and hence realizing itself through the metastabilization of knowledge as the common ground of retentions and protentions unified by the prospect of a possible future within becoming.

The destruction of processes of transgenerational identification and idealization, however, is the basis of *consumer capitalism*. When information and communication technology combines with the digital reticulation of relational technology, it becomes disruptive, producing an industry of 'lifestyles' [*industrie des modes de vie*], in the sense given to this expression by Mark Hunyadi.[29] As the liquidation of what, after Groddeck, Freud called the id,[30] then as the liquidation of psychosocial protentions by automatic protentions,[31] this lifestyle industry engenders immense suffering – symbolic, affective, spiritual and intellectual – which is bound to provoke in return *a literally dia-bolical unleashing of* ὕβρις: to destroy the sym-bolic is to install the dia-bolic.

At the end of the process of disinhibition that began in the fifteenth century, became industrial in the eighteenth, and since that time has engendered the Anthropocene, *disruptive nihilism has radicalized disinhibition by nipping all idealization in the bud*. In so doing, it infantilizes all psychic individuals, who, becoming powerless, lose all ascendancy over their descendants, and no longer transmit to them any knowledge, *so that they become little more than illustrations of counter-examples not to be followed* – consequently arousing *disgust, contempt and despair*.

So it is that nihilism succeeds in disindividuating individuals, while claiming rights that it renders inaccessible to them. It makes them minors for life, and hence makes them irresponsible, either as liars, simulators and shameless sneaks, or as deniers of their own condition, which they want to know nothing about so as to sleep the bad sleep deserted of dreams. It is this process of the liquidation of the *différance of generations* that leads to the ordinary, extraordinary and reflective forms of contemporary madness – from 'France from below' [*France d'en bas*] up to the highest level of the Republic and international institutions, by way of a crooked surgeon who became minister of tax evasion.[32]

105. Idealization, dream, transition

Idealization must be satisfied: it is through idealization that, under the pressure of the successive shocks provoked by exosomatization across the ages, whose effects on moral being are recapitulated through knowledge and its transmission, the new generation, in thereby becoming adult (adolescent in this sense), bears within it a *critique of the super-ego* that amounts, not to a rejection, but to a transformation.

It is the transformation of the common rules of life by the new generations that enables new circuits of transindividuation to be constituted. These new circuits reformulate knowledge, that is, the collective rules without which it would be impossible to live in a society founded on the organic solidarity described by Durkheim.

The formation of such rules establishes prohibitions – on killing, stealing, raping, lying, fraud and all kinds of other *manners* of living, as Lordon calls them[33] – which establish the frameworks within which limits are imposed on ways of life, as Freud says in the following passage:

> If one imagines [civilization's] prohibitions lifted – if, then, one may take
> any woman one pleases as a sexual object, if one may without hesitation

kill one's rival for her love or anyone else who stands in one's way, if, too, one can carry off any of the other man's belongings without asking leave – how splendid, what a string of satisfactions one's life would be![34]

But such limits (which are here presented according to an order of values typical of the West) are only the most immediate and constantly perceptible echoes of the ὕβρις unleashed by the default of origin – that is, *delinquere. Theft* is possible only because *removable organs* are *detachable* from noetic bodies, murder transforms the tool into a weapon, rape stems from the ὕβρις that turns instinct into drive – and the latter is itself detachable, which is also to say perverse,[35] which distinguishes it radically from instinct.[36]

In the *transformation of rules provoked by the unfolding of the generations* in exosomatization, the production of *transitional objects* is the key factor in the production of new forms of knowledge – from the infant's toy or teddy to 'spiritual works' [*oeuvres de l'esprit*], via instruments of worship [*culte*]. If the adult must leave adolescence behind, its neganthropic capacities are commensurate with the opportunities it provides to *transport, into the adult age of ascendants, the virtue of transitional objects that are the prerogative of childhood* – objects that, by opening the transitional space between the child and the 'good enough mother'[37] (where 'transitional' means that *dream and reality are still indistinct*), inscribe *possibilities of dreaming* into the *social realization of the noetic real*, that is, of the *desiring* being that is the *moral* being. The infantilization of adults keeps their sights set on an immature age that has yet to recognize that the transitional object is a *pharmakon*. In fact, however, it is as a *pharmakon* that knowledge can enter adults into a mature relationship to the transitional objects that are objects of knowledge in general, insofar as they *remain objects of idealization* well *beyond* adolescence. It is in this sense that Alain Bergala refers to *child's play* [*enfance de l'art*] with respect to the work of Victor Erice.

The transitional object is the matrix through which, in the *native* relationship of *coming into the world* into which the mother (who can be the father) introduces the child, the noetic soul *existentially experiences* exosomatization throughout noetic life inasmuch as it consists in generating *transgenerational transitional objects* that give exosomatization its meaning (its direction), thereby becoming capable of *inscribing bifurcations of the desired future into blind becoming*.

Ever since exosomatization began, the irreducibly moral dimension of the noetic being has, *in this way* – that is, *transitionally* – proceeded from the αἰδώς that, because of ὕβρις, must be concealed within neganthropogenesis. And it is *in this way* that therapeutic prescriptions are both necessary and possible. It is this necessity and this possibility that, in their unity, form the object of moral philosophy.

The moral dimension of any noetic existence must regulate behaviour precisely because – in addition to the fact that social cohesion is not conceivable without *shared rules of social affinity*, which are not just laws, but arts of living, which, as forms of knowledge [*savoirs*], 'manners', ways, fashions, confer flavour [*saveurs*] upon any existence worthy of the name, and thereby *fashion* it – all these behaviours themselves *contain* ὕβρις within them, ὕβρις as the echo of the default of origin (*delinquere*) of which exosomatization is the *fact*.

106. The liquidation of *Sittlichkeit* by the ethics of 'lifestyles'

As in Frédéric Lordon's reading of Spinoza, and in particular his reading of Chapter 3 of the *Ethics*, where Spinoza sets *desire and the processes of evaluation that constitute it* at the heart of the process of transindividuation, one can indeed say that the latter amounts to 'a convergence of manners [that] is, by the very effectiveness of the mechanism of mimetic confirmation, an unparalleled power of collective coherence, and that it is so insofar as it is a power to appease axiological disquiet through the strength of the collective'.[38] A society produces itself through the arts of living thereby engendered, which amount to so many forms of knowledge supported by what Spinoza calls *imperium*, and which they support in return: '*Imperium* is not the law, nor the instituted State, but it is in the first place the force of autoaffection of the multitude, and, as a result, but only as a result, the principle of the efficaciousness of the law, if it is a written one.'[39] In fact, however, the therapeutic prescriptions that are, in this way, *self-produced through social cohesion* have now *completely disappeared*: they have been replaced by what Mark Hunyadi calls 'lifestyles' [*modes de vie*].

It is these *modes* of life that, insofar as they are no longer mores concretizing *rules* of life, but, precisely, styles, fashions [*modes*], destroy social systems and the psychic individuals who conform to them *without identification, without idealization*, without taking part in their genesis: they are psychically as well as collectively *disindividuating*. 'Lifestyles [*modes de vie*] refer to long-term behavioural expectations imposed by the system on individuals and groups, and which are imposed independently of the will of the actors.'[40] Once we enter the twenty-first century, once we enter reticulated society that implements algorithmic governmentality, these expectations come to be generated by processes that automatically capture retentions in the form of data, and processes that computationally produce the protentions deduced from this data by systems of intensive computing capable of operating four million times faster than noetic souls.[41] The 'tyranny of lifestyles' thereby turns into the tyranny of the data economy, and it is precisely in so doing that it becomes disruptive.

It is starting from the *hegemony of strategic marketing*, in particular, which established itself in the 1980s[42] by replacing entrepreneurship, that 'lifestyles are imposed upon us in an almost autonomous way, without anyone having wanted them, and without anyone being able to oppose this'.[43] Lifestyles, modes of life, then, come to replace the mores and knowledge that result from the autoaffection described in Spinoza's *Political Treatise*.

Hunyadi shows that ethics committees, which then begin to proliferate, exist in order to conceal and deny this liquidation of moral being, replacing 'morality' with the defence of 'the rights of the individual':

> Even though ethical rules continue to multiply, we are no longer capable of treating the fundamental ethical question, the question of knowing whether this is the world we want. [...] In the ethical respect for the rights of the individual, we are paving the way for a world that may turn out to be ethically detestable.[44]

These 'individual rights' are nothing but the empty rights of disindividuated 'dividuals'[45] – that is, psychic individuals deprived of the knowledge in which collective individuations *consist*, and so *incapable* of exercising their rights. We will see in the following chapter that this incapacitation is induced by the generalized proletarianization that disruption and transhumanism are now taking to extreme levels.

Ethics committees protect 'individual rights' only by *legitimizing the liquidation of common rules* insofar as they are *communally* produced and *practically* consented to (that is, practised through experience, and not followed like fashions [*modes*]):[46]

> Omnipresent ethics is in reality an ethics restricted to the defence of a few principles that, once satisfied, leave the world free to pursue its course. This restricted ethics [...] is [...] fundamentally a-critical: under the guise of ensuring the highest level of respect for individual rights, it serves in fact to ethically whitewash practices whose general ethical character it is careful not to question.[47]

Conceived by design and marketing – which thereby provide the interface between the technical system and psychic individuals, and which do so by short-circuiting the social systems, and, along with them, the mores and rules that, collectively individuating (in the sense of both Simondon and Spinoza) what the Greeks called αἰδώς and δίκη, have hitherto constituted the *social cohesion of the moral being* – 'lifestyles' install a bottomless *moral poverty* [*misère morale*] resulting from what Hunyadi describes as a *tyranny*, and *impose anomie*, the threat of which Durkheim felt coming in the late nineteenth century, and which the libertarians now claim as such.

'Ethics', understood in this manner, liquidates the moral being and his or her knowledge, insofar as the latter forms the arts of living. Such an 'ethics' is what, through the globalization of consumer capitalism, and increasingly as disruption, continues the process of disinhibition by replacing ways of living with technological systems and apparatus that are purely and simply computational. Through this, hyper-control 'societies' are imposed, forming what Berns and Rouvroy describe as algorithmic governmentality.

These 'societies', however, are nothing of the sort: an anomic society is *no longer* a society – except insofar as it tends to become an arthropod society, which, as I have previously recalled, was Leroi-Gourhan's dystopian hypothesis, and which, as Freud already concluded, is unbearable for those who, bearing libidinal energy, are irreducibly *constituted by the* ὕβρις that it is a matter of *containing* – and of containing precisely through *nomos* and *Sittlichkeit*, whatever may be their peculiarities.

This is why hyper-control societies are bound to engender *dissociety*, through which they are in a certain way collapsing in advance. The most glaring symptoms of this collapse are the transgressive behaviours of all kinds that increasingly commonly emerge from the *submission* in which all this consists – from lying to committing massacres, via indiscretions and frauds, whether financial or otherwise – while *demoralization crushes the vast majority, pushing them to the brink of despair.*

14

Nonconformism, 'Uncoolness' and *Libido Sciendi* at the University

107. Conformists, 'petits-bourgeois' and the 'uncool'

In the 1950s and 1960s, prior to May 1968, the adjective 'conformist' was used to describe 'bourgeois morality', and 'nonconformist' [*anticonformiste*] was used to describe the new attitude emerging from the 'younger generations' – through which Isidore Isou, one-time companion of Guy Debord, foresaw the coming of a 'youth uprising'.[1] *Conformation* and *conformity* were the questions and the challenges of a time when the *arbitrariness of agreed things* became confused in young minds (including my own) with the *emerging standardization of 'lifestyles'*.[2]

The bourgeois and Gaullist conformism of that time was indeed suffocating: it thoroughly contradicted the 'values' claimed by 'Christian morality', which was nevertheless its reference point – and Charles de Gaulle never failed to have himself filmed leaving church on Sunday mornings. The contradiction between, on the one hand, the organization of the process of *standard disinhibition* by the bourgeoisie via the culture industry, and, on the other hand, its moral conformism, was experienced as an *intolerable hypocrisy*. Hence was already foreshadowed the absenting of epochs in the disillusioned West.

The 'lifestyles' then advocated in France, principally through the radiophonic and 'generational' mass media manipulation of the counter-culture coming from across the Atlantic (television was at that time still national and free of advertising; it still 'belonged to the "General"'), would give rise to a profound discomfort with respect to forms of knowledge more generally.

As readers of *Eros and Civilization*,[3] *One-Dimensional Man*[4] and *Society of the Spectacle*,[5] the future 'sixty-eighters' took aim at the 'arts of living' emerging from 'tradition' and from the 'savoir-vivre' of bourgeois society just as much as they did at the 'lifestyles' [*modes de vie*] heralding the 'new form of barbarism' redefined by Marcuse and Debord after the initial analysis put forward by Adorno and Horkheimer.

All of this was essentially based on 'Marxist thought', which at that time occupied a dominant place on the horizon of idealities, and did so as 'materialism': idealities themselves were therefore understood as lures of 'petit-bourgeois thinking', which was for many years referred to, after Marx and Engels, as 'German idealism', by those who may not even have read the authors (as was the case for me).

This dynamic was already thoroughly related to the doubly epokhal redoubling, which, with the newly emerging information management that by the end of the 1970s would become telematics, was heading towards computationalism and disruption as what would no longer allow the second moment of the redoubling to metastabilize.

It was this emerging context that saw the appearance of what Luc Boltanski and Ève Chiapello would later call 'artistic critique'.[6] Emerging from the May '68 movements in France, artistic critique would go on to reshape – in particular through the discourse of Yvon Gattaz, father of the current president of the business group Mouvement des Entreprises de France (MEDEF) – the discourse of a capitalism that was less and less bourgeois and more and more 'unburdened' and 'uninhibited' [*décomplexè*], where to be moral became *inherently 'uncool'* [*ringard*].

108. Understanding, reason and disinhibition

With the occurrence of the 'events' that began to unfold on 3 May 1968, following the evacuation of a Sorbonne courtyard where students had gathered, the university and its role became the focus of general political debate – and, through this question of the university, so too did questions of reason, rationality, imagination and the relationships between power, knowledge, alienation and emancipation. After 1968, however, with Foucault and Deleuze as well as with Derrida and Lyotard, the questions of alienation and emancipation, as these were understood by the students of 1968, were themselves placed into question.

We have seen how Sloterdijk insists on the fundamental (in the strict sense of *foundational*) contribution of 'classical reason' to disinhibition. For in fact, *academic activity in itself essentially amounts to a process of disinhibition*, given that it is above all unthinking and automatically reproduced behaviours that inhibitions manifest as forms of reserve, limits and prohibitions, and given that thinking is already a matter, if not of dis-inhibiting properly speaking, at least of dis-automatizing, which generally leads to a critique of these ways of living and to their transformation in the course of what amounts to the second moment of the doubly epokhal redoubling, through which a new epoch is constituted.

It is a matter, therefore, of correlatively analysing the relations between *two distinct but inseparable processes of disinhibition*.

1. On the one hand, there is the *always transgressive* process of disin-hibition that results from the *intermittent critique of foundations and dogmas by the disciplines* and through the rational activity of the mind as it relates to the *libido sciendi,* and of which Cartesian modernity is obviously a new stage, expressing a *new critical radicality* – against the School, of course, but also against everything that has been passively received 'into my set of beliefs' [*en ma créance*].[7]

The Enlightenment will be the next stage, deepening this radicality, and, after that, Freud will constitute a new critical threshold, because he will make it possible to *think transgression itself* – while Marx and Engels themselves analysed *disinhibition as the revolutionary process* accomplished by what they call both *capital* and the *bourgeoisie,* a discourse that in the following instance we find coming from the pen of Engels:

> It was the greatest progressive revolution that mankind had so far experi-enced, a time which called for giants and produced giants – giants in power of thought, passion and character, in universality and learning. The men who founded the modern rule of the bourgeoisie had anything but bourgeois limitations. On the contrary, the adventurous character of the time inspired them to a greater or lesser degree.[8]

2. On the other hand, there is the process of disinhibition that results from *conquest* – firstly of territories, then of markets – which concretizes the *emancipatory dreams of the bourgeoisie* that give rise to modernity just as science emerged from the humanism of the Renaissance, and which, installing capitalism and its process of fundamentally computational generalized disinhibition, and thereby becoming the Anthropocene, eventually comes into conflict with intermittently noetic disinhibition.

At the end of this 'great transformation'[9] that is also a 'chain reaction',[10] the Anthropocene – by enlisting science in order to transform it into technology (which is a specific epoch of noesis as realizable dreaming that turns into the disruptive nightmare) – comes to struc-turally short-circuit it by radically distorting it. Or, in other words, it comes to *replace noetic activity,* understood as the *passage to the trans-gressive act of reason,* with an *automatic understanding* that functions *without any reason,* regardless of whether we are talking about 'big data' or 'deep learning'.

Here, it is *reason as a process of disinhibition by synthesis* that is short-circuited by the *understanding as a process of disinhibition by analysis* – the first being a process of *dis-automatization,* the second being the *generalization of automatisms.*

In addition to the *critique of those beliefs* that support the apparatus confining us within what Kant called 'minority'[11] (beliefs in relation to which the 'great confinement' described by Foucault in *History of Madness* is the guardian – which is also to say, the repressive system), one of the modalities of which is 'bourgeois morality', as are, too, the forms of *tradition* that are shaken up by the bourgeoisie through the epokhal redoubling of which it is the epoch – in addition to all this, the passage to the act of reason *always* transgresses circuits of transindividuation that have been fixed in place and socialized in highly diverse forms and supported by all manner of *technologies of power* that enable the *exosomatization of the understanding.*

Through *grammatization*, which, as I have explained elsewhere, begins in the Upper Palaeolithic,[12] it is always the *relationship between understanding, reason, imagination and intuition* that comes to be transformed. Every Western epoch stems from such a transformation, which engenders a 'regime of veridiction' or 'truth'. Over the course of these new arrangements of the 'faculty of knowing' – a faculty *conditioned* by tertiary retentions, that is, by *pharmaka* – new circuits of transindividuation accumulate, forming knowledge and installing eras, and, within these eras, successive epochs.

It is an objective critique that operates in advance through the advances of what Hegel already saw as an *exteriorization of the Spirit*, and that amounts to an *exo-somatic* and as it were *exo-psychic discernment* of discrete elements. These elements are discretized through that exteriorization and reproduction of living flows that is *grammatization* – whether of bodily movements, gestures, speech, images, calculations or dreams. And this is something that, ultimately, Hegel did *not* think, because he believed these elements were themselves *soluble* into the *grand synthesis* of absolute knowledge, and because he was *completely unaware* of the *indissoluble* play of entropy and negentropy.

109. Disinhibition as the revolutionary power of the bourgeoisie, proletarianization as demoralization, the new prospects opened up by the general intellect, and the question of entropy

Towards the end of the nineteenth century – after Marx had raised the question of the historical role of the bourgeoisie, both as process of disinhibition and as the ideological and contradictory domination of *bourgeois* (formal) *law* and of *bourgeois* (idealist) *morality*, law (νόμος) and morality (αἰδώς) thereby concealing the process of exosomatization, that is, the ἦθος in which man's production of his artificial organs consists, and as the 'means of production' of man by man in the service of the exploitation of man by man – the *genealogy of nihilism* undertaken by Nietzsche and the *economic analysis of the production of libidinal energy* undertaken by Freud allowed all this to be seen as the domination of 'instrumental reason' and of moral and spiritual values made to serve the domination of labour by capital.

This description and its objects are what it is now a matter, not just of offering a new critique, but of 'transvaluing'. In what way must this transvaluation beyond critique be an 'active forgetting'? And how to *actively forget* in order to be able to remember otherwise, completely otherwise, the originary default of origin beyond the Anthropocene, in the Neganthropocene, and contrary to transhumanist free trade? We will return to this in *La Société automatique 2. L'avenir du savoir.*

Dealing with such questions, which are also experiences and challenges that *exceed the understanding*, requires a better comprehension of the process of disinhibition. Marx and Engels frequently declare their admiration for the historical process of modern and bourgeois disinhibition, especially in *The Communist Manifesto*: in it, they see the *dialectical necessity* of the great historical movement of emancipation:

> The bourgeoisie, historically, has played a most revolutionary part.
>
> The bourgeoisie, wherever it has got the upper hand, has put an end to all feudal, patriarchal, idyllic relations. It has pitilessly torn asunder the motley feudal ties that bound man to his 'natural superiors', and has left remaining no other nexus between man and man than naked self-interest, than callous 'cash payment'.[13]

Here, disinhibition is a process that reveals the reality of social relations – as relations of force between classes that have unequal relations to the means of production, that is, to exosomatization, and into which *bourgeois time-measurement will, in unvarnished fashion, introduce the law of calculation, thereby dissolving the moral values of former times*:

> It has drowned the most heavenly ecstasies of religious fervour, of chivalrous enthusiasm, of philistine sentimentalism, in the icy water of egotistical calculation. It has dissolved personal worth into exchange value, and, in place of the numberless, indefeasible chartered freedoms, has set up that single, unconscionable freedom – Free Trade.[14]

There is no doubt that, *just as Nietzsche* sees in nihilism the promise of a new belief and a new sensibility that will pass through the transvaluation of all those values devalued by nihilistic disinhibition, whether active or passive, *Marx and Engels see, in the revolutionary dissolution of social ties by calculation, the promise of a reversal*, which for them is 'dialectical', and where it is no longer a matter just of 'fighting against the bourgeoisie', but of *overcoming the contradictions induced by the work it does to undermine the orders of former times.*

It is firstly the *surpassing of the bourgeoisie by itself* that, in the eyes of Marx and Engels, is *intrinsic to the activity of capital* – until it *destroys* the bourgeoisie itself. But what is less certain is if the authors of the *Manifesto* and *Capital* ever took the measure of the consequences stemming from this fact:

> The bourgeoisie [...] has been the first to show what man's activity can bring about. [...] The bourgeoisie cannot exist without constantly revolutionizing the instruments of production, and thereby the relations of production, and with them the whole relations of society. Conservation of the old modes of production in unaltered form, was, on the contrary, the first condition of existence for all earlier industrial classes.[15]

What Marx and Engels see coming is what we are here calling disruption – but perhaps without truly taking stock of the unendurable consequences on the planes of the ecologies of nature and spirit, and on social life: 'Constant revolutionizing of production, uninterrupted disturbance of all social conditions, everlasting uncertainty and agitation distinguish the bourgeois epoch from all earlier ones.'[16] The bourgeoisie therefore constitutes an *epoch* precisely inasmuch as it gives rise to an epoch that does not cease to destabilize itself, through which it seems to turn into a perpetual doubly epokhal redoubling, but a redoubling that is less and less stable. Here, metastability, which is always contained within its limits by the bipolarity of synchronization and diachronization, moves *ever closer to the brink of its actual destabilization*, to the edge of *ruining* its metastability, which, in the twentieth century, already gave rise to two world wars that Marx and Engels would never know:

> All fixed, fast-frozen relations, with their train of ancient and venerable prejudices and opinions are swept away [...]. All that is solid melts into air, all that is holy is profaned, and man is at last compelled to face with sober senses, his real conditions of life, and his relations with his kind.[17]

What relation does the *coldness* with which the bourgeoisie 'draws all, even the most barbarian, nations into civilization', which it does 'by the rapid improvement of all instruments of production, by the immensely facilitated means of communication',[18] *what relation does this coldness have to the cynicism of the new barbarians and to the calculations involved in the pseudo-jihadism born of the second war against Iraq* – not to mention the cold calculations carried out by puppets in order that they may remain in power, even though they are devoid of knowledge and therefore do so at the cost of seeing their noses constantly lengthen?

To pose this question correctly, which also concerns the organological genesis of the spiritual and intellectual misery provoked by calculation's denoetizing effect, it is necessary to take account of what it means to say that the *bourgeoisie destroys itself*: it means *that the proletarian power of the negative fails to 'sublate' the thetic moment in which the bourgeoisie consists in the materialist dialectic.*[19] And this is so firstly because, in fact, proletarianization spreads throughout 'all classes of the population', as Marx and Engels themselves predicted in the *Manifesto*:

> The lower strata of the middle class – the small tradespeople, shopkeepers, and retired tradesmen generally, the handicraftsmen and peasants – all these sink gradually into the proletariat, partly because their diminutive capital does not suffice for the scale on which Modern Industry is carried on, and is swamped in the competition with the large capitalists, partly because their specialized skill is rendered worthless by new methods of production. Thus the proletariat is recruited from all classes of the population.[20]

Here, we must emphasize the following:

1. What above all constitutes the process of proletarianization is the *loss of knowledge* (even if it also has causes stemming from the lack of capital held by the small entrepreneur), *proletarianization being itself described as the exosomatization of knowledge in the means of production*, which Marx and Engels call 'the extensive use of machinery and […] division of labour'.[21]
2. The generalization of proletarianization does not occur only on the side of the 'means of production': it is also what is generated by 'lifestyles', which, as behaviours 'expected by the system', as Hunyadi says, destroy life-knowledge [*savoir-vivre*] and the social systems that cultivate it. But this will occur during an epoch that Marx and Engels will not experience.
3. Intensive computing and 'deep learning' lead to the proletarianization of intellectual and scientific work, and, more generally, to the proletarianization of conception, thereby constituting the *general intellect* referred to by Marx in the *Grundrisse*.[22] This invites the possibility of a critique of *Capital* on the basis of this posthumously published work,[23] the remarkable foresight of which demands *a general reinterpretation of the prospects opened up by the philosophy of Marx and Engels*.

We now know that proletarianization extends to all classes of the population *as demoralization* (and without any 'revolutionary' prospect borne by the 'power of the negative'[24]). But this is what Marx and Engels were not yet able to see, nor were they able to draw all the consequences, including, in particular, the fact that, before anything else, *proletarianization is a process that entails a colossal increase of entropy*.

The theoretical formulation by Clausius of the physical law of entropy, which in the twentieth century will with astrophysics become 'cosmic', occurs contemporaneously with the writing of *Capital*, but

Marx and Engels were destined to remain ignorant of the issues raised by this formulation, as attested by Engels' analysis, in *Dialectic of Nature*, of the cosmological situation:

> Hence we arrive at the conclusion that in some way, which it will later be the task of scientific research to demonstrate, it must be possible for the heat radiated into space to be transformed into another form of motion, in which it can once more be stored up and become active. Thereby the chief difficulty in the way of the reconversion of extinct suns into incandescent vapour disappears.[25]

110. Transvaluation without concessions

This ignorance of the fate of the bourgeoisie, itself destroyed not by negative proletarian power but by the anthropic activity that it itself generates as the dominant force within the Anthropocene, as the *speculative shortening* of the timescales required for a *return on investment*, which thereby turns into destructive counter-investment because it renders structurally insolvent both exosomatization and the organological cohesion of societies, and as an increase in the rate of entropy inherently contained in calculation – all this is what forms the question of a *new critique of political economy capable of taking full responsibility for the question of exosomatization from a neganthropic perspective*.

After Hegel posed the question of exteriorization as a *moment* of Spirit, and hence of noesis as historical, the question of exosomatization was opened by Marx and Engels themselves. After that, it was taken up by Nicholas Georgescu-Roegen (who was at one time the assistant of Joseph Schumpeter, thinker of 'creative destruction') in the context of the new questions posed to economics by entropy and negentropy, as that *neganthropological imperative* which Marx and Engels were not themselves able to formulate: just like Nietzsche, ultimately, they rejected the new problem that entropic fate poses for any economics – whether it is an economy of capital or of libidinal energy.

On the other hand, the questions raised by Freud in *Beyond the Pleasure Principle* inevitably lead to the question of how the *death drive*, also called the drive to *destruction*, involves a *fundamentally entropic tendency of unbound desire*. And this entropic tendency is what desire binds – *bound by rules of life held in common and formed in the course of transindividuation, and as knowledge* (of how to live, do and contemplate) – *containing* this entropic tendency as ὕβρις itself, *where the death drive and the life drive merge*: they are *distinguished* only through the fact of being *bound*, that is, *transductively deferred and differentiated* [*différées*]. Such is the *economy of noetic différance*.

This complex heritage – Cartesian, Kantian, Hegelian, Marxian, Nietzschean and Freudian – as it has been described in the preceding paragraphs, will provide the beginnings of what will become, for Derrida, the 'deconstruction of metaphysics', or, for Foucault, the 'archaeological' analysis of technologies of power and techniques of the self, and that will also become, after Freud, the 'materialist' attempt to think desire, and, for Deleuze and Guattari, to think this as the primary question.

These elements of the critique of values and beliefs, however, thus revisited by 'poststructuralism', will participate, through American and European counter-culture, then through the new libertarian ideology and the ideology of northern California in particular, in *legitimizing disinhibition in the social sphere*. In *States of Shock*, I have tried to show why this inheritance must for all that not be rejected, and must on the contrary be *re-evaluated and 'transvalued' without concessions*: if we must deal with the great trial and ordeal entailed by the bifurcation that has begun to unfold, then it is a matter of doing so *through* the 'transvaluation' of this 'active nihilism'.

Such a *theoretical re-evaluation*, leading ultimately to a *practical transvaluation*, equally involves a re-evaluation of the *faculty of knowing in general*, and, along with it, a re-evaluation and transvaluation *of reason itself* – both as theoretical reason and as practical reason constituted as the motives of desire. Noesis struggles against what, in the faculty of knowing, stems from rules that tend no longer to be applied other than to preserve themselves by pandering to mental laziness that goes around in circles and thereby turns knowledge into the 'well-known', which is to say into stupidity (which is the rule of the pharmacology of spirit, and the subject of *Bouvard and Pécuchet*). Hence it may abandon itself to that vain *libido sciendi* by which 'men are led to investigate the secrets of nature, which are irrelevant to our lives, although such knowledge is of no value to them and they wish to gain it merely for the sake of knowing'.[26] Noesis is, then, the struggle against this entropic tendency that belongs to it, and that belongs in particular to its rational form, that form which, in principle, the university shelters and cultivates. It is a struggle that consists in *fighting against all fixity* that creates obstacles to the realization of the rational dreams in which exosomatization consists.

Driven by the *libido sciendi*, reason and the faculty of knowing in general were, before long, channelled into the industrial economy and have today become the *main production function*. But in this process, reason finds itself degraded, turned into rationalization, into the analytical specialization of industrially divided intellectual labour, into automation by calculation, and, finally, into the destruction of reason as a synthetic function of the generation of bifurcation by the projection of ends – that is, of protentions – conceivable as the *social as well as cosmic coherence of affinities*. This outcome amounts to a *profound demoralization of researchers, teachers and students*, which is one of the most serious concerns of our time.

The understanding, to which reason thus submits, is what delivers the faculty of knowing wholly and without limits over to the drive of destruction – legitimized by 'creative destruction' as generator of employment. *Such employment generation by industrial exosomatization, however, no longer has any reality behind it whatsoever.*[27] What remains of the *libido sciendi* then becomes a pure activity of *disinhibition without ends*, totally submitted to and structurally devoid of any transgressive capacity other than that which legitimizes the *senseless* and fundamentally irrational transgressions of capital.

As for the entrepreneurial literature of Michel Houellebecq, which submits to this general abandonment by exploiting the ploy of the scapegoaters, it presses itself into the service of computational entropy, simply in order to earn a little more money and to increase a little further what this oh-so-fashionable writer believes to be the 'recognition' that he so desperately seeks – to the point of looking for it right down the bottom of his garbage chute.

111. Organology of exemplarity

Throughout the succession of epochs resulting from the epokhal redoublings provoked by the pursuit of exosomatization, the latter being considerably intensified and accelerated since the advent of modernity and via industrial production, processes of transindividuation amount to so many successive forms of libidinal economy, each one characterized by original identifications, idealizations and sublimations that are different in each case. It is through these links that the economic processes of noetic *différance* are woven, and this includes the forms of sacrifice and sumptuary behaviour that Georges Bataille, on the basis of Marcel Mauss, described in terms of 'general economy'.

As *very bad examples* imposed upon the generation deprived of the faculty of dreaming by the absence of epoch (which Houellebecq reveals, and which he legitimizes), Strauss-Kahn and Hollande are, in France, each in their own way, and along with Cahuzac and after Sarkozy, the most glaring symptoms of a generalized demoralization that finds expression even in Juppé, and in his claim that we must renounce exemplarity.

Contrary to this tired and disillusioned discourse [*discours usé et désabusé*], we must reaffirm that *only exemplarity* can open up the future and draw from out of despair the possibility of *inscribing something new in becoming* – exemplarity can, like madness, be ordinary (goodness), extra-ordinary (heroism) or reflective (wisdom).[28] Exemplary does not mean perfect: on the contrary, what any noetic exemplarity manifests is the quasi-causal adoption and transformation of the default of origin – irrespective of whether that exemplarity is religious, artistic, scientific, military, maternal, paternal, athletic, scholarly and so on.

To be an 'exemplary' figure is not to strive for perfection, but to *singularly* embody – whatever that singularity may be, and as modest or invisible as it may seem in the eyes of many others – the possibility of turning exosomatic incompleteness into a future in becoming for any descendants there may be, that is, in the succession of generations. Such is, for example, the art of being a grandparent.

Such an exemplarity is *organological*:

- given that, on the one hand, it stems from an exosomatization that always leaves the moral and noetic being – that is, the ethical being in the sense of the being bound to its ἦθος, within αἰδώς and δίκη – *without a model*, and forces the *noetic imagination to invent, with reason and understanding, new dreams*, forming the *motives of new forms of life* that *cannot be models* because they *themselves remain to be interpreted more or less exemplarily*;
- and given that, on the other hand, *this exosomatic organogenesis always introduces the pharmacological threat of a self-destruction that seduces the drives* – a seduction that monotheism, mistaking the default for a fault, will present as the Devil opposed to the Good.

112. The pleasure principle, the reality principle and the drives in capitalism, according to Marcuse, Dostaler and Maris

That this seduction lies at the heart of contemporary capitalism is plain to see. That this incites some to move regressively towards demonization [*diabolisation*] and scapegoating is clear. And this is what installs the barbarism of the neo-barbarians and of those who hate by looking for scapegoats instead of thinking or taking care of what unfolds with disruptive exosomatization.

As the dominant *Stimmung* of the absence of epoch that destroys the energy that Bergson called spiritual, and which Freud showed to be libidinal, and which must be made into the object of an economy, there is no way of overcoming demoralization other than by initiating the turn and the leap in which consists the possibility, not just of a new epoch, of another epokhality or of a new epokhal redoubling, but also and firstly of a *new era*.

If this new era is an *exit from demoralization*, should we then conceive it as a 'remoralization'? And what is its relationship with 'ethics'? Note that in this regard Mark Hunyadi's proposals evince a certain ambiguity, in particular when he writes that 'never before have moral rules been so present or so constraining in our society. [...] Ethics is the key word, the word on everyone's lips.'[29] The morality of the moral being studied in moral philosophy is not ethics – nor is it the 'ethics' of 'ethics committees', or that of Spinoza, for example, who obviously refers to ἦθος – and confounding them impacts on the analysis of 'lifestyles'. Morality prescribes rules that are more or less positively stated or manifested, whereas ethics confronts the singularity of each situation: there where there are no rules, and where we must imagine, dream and invent a new exemplarity – a new *motive* for living.

Now, it is true that the *singularity* of each situation provoked by advances in exosomatization is the very reason given by ethics committees to justify their own existence in the face of technology that proves to be increasingly disruptive and that short-circuits all rules. The problem with this, as Hunyadi rightly emphasizes, is that, in so doing, such committees tend to legitimize both:

1. the short-circuiting of social systems, in the absence of which there can be no collective individuation, and hence no psychic individuation, a short-circuiting that is provoked by this disruptive innovation that outstrips and overtakes individuation, committees of all kinds arriving always too late, and society even more so;
2. the individualistic ideology that legitimizes this strategy by affirming individual rights and uprooting individuals from their communities [*groupes d'appartenance*], which in reality disindividuates them (and on this point, it is clear that Spinoza should be mobilized in a systematic way, which I will do elsewhere[30]).

To confront contemporary ethical questions, and the issue of demoralization within which these questions arise in new ways, we must return to the *drives* and to their *economy* inasmuch as it is becoming a *diseconomy*: the *functional unbinding of the drives* leads both to the *disintegration of social systems* and to the *generalization of capitalism as the drive to death*, the latter now being computationally effected through machines, as an operation of the understanding without reason.

This is why, in *Capitalisme et pulsion de mort*, published in 2009, Gilles Dostaler and Bernard Maris can quote John Maynard Keynes: '"Flight towards liquidity", "unquenchable thirst for liquidity", "morbid desire for liquidity", [...] the expression "morbid desire for liquidity" is not ours, it was invented by Keynes, and it refers to the death drive discovered by Freud.'[31] If this initial hypothesis is one I share, I cannot, however, agree with the consequences they draw from it. They offer what amounts to a Marcusian reading of Freud, as can be seen, for example, in the following quotation:

> Man, who under the law of the pleasure principle is merely a set of animal drives, collides with necessity, or with the exterior world. He learns to repress pleasure, the pure overflow of which would lead to death, and discovers utility and reason. He discovers the reality principle.[32]

The pleasure principle, in Freud, has nothing to do with a 'set of animal drives'. And this is so because it is *constituted* by its transductive relation to the reality principle, to which it is therefore not opposed.

In *The Lost Spirit of Capitalism*, I have tried to show why Marcuse, *confounding libido and drive* (and we saw in the preceding quotation from Dostaler and Maris that this rests on the confusion of drive with instinct, with 'animal drive'), and *opposing* the pleasure principle to the reality principle,[33] takes no account of Freud's 'second topography' as outlined in *Beyond the Pleasure Principle*.[34] It is precisely here, however, that the theory of the drives is reformulated, calling his earlier account into question, as Freud states very clearly,[35] in so doing elaborating the new concept of the 'death drive', or, again, the 'drive of destruction'.

The Marcusian reading on which Dostaler and Maris rely in the preceding quotation does not enable a struggle to be engaged against the situation described by Mark Hunyadi: it is on the contrary and precisely *because* he *celebrates* the *unbinding of the drives* and libidinal *diseconomy* that Marcuse himself was able to be pressed into the service of disinhibition via 'lifestyles' – on the basis of the demands that emerged from May '68 in France, and as was the case for Californian counterculture. And where this was also a situation in which Marcuse was himself, during this period, teaching in California.

Before expanding on this point, let us see how, according to Dostaler and Maris, capitalism stems from the death drive 'discovered by Freud'. They write: '[Capitalism] has liberated what lies buried deep within it and [that] moves it with all its energy: the death drive. What we believed to be "happy globalization"[36] was nothing but the excesses of mad money and its destructive drive.'[37] It would therefore be through a *fundamental transformation of the meaning of science and technics* that capitalism is able to fundamentally damage the vitality expressed by the life drive: 'Capitalism is a particular moment of human history in which technics and science are perverted towards the over-productivity of labour.'[38] Let us note here that this perversion is also that of reason, and that reason is in no way that towards which the reality principle is directed, which is on the contrary to serve utility, whereas Dostaler and Maris confound reason and utility when they write that the reality principle 'discovers utility and reason': reason is not on this plane. It is with this that Aristotle begins his *Metaphysics*, and this is the basis of all philosophy.

Far from being the actuality of the reality principle, which is itself *what ties the plane of subsistences to the plane of existences*, reason, which is *what ties the plane of existences to the plane of consistences*, is, on the contrary, the *function of idealization* insofar as it aims at consistences that are anything but useful – since they do not exist, no more than does the teddy bear [*doudou*] that the young child raised by its mother in transitional space sees in the dirty, smelly fabric of the toy: the teddy does not exist, but it consists in this space that is precisely that of reason, and which is the space where noetic dreams and desires are formed.

113. The market as catalyst of the death drive and Keynes' dream

The perversion [*dévoiement*] of technics and science by industrial capitalism, through which the Anthropocene is installed, enables exosomatization to be intensified, and, *through that*, it enables control to be taken over exosomatization via the application of criteria that are ever more exclusively those of the market. In other words, it enables industrial capitalism – and technology along with it – to be 'disembedded' from society and from social systems: such is the 'great transformation'. And, inasmuch as it stems from the submission of *otium* to *negotium*, this perversion relates to the secularization that Dostaler and Maris here refer to as laïcization: 'By laïcizing time, by distending it and making it the object of accumulation, man has recuperated, monetized and exchanged what belonged only to God, breaking the religious prohibition on charging interest on loans.'[39] It should be emphasized here that this 'interest' is itself the condition of industrial investment, and hence also of the financing of research, which in this way becomes 'technoscientific'.

This so-called technoscientific research is, therefore, *essentially* placed into the service of the acceleration of exosomatization, which leads to the age of disruption, of which it is the main factor: in this new age of research, in which science and technics are brought together to form industrial technology, *developmental opportunities, possibilities of becoming, are systematically explored and selected according to the criterion of greatest calculable profit.*[40]

Calculation thereby becomes the criterion by which to differentiate the future [*avenir*] from within becoming [*devenir*]. The problem, however, is that calculation cannot satisfy the criterion of improbability that is required by any negentropic evolution or any neganthropic evolution. And this also means that the *exploration of possibilities by calculation comes at the cost of the neutralization and systematic denial of the negative externalities to which this exploration may give rise.*

What, then, is *systemically and epistemically disavowed* is this *pharmacological question inherent to exosomatization.* And it is to overcome this denial and its disastrous consequences that a new *epistēmē* must emerge – amounting to what is here being called the Neganthropocene.

To the extent that man, through this 'perversion', according to Dostaler and Maris, 'through technics, [...] thinks to touch the divine',[41] this 'laïc' practice of technics – a practice that the religious would call 'impious' because it is, *par excellence, the diabolical dream of omnipotence* – is in reality and above all a *colossal process of destruction*, which 'is nothing but the death drive. To destroy, and then to destroy itself and die, too, amount to the spirit of capitalism. [...] The market [...] is an incredible catalyst of the death drive at work in accumulation.'[42] Hence Dostaler and Maris introduce the terms of an *alternative*:

> By developing technics, man has not increased his happiness, but released a force that will lead we know not where. Towards a new society of abundance, finally putting an end to the economic problem (scarcity), towards the arts and friendship, as Keynes dreams in 'Economic Possibilities for Our Children'? Towards the termite mound, evoked by Freud in his book,[43] and the abolition of individual will, as the Nazis wanted? Towards the apocalypse?[44]

We cannot be satisfied, however, with an alternative formulated in such a manner: man has not 'developed technics'. Man *is* technics. And the force that he always releases in developing himself is ὕβρις. Man could not have developed otherwise than by 'developing technics', and in the ὕβρις that he has therefore constantly produced and renewed, and contained, weaving the life drives and the death drives. This mixture, unbound, destroys both man and technics, whereas, bound, it constitutes, *through exosomatization's formation of a new organological milieu*, the neganthropological power that is formed as libidinal economy, that is, as circuits of transindividuation, giving rise to epochs that themselves belong to an era of moral being.

By referring to man's 'perfecting his own organs'[45] in *Civilization and Its Discontents*, a text that Dostaler and Maris cite extensively, Freud ties the 'fate of the drives'[46] to this improvement that is, not just organic but, precisely, organological. And in the current stage of the development of capitalism as the concretion of a drive to death, and as force of destruction, according to Dostaler and Maris – and on this point I am in agreement with them – an alternative presents itself, which for them is between the anthill on one side and the 'dream' of Keynes on the other.

It will be obvious that, given my quite different interpretation of the drives – whether of life or of death – I do not draw the same conclusions as Dostaler and Maris with respect to the outcome of this alternative.

114. Collective suicide in the face of what capitalism no longer succeeds in containing, and the question of investment

The death drive, which Freud also names Thanatos and which culture constrains, lies *within culture*, and hence constitutes *one of the faces of* ὕβρις – of which the life drive, Eros, is the other face. Dostaler and Maris affirm the *probability of a coming unbinding of* ὕβρις *as death drive*: 'Every bad omen suggests that the death drive is merely waiting to overwhelm the capitalism that contains it. Until when?'[47] It is quite striking and affecting to read this question posed by Bernard Maris, as if he anticipated his own assassination on 7 January 2015 by the unleashing of what is, indeed, a death drive that capitalism no longer contains – that it is no longer *able* to contain – *and that it provokes*: this was indeed Maris' own analysis.

Dostaler and Maris also recall that the death drive lies, as well, *within* the life drive, as for example in the form of sadism. They then describe drive-based excess, the overflowing of the drives, as a true *collective suicide*: 'Like lemmings who, pushing and shoving in large numbers, fall off the edge of cliffs [...], is humanity itself in the course of unconsciously rushing towards death?'[48] As striking as this analysis may be, it is not really convincing – and it does not make it possible to think a true alternative, for it confuses the accursed share analysed by Bataille with consumerist waste: quite unlike the unbinding of the destructive drives by calculation, which is characteristic of the unbridled capitalism criticized by Keynes and described by Dostaler and Maris as constituting our mortiferous fate, Bataille's general economy affirms *incalculability* – where the surplus, instead of submitting to calculation in reinvestment, is *sacrificed*, and not simply destroyed.[49]

Here, the references to Marcuse and Norman O. Brown[50] (as well as to Reich and Fenichel) somehow neutralize the hypothesis that Dostaler and Maris are trying to put forward, and lead them into a dead end. For after recalling that Keynes established (150 years after Adam Smith) an *essential relationship between economics and morality*,[51] and 'made the link between the death drive, the preference for liquidity and the tendency towards rentier economies',[52] they point out that man, caught up in the morbid tendency of capitalism, 'does not just accumulate wealth, but also and above all negative goods, waste [...], destroying more than he accumulates'.[53] They then relate this destructive accumulation to the reality principle (in the sentence already quoted), which they equate with investment.

In so doing, they posit that in economics, as in the analysis of the reality principle by Freud according to them, investment 'requires the abandoning of immediate destruction, consumption, and allows deferred action, greater future consumption'.[54] They insist here that, through this 'différance', 'it is indeed a question of deferring a present destruction *in favour of a much greater future destruction*'.[55]

Two points, however:

1. On the one hand, we have reason to doubt that what Freud, too, theorizes as investment, under the name of *Besetzung* (which James Strachey subtly translates into English as *cathexis*) – which stems from the différance evoked with respect to the reality principle,

but such that it leads *well beyond a mere principle of reality* and equally *beyond a principle of pleasure*, as Freud states explicitly in *Beyond the Pleasure Principle*, where he develops his theory of the death drive – we have reason to doubt that this investment consists simply in deferring to 'a much greater future destruction', for *it is the complete opposite*. It is, in fact, this *différance of the satisfaction of the drives* that, in 1923, will lead Freud to introduce idealization and sublimation as conditions of this investment (*Besetzung*) in an object of desire, which is defined as a 'diversion of aims'.[56]

2. On the other hand, we have reason to doubt that this, which is Freud's second conception, presented as *that which leads beyond the pleasure principle*, and as the libidinal *economy* of the drives, is reducible to what these economists present as an economic investment that more closely resembles speculation about the future than investment. For in fact, *it is highly debatable whether investment is undertaken for the purpose of future destruction* – in economics or in Freudian theory, *which refers to economics, here, for precisely this reason.* Investment – as *Besetzung*, as cathexis – is, rather, undertaken in order to *construct the future*.

Cathexis (κάθεξις) means 'the act of arresting, from which: the act of retaining, of conserving [...], the act of intercepting [...] or of containing, of repressing'.[57] This *retaining that contains* is a kind of retention that bears a protention, and it is obviously also, as an arrest, a kind of suspension, an *epokhē*. Investment, if we wish to keep this translation of *Besetzung*, is one dimension of epokhality as the initiation of a process of transindividuation in the doubly epokhal redoubling.

While it is obvious that Freud does not at all say this here, it is equally clear that this investment participates, as an economy of the drives, in the construction of processes of transindividuation in general, and in an essential way, just as it is obvious that *economy in general is the materiological dimension of transindividuation in general* – some of whose theoretical elements I have outlined in *For a New Critique of Political Economy*.[58]

As such, the capitalist economy, *today subject to a mortifying computational totalization that makes idealization impossible*, ruins transindividuation, which it replaces with transdividuation, just as entrepreneurs are replaced by shareholders, then development projects by takeover bids and later by the short-circuits to which the strategies of disruption give rise. It then becomes possible to accrue huge profits, but at the cost of destroying what they exploit as surely as the barbarians who formerly invaded civilized territories would ruin them by plundering them.

Dostaler and Maris are no doubt right in wanting to point out that the construction of the future also frequently leads, and especially today, to

destruction, in particular if we take into account the negative externalities generated by what is ultimately and above all, in the psychoanalytic
sense and in the economic sense, the *realization of a dream*. But this issue
does not just concern capitalism: it involves the question of entropy and
negentropy, which after the fact of exosomatization become neganthropology, organology and pharmacology.

Precision with respect to the *question of investment* is fundamental,
given that – in addition to the fact that this question arises in Freud
after he *puts back into question* his theory of the pleasure principle and
reality principle, which, in any case, he never placed into opposition – *it
thoroughly conditions the Keynesian discourse on what Keynes dreams
of* as a *society founded on arts and friendship*.

115. 'To think otherwise': it's always about Keynes' dream

Art is an activity of sublimation that gives full meaning to the notion
of what Freud refers to with the name *Besetzung*. And friendship is
obviously a highly significant form of sublimation and investment
insofar as these involve a fundamentally *free share* [*part foncièrement
gratuite*] that is the object of a *sharing* [*partage*], that is, of a *transindividuation* forming a *moral being*, and that relates to the *accursed share*
referred to by Georges Bataille – that great friend of Maurice Blanchot.

This is why we cannot agree with Dostaler and Maris when they
write, relying on Freud but seeming only to have read Marcuse, that:

> the more the death drive is repressed, the stronger the sublimation of
> the libido in the work, the greater the repression exerted on the morbid
> tendencies, the more technics is powerful and accumulation important
> [...] or, as Marcuse puts it: 'Then, through constructive technological
> destruction, through the constructive violation of nature, the instincts
> would still operate toward the annihilation of life.'[59]

At the end of their work, they do quote from Freud's *Civilization and Its
Discontents*: 'The fateful question for the human species seems to me to
be whether and to what extent their cultural development will succeed in
mastering the disturbance of their communal life by the human instinct
of aggression and self-destruction.'[60] But here it must be remembered
that, prior to this statement, Freud stresses that man is constituted and
moved by the process of constantly perfecting his organs,[61] that is, by
exosomatization, which in the twentieth century reached such a degree
that technics, which is its fruit, and which characterizes humanity as
a form of life, presents itself in the way that the tragic Greeks (which
the monotheism of god-the-creator had erased) had already viewed it,
namely, as a *pharmacological fate*.

I have argued elsewhere[62] that Freud, in rediscovering the *tragic* bearing of the libidinal economy inasmuch as it is constituted by the *pharmakon* emerging from 'organic perfection', here takes note of that to which Valéry in 1919[63] (then again, in another way, in 1939[64]) and Husserl in 1934[65] drew attention, each in their own way, namely, that the life of the mind, spiritual life [*la vie de l'esprit*], insofar as it is produced socially as *moral virtues, science, arts, skills and technics*, must be completely reconsidered from the perspective of what, in 1919, Valéry called the mortality of civilizations, and of what, in 1939, he called the political economy of spirit, leading to a 'fall of spirit value'.[66]

Dostaler and Maris note that 'the collapse of those regimes that claimed to be Marxist was accompanied by a return of the religious, not only in Muslim countries but in the United States and other Western countries'.[67] Besides the fact that it remains to be verified that this is indeed a question of a 'return of the religious', and not just compensatory fantasies mistaking religion for the most regressive and herd-like aspects contained within it, such a 'return' could be confronted only by *taking seriously into account what, in the death drive as in the life drive, requires them to be bound together* by a libidinal economy that is constituted, precisely, *by investment*, and by an investment that is irreducible to the mere différance of a destruction aiming at a 'greater future destruction': it is the *complete opposite*, being first and foremost *the protention of a construction.*

It is in this sense, as the *power to bind the drives to one another*, and as what makes it possible for *investment to contain* ὕβρις, that we must understand the statement with which Freud follows up his previous remark, when he says that 'it is to be expected that [...] eternal Eros, will make an effort to assert himself in the struggle with his equally immortal adversary'.[68] And here, it is interesting to quote Keynes, whom Dostaler and Maris themselves quote, just after they refer to Freud:

> I see us free, therefore, *to return to some of the most sure and certain principles of religion and traditional virtue* – that avarice is a vice, that the exaction of usury is a misdemeanour, and the love of money is detestable, that those who walk in the paths of virtue and sane wisdom will take least thought for the morrow.[69]

Can we, however, remain content with these remarks that date back to 1930? Surely not. If, indeed, 'it is no longer a matter of refounding, but of overcoming in order *to think otherwise*',[70] and if the question is either to go *beyond capitalism*, as the epilogue to *Capitalisme et pulsion de mort* proposes, or to go *beyond the Anthropocene*, as seems to me to be the real question, that of capitalism coming *after* that of neganthropy in the *order of questions*, in all cases the context within which we speak is that of disruption[71] inasmuch as it has led dissociety to the brink of madness, to become murderous in actuality – for which Bernard Maris himself paid dearly, with his life.

Earlier, I asked whether, in order to combat demoralization and the absence of epoch, and to enter a new era, it is a question of enacting a 'remoralization'. And after this reference to the call by Keynes for a return to 'certain principles of religion and traditional virtue', he who did not yet know demoralization as the effective accomplishment of nihilism, I repeat that this is a question of *transvaluation*, that is, of redefining value in all its forms, in both restricted and general economies, or, in other words, by reintegrating the restricted economy into a libidinal economy reinventing investment as investment *in these transvaluated values*.

From this point of view, then, it is, necessarily and inevitably, a matter of moral and ethical questions conceived through the lens of a new moral philosophy, as well as being a matter of manners and mores in the sense of Kant's *Metaphysics of Morals* – all of which requires revisiting Spinoza, Nietzsche, Freud and beyond. Here it is indeed a question of quasi-causality, that is, of worthiness, of dignity: quasi-causality as the condition of dignity in the organological and pharmacological condition, where the question of dignity is not merely a 'moral question', but a *vital* question.

Dignity and its cultivation by a moral philosophy taking care of morale is the condition of any neganthropological bifurcation. Dignity is no more and no less than noesis in all its forms: noesis is not simply abstract thought, rational thought, logical thought. It is the *manner* in which the noetic body moves itself through its life-knowledge [*savoir-vivre*] and work-knowledge [*savoir-faire*], as well as its intellectual knowledge and spiritual knowledge.

Now, demoralization is a denoetization – and it is on this basis that we must rethink moral philosophy starting from the fact of generalized proletarianization and the obligation (in law and of *dikē*) to de-proletarianize. To revisit all these questions in the wake of Foucault, Derrida, Deleuze and Guattari, and to do so with the ecological question that is really opened up only after Deleuze produced his 'Postscript on Control Societies', we must take note of what we have learned, for example, of the conditions of the constitution of the psychic apparatus, and of the fact that, more precisely, there is a libidinal *economy* within which this psychic apparatus, which is not simply cerebral, becomes noetic in its effects.

Such noetic effects make the pursuit of exosomatization neganthropologically possible. This is done by minimizing [*économisant*], through this noetic différance, its entropic consequences, insofar as this is possible – it being taken as given that it is impossible to *eliminate* these consequences (contrary to what Engels postulates in his dialectics, in a manner somewhat similar to the Extropians[72]), and that this différance is therefore ineluctably and inevitably temporary.

The noetic cerebral organ is immersed in the libidinal economy of this noetic *différance* that forms it as an organ of an individual who is noetic only insofar as he or she is strictly social – since this noetic organ is constituted through identification, idealization, sublimation and the super-ego, which are not, properly speaking, cerebral, even though they pass through the brain. This is what Catherine Malabou fails to under-stand.[73] And this is why idealization and sublimation, as productive economic functions of this form of energy that is the libido, transindividuate the drives by *binding* them – which is possible only because the drives are in no way instincts.

Denoetization, on the other hand, is the unbinding of the drives, that is, the process of demoralization provoked by disinhibition that here approaches its limits. And it is at this point that the programme of transvaluation becomes necessary, transvaluation as a collective protention – *that is, as affect* – of de-proletarianization as the condition of possibility of the Neganthropocene.

15

The Wounds of Truth: Panic, Cowardice, Courage

116. A reminder concerning questions of denial and disavowal

The knowledge of death[1] is the archi-protention that constitutes Dasein: death is *what Dasein expects* throughout everything it does – but *most of the time without being aware of it*. This primordial protention conditions and configures all other protentions, that is, the entire relationship to the future. It inscribes them on the path of the vanishing point that it constitutes as this irreducibly thanatological perspective, but this knowledge is also and constantly *concealed by processes of denial of all kinds*, where all of this falls under what Heidegger calls *Besorgen* (busyness, pre-occupation).

The whole existence of Dasein is a way of both knowing this thanatological perspective and refusing to know it, that is, of disavowing it [*denier*]. All of Dasein's knowledge is *in this sense* a *différance*. It is the existential analytic of *Being and Time* that inscribes this knowledge of the end – being towards death, *Sein zum Tode* – at the *origin* of the noetic soul, and as the *default* of origin, that is, as *denial* [*dénégation*].[2] By returning to the myth of Prometheus and Epimetheus, I have endeavoured to show why and how this default of origin that is mortality is constituted as prostheticity, that is, exosomatization.

The fruit of this exosomatization is tertiary retention in all its forms, that is, technics in general inasmuch as it conserves and materializes an individual's time in the form of traces – a spatialization of time, which is a *différance* and which allows the constitution of what Heidegger described as the past of Dasein, a past that Dasein has never lived but which it inherits as its 'already there', which also amounts to what Simondon called the preindividual funds [*fonds préindividuel*] of psychic and collective individuation.

It is in terms of the archi-protention of the end that Dasein interprets the tertiary retentions accumulated as the already there – even though most of the time Dasein does *not* interpret this already there: it remains content merely to reproduce it through its 'busyness', its pre-occupation, its *Be-sorgen*, and to do so as the practical denial of this end. Nevertheless, intermittently, and in going over tertiary retentions that, instead of repeating, it interprets, on the basis of which it inscribes a difference, Dasein projects its protention of all protentions, *its* end, and its end insofar as it is indeterminate – insofar as it is incalculable and where, reaching it, no one can take Dasein's place.

This path towards its end is therefore what generates a bifurcation – that is, what inscribes a negentropic difference beyond pre-occupation, *Be-sorgen*, and does so as the singularity of a care, *Sorge*. Such a care is a knowledge, which means that it itself becomes transmissible, interpretable and fruitful through new negentropic différances that we must here also and more specifically describe as neganthropic.

Nowhere does Heidegger integrate the negentropic conception of life. Nor does he explore the metaphysical consequences of the new consideration of the cosmos to which the theory of entropy gives rise – even though the end of Dasein, its death, is the return of its body 'to dust', that is, to a mineral realm itself overwhelmingly subject to entropic becoming.

On the other hand, and above all, Heidegger does not investigate the protention of what happens beyond the end of Dasein in terms of what it means for the way that Dasein relates to this end: he does not question the care for what comes after the end, which, for example, makes us wonder about the future becoming of the universe, the solar system and the biosphere *beyond* our own end.

Until quite recently, *beyond* this archi-protention that is *Sein zum Tode*, there was, rooted in every noetic soul – that is, in every form of exosomatic, organological and pharmacological life – a more archaic and, as it were, 'more primordial' archi-protention than the thanatology of being for the end. This protention was *being for life*, and it posited that beyond my death, beyond the 'instant of my death', *life would continue*, that there would be descendants who succeed me, that a legacy would continue on that would make fruitful what I would myself have made fruitful and thus brought to life – so that I would, then, not have lived for nothing and not have died *in vain*.

In our time, at the end of nihilism, of which the capitalism of the death drive is the fulfilment in and through disruption as the radicalization of the Anthropocene, this *absolutely positive* archi-protention that is the *affirmation of life itself* – and the one that variously but jointly animated Schopenhauer, Nietzsche, Bergson, Freud (for whom the life drive expresses and constitutes the first and last meaning of Eros, which incessantly composes with Thanatos in exosomatization) and, indeed, many others – this absolutely positive archi-protention has turned into an archi-protention that, if not purely negative, is at least tragically interrogative: 'Will life continue after I'm gone? For how long?'

Such a reversal is what occurs when *Neganthropos* discovers that, as *Anthropos*, and through its anthropic activity, it massively and irreversibly increases the entropy that, in the biosphere, it had been concerned with minimizing [*économiser*], that is, reducing, deferring [*différér*], and doing so through noetic différance, that is, as knowledge – of how to live, do and conceptualize. It discovers that it has destroyed its knowledge [*savoir*], that its life has thus lost all flavour [*sans saveur*], and that, by this very fact, the future of life in all its forms has been jeopardized.

It is this discovery – which is the result of both the so-called 'death of God', that is, the domination of calculation, and the discovery of entropy made possible by the steam engine – which has precipitously increased the rate of entropy in the biosphere, leading Lévi-Strauss to refer, at the end of *Tristes Tropiques*, to 'entropology'.[3] And this is *what we think, along with Florian, and in fact all of us, whether we want to or whether we don't*: we have compromised the future of life and we are threatening our descendants.

But at the very moment when we think this, when we say to ourselves that this positive archi-protention that we all share of a life after ours has become the negative protention of the end of every form of life after our death, *we all, all of us except Florian*, we all at the same time continue to deny this thought, which thus becomes our unthought. But if we thus deny and 'unthink' this thought, it is because such an archi-protention is literally *unbearable* [*insupportable*]. And as long as it is unbearable, as long as it remains *unsupported*, it cannot *support* our existences by constituting the ὑποκείμενον πρῶτον (*hypokeimenon prōton*).

117. Knowledge, thermodynamics, philosophy and economy

Such an archi-protention, however, is what results, on the one hand, from the second law of thermodynamics, and, on the other hand, from the fact that the Anthropocene has amounted to an astounding concentration and acceleration of the local effects of this second law within the biosphere. The anticipation of these consequences necessarily leads to a negative collective protention. Unless it becomes the object of new cares, that is, new knowledge, both theoretical and practical, such negative collective protention can only drive us mad.

Without such knowledge, this negative protention – whether we deny it, as most of us do most of the time, or whether we declare it, as Florian exemplifies, but where this is also expressed through forms of behaviour that stem from negative sublimation, which is itself the outcome of this negativity – is bound to produce a kind of cosmic panic at the scale of the biosphere, a biosphere that is suffering the consequences of the entropic becoming of the universe in the Anthropocene era, that is, the era of negligent carelessness [*incurie*], the era of the absence of care. On the horizon of this negligence, and without a highly improbable jumpstart [*sursaut*], many scientists predict that the collapse of the Anthropocene may only be decades away.

It is striking to note here that, with the exception of Bergson, the consequences stemming from the law of entropy – as revealed by thermodynamics, itself emerging from the development of the machine, which produced a sudden increase in the local entropy rate in the biosphere – failed to be drawn by so-called 'continental' philosophy, and more particularly 'French' philosophy. In addition to the silence of phenomenology on these questions, Derrida says nothing about it, and nor does Foucault, while Lyotard may occasionally do so but only in passing. In some respects, Deleuze and Guattari do refer to these questions, but via dissipative structures and chaos theory.

The forgetting of these questions clearly stems from the denial highlighted by the existential analytic as constitutive of what I myself analyse in terms of the intermittent condition of noesis – something that afflicts this existential analytic itself. We will see that this forgetting stems more profoundly from a forgetting of the *pharmakon* and the consequences of the pharmacological situation of exosomatization in which noetic life consists, and that this strikes Derrida himself, at the very moment when he tries to think forgetting as such by deconstructing the deconstruction that Heidegger called the 'forgetting of being'.

We have seen[4] that the madness to which today's extreme demoralization leads is expressed by denial of all kinds, cloaked in statements either *pessimistic* or *optimistic*. We see now that all this stems from a more profound denial, on which neither philosophy nor psychoanalysis has ever really reflected.

The positive collective protentions from which epochs are formed are conditioned by the archi-protention of a life yet to come for descendants: this is the unconditional affirmation of a life after my death, not as immortality, but as the preservation of negentropic possibility. In exosomatization, it is knowledge that constitutes such a possibility, but where this knowledge is pharmacological, that is, where it can turn into its opposite. And in the Anthropocene, the highly entropic automatisms this knowledge generates lead to its disintegration.

This disintegration is the vanishing point of the moral philosophy that we are trying here to conceive as a way of responding to questions raised by Keynes and so many others (including Georges Friedmann and Nicholas Georgescu-Roegen), and as articulations between the market economy and the libidinal economy. The moral dimension of economics and the economic dimension of morality lie on the horizon of works as diverse as those of Adam Smith and Karl Polanyi – provided that we are willing to reinterpret them from the perspective of the only noetic dream possible today: the Neganthropocene whose possibility arises from out of the nightmare of the Anthropocene and disruption.

It is a question of revisiting the questions of morality and exchange by reconsidering the market and libidinal economies from the perspective of a life drive that, economically *bound*, becomes the key dynamic element, not just of a bio-economy, which would be an economy of negentropy, but of a neganthropology, that is, an economy of exosomatization – as Georgescu-Roegen allows us to think it.

118. State of emergency and splitting: courage, object of moral philosophy in the twenty-first century

Faced with the fears expressed by Keynes, Freud, Dostaler, Maris and so many others before[5] and after[6] them, faced with the terrible 'negative collective protention' expressed by Florian, the overwhelming tendency is towards denial: it seems *practically impossible* to live in awareness of the immense dangers that humanity is now courting – we find ourselves paralysed. We try to think of something else, to hold onto that portion of energy we need just to live our everyday life.

Just as after Fukushima a syndrome *erasing the memory of the catas-trophe* appeared in Japan, increasing the likelihood of its repetition, or of something even worse,[7] so too deniers of all kinds proliferate through the effect of this psychic defence mechanism that is self-deception and a kind of lying to oneself, which the immense pressures of ideological bombardment attempt to consolidate and reinforce, by which it is turned into diverse forms of cowardice.

In the absence of epoch, in the absence of positive collective protention, everyone more or less effectively cultivates this kind of lying to oneself. This gives rise to a particularly pathetic symptom of demoralization: *denying demoralization itself*. This was already dramatized by Alain Resnais in *On connaît la chanson* (*Same Old Song*, 1997), where all the characters are depressed and all of them deny being so, such as Camille: 'Depressed, me? Pfff...anything but!' All but one, Camille's interlocutor Simon, who is in love with her *because* she is depressive *like* him, that is, for the same reasons as him, and who tries to make her recognize these reasons, if not know them.

This denial of depression from within depression is both psycho-logical and socio-economic, and believes it can justify itself from the side of economic and political power by declaring that it has the goal of 'supporting the morale of markets' and 'investors'. This denial, however, is the very thing that continues to feed depression, and to feed it *more than ever* – and to do so on the scale of whole countries, about which we act as if 'everything is going fine, Madame la Marquise',[8] which *nobody believes*, but which many want to hear, while nevertheless finding these lies shameful and scandalous.

This is so because both psychic individuals and collective individuals are *split*: one part wants to hear the *opposite* of what the other part knows and tries to say and to make heard. This is what since Freud has been called *Spaltung* (splitting), which opens up the possibility of lying to oneself, that is, of the denial by one part of the self of what another part of the self knows.

Such splitting is also the starting point of any moral philosophy, since it establishes the bipolarity that cuts across both psychic and collective individuation, binding them to one another. Because it may lead to denials that threaten the unity of the psyche, this splitting can also drive us crazy: it was by starting from splitting that Freud tried to think psychosis.

As the deniers that we all tend to become in order to defend our *psyche*, and so as not to completely exhaust our 'morale', what we do consists in *also and firstly denying the absolute need to think afresh* – this necessity to which Dostaler and Maris invite us at the end of their work.[9] For thinking is painful: when it involves an authentic noesis, which happens only intermittently, it inevitably leads to *questioning the foundations of one's own life.*

By *refusing to think afresh (anew, otherwise)* in the Anthropocene, and through the *denial* that this contains and that *justifies* this refusal and leaves us lost in disruption, we are courting disaster. To think in such conditions, and to think such a contemporary condition, inasmuch as it sets up a vast denoetization, is therefore and above all to think [*penser*] and to think care-fully [*panser*] these conditions themselves: it is to try to think, and think care-fully, the *necessity* of this process of denial.

Denial – which organizes noetic life in general as it confronts the fact of the relation to the end, and which is borne by a *more profound denial* of the exosomatic and pharmacological situation as it poses a specific problem of 'entropology', and such that it requires a neganthropology – is what *professional deniers* systemically and systematically exploit. And they do so, at least in the short term, inasmuch as they are caught up in their own denials about *their own long term*, believing that it is wholly within their interests to *stir up obstacles to thinking and caring newly and otherwise.*

To think and to care today, in the *exceptional* circumstances that have been imposed *precisely in this regard*, is to try to combat at all costs the deeper *disavowal* on which these denials are built, and which is in truth constitutive of ὕβρις itself (and as the denial of ὕβρις itself). And this is the way to combat the consequences of disavowal – which are the denials themselves, inasmuch as they are all tied to *cowardice*. As such, this is a question of the moral philosophy of the 'moral being', who possesses a noetic soul insofar as this refers to the soul of a non-inhuman being.

To think by caring [*penser en pansant*], in other words, is to have the *courage* to think, to counter this cowardice that prevents thinking – and that does so within thinking itself, which means that it is firstly *to think (and to think care-fully) cowardice as such, inasmuch as it is something from which there is no escape*. It is to conduct, by all means possible, this fight that is firstly with oneself – and by all means *that can re-establish desire, hope, trust, courage and therefore reason (for hope)* in the face of what seems above all to make all of these impossible.

This impossibility that must be overcome, this impossibility in which the only possible possibility is the impossible, is the *bifurcation* that must give rise to the Neganthropocene from out of the Anthropocene, and it is a question that is simultaneously and *inextricably political, economic and epistemic*. It is *epistemic to the highest degree*, that is, *to the point where a new era is initiated*, bearing other epochs, themselves emerging from *new conditions of epokhality* established by this era, and hence discovering new consistences, Inshallah.

This is what is most urgent, the real 'state of emergency', and it is what is denied at the highest levels, in particular as the denials and lies of the state – which everyone feels but without having a clear idea of how to describe this feeling – and which amounts to the height of carelessness.

The process of disinhibition began from the very first moments of modernity, and amounted to an incredible dream, but the effects of this dream would begin to accumulate and be systematically realized from the onset of the Anthropocene, only to be radicalized by disruption. At the end of this process of disinhibition, the dream seems to have turned into a nightmare, and so the defence mechanism of generalized denial tries to blind itself to the obvious, from high to low, both in 'France from below' [*France d'en bas*] and at the 'highest levels'. Those on these 'highest levels' believe they will be able to seduce 'France from below', but the latter, on the contrary, and for this very reason, despise them even more. And so this generalized denial continues, despite the *parrhēsia* of a few.

Through denial, we, the demoralized that we all more or less remain, nevertheless try to keep sleeping – and to do what accompanies it: to keep dreaming – but we find ourselves sleeping and dreaming less and less. Meanwhile, other deniers, the professionals, *maintain* this bad sleep haunted by dreams that are less and less noetic: the professionals of denial deny themselves the noeticity of dreams in order to *keep* economic or political *power*, and, as the power over the power of dreaming, to pervert it, and, ultimately, to annihilate it by disintegrating it – and, along with it, any horizon of promise worthy of hope.

It is within this state of negligent carelessness that Florian and his generation find they have become incapable of dreaming: deprived of a place in the future, deprived of a right and an ability to create a place for themselves, deprived of the knowledge that we no longer know how to transmit to them because we no longer know how to live, deprived of the possibility of *being the future* that they alone can *incarnate*, they can no longer undo these denials upon denials that may be involuntary for some but are nevertheless programmed by others – making perfectly visible to them the negligence involved in what some claim to be untreatable [*impansable*].

Engendering a generalized cowardice from which *no one* completely escapes – it functions both as an *epidemic* and as an *omerta* – this generalized denial raises, as never before, the question of courage: of the *courage of truth*, of which *parrhēsia* is thus, according to Foucault, the first resource, and of the courage of truth that is not just being-towards-death as the archi-protention of the existential analytic, but the absolute danger that *Anthropos* has become in the cosmos, whose becoming is the slope on which courage becomes the *force affirming a future that pushes back up this slope* – against becoming, right up against it [*tout contre lui*].

119. Muddying the waters: denial, regression, democracy

It was Nicolas Sarkozy who installed government by systemic stupidity, and he did so at the level of the presidency of the French Republic – a governance consisting essentially in encouraging regressive processes, of which his successor's governance by lying was a frightening escalation. Lying is obviously a regression – towards minority, that is, into infantilization. We have already seen how much this has to do with the promise and with the dream. It is this obviously regressive character of lying that lies on the horizon of Kant's famous essay on the 'right to lie'.[10]

Lying introduces into the noetic soul – which must always *produce* its unity – not just a split, but a divorce with itself.[11] When heads of state indulge in such cowardice, they inevitably generate a collective madness that usually begins by turning against the heads of state themselves. Given the position these figures occupy within transindividuation, such cases are particularly harmful. This harmful power then appears to be the expression of power itself, which thereby finds itself inherently discredited.

Analysing and describing individual and collective regressive processes is always a delicate matter: when people who are regressing are told about their regressive tendencies, they generally regress *even more* – and they become *aggressive*. Moreover, this is a process from which no one escapes, precisely because it lies *at the root of denial itself, which is also the ordinary mode of functioning of transindividuation*.

Anyone who tries to analyse regression and its denials may be tempted to believe that doing so shelters them from its effects, that it sets them apart from the crowd. If this is what they believe, it is because they do not see that, in the *no-win scenario* in which exosomatization ultimately consists, and the ὕβρις that it provokes and that provokes it in return, *regression is constitutive of noetic life*: it inhabits it, and hence it is never simply a matter of denouncing it. It is a matter of thinking and caring about it [*panser*], and firstly of thinking care-fully of and by oneself.

Because it constitutes the noetic life of which it is in this way a 'condition of impossibility', regression *always ends up denoetizing noesis itself*, if only by turning it into the 'well-known' through which knowledge [*savoir*] and savours [*saveurs*] become clichés and hackneyed expressions. Hence it is that noesis is intermittent, and that education, attention-formation, culture and civilization all amount to systems and apparatuses [*dispositifs*] whose aim is to struggle against this regression. The latter, however, *always* returns, and accompanies the 'revenance' of the spirit like its shadow, to the point that education, attention-formation, culture and civilization can be made to serve regression at the very moment when they believe they are fighting against it: such is the 'mortality of civilizations' upon which Valéry meditates in 1919 in 'The Crisis of the Mind'.[12]

It was the *denial of regression itself* and of the *complicity that it so often induces among those it affects* – as an anti-noetic, highly mimetic epidemic – that drove Marie-Noëlle Lienemann and Paul Quilès to take France Télévisions to court over a programme called *Le Jeu de la mort*. This show recreated the Milgram experiment[13] with the goal of *highlighting the degree to which television demoralizes viewers*, not just by showing them degrading images but by leading them to identify with the televisual apparatus itself (thereby pursuing Milgram's hypothesis a step further), and, through it, with the viewers selected as part of the 'games' of reality television – where they lose their dignity.[14]

In March 2010, Quilès defended their lamentable action, which he and Lienemann felt was necessary, by claiming that with this programme public broadcasting had gone a step too far, that it had crossed boundaries: 'There are limits to what should be shown – even in the name of worthy intentions – on this powerful medium that is television.'[15] On the basis of such a placatory discourse, which consisted in concealing the fact that television is *constantly crossing boundaries* (especially when it is not part of the public service) without Quilès or Lienemann or anyone else being bothered by it, these two 'elephants of the PS'[16] brought to an end, through the threat of a trial, and therefore *by intimidation*, the public debate that France Télévisions had been courageous enough to want to open up.

In so doing, they intimidated or intended to intimidate all those who would like to combat the degrading state of affairs that offends youth who have themselves become incapable of dreaming, and incapable of being attracted by the solicitations of false 'idealizations'. Quilès and Lienemann did not allow them to occupy the position that the programme was attempting to set up so as to provoke a shock in public opinion through a kind of televisual *parrhēsia*. They failed to recognize the *courage of these filmmakers* who were creating this dream of a public television service that would be thought-provoking and that, in so doing, would be care-ful [*panserait*]: they took advantage of what were no doubt the programme's weaknesses in order to muddy the waters, or, as we say in French, *pour noyer le poisson*, 'to drown the fish'.[17]

120. The powers that be in the face of the parrhesiasts of our time

Bullying and intimidation is what, by means of fear – and through the cowardice it provokes – accompanies and reinforces the denial that afflicts the state or the party.[18] We have witnessed the way in which pressure and intimidation continue to be used against all those who would dare to follow the examples of whistleblowers such as Julian Assange, Edward Snowden and Denis Robert,[19] and the way in which Jérôme Kerviel was turned into a scapegoat, which in this case means a 'fall guy',[20] after having been chewed up and spat out by his superiors. We have also seen how laws have been implemented to reinforce this power of intimidation against the parrhesiasts of our time.

In fact, over the course of history, *parrhesia* is reconfigured as the stage of a regime of veridiction according to the *configurations of denial* to which it is exposed – and which are conditioned by the state of exosomatization, the organological arrangements produced by this state and the positive and negative pharmacological effects that unfold as a result. An organology of *parrhēsia* in the epoch of networks, platforms and algorithms is clearly required, and a parrhesiastic right should be allowed to emerge.

Today, all this must be situated within the cosmic locality that is the biosphere, where the 'struggle' against entropy occurs in an Anthropocene in disruption that is itself highly 'entropogenic' and where denial strikes the industrial democracies, and particularly France, with unprecedented vigour. Consequently, the regressive tendencies that democracy fights against, and that constitute it as the problem of noesis itself, seem to be set free thanks to the effects of panic, the symptoms of which are all those forms of madness that we have referred to as ordinary, extraordinary and reflective.

When, on the evening of the first round of the 2012 presidential election, Eva Joly was overtaken by Marine Le Pen, leading the former to declare that the result achieved by the National Front represented an 'indelible stain on democratic values', her reaction stemmed from a fundamental denial with respect to the real: it stemmed from *not wanting to know* what was going on in the world in which she lived, or the immense suffering provoked by this denial. Since she did not know how to respond, she turned those who voted for the National Front into *scapegoats* for democracy, responsible for its own negligent carelessness.

This practice of scapegoating, in which she claims to be the one denouncing the scapegoating of the opponent who beat her, is one case among many of the general regression afflicting a French democracy that is running out of steam.[21] This case raises the question of democracy as such, of the regressive temptations in the name of which it is always attacked by its opponents, as well as the question of the status of courage and truth in the democratic exercise in general, and in our absence of epoch in particular.

In *The Courage of Truth*, Foucault recalls that Plato effects a reversal in the sense of *parrhēsia* as true discourse 'in the form of philosophy as the foundation of the *politeia*, [which] can only eliminate and banish democracy'.[22] 'Democracy cannot appeal to true discourse.'[23] In the Greek age, democracy – of which Solon, along with Cleisthenes, was one of the founders, Solon whom the Athenians treated as mad when he denounced Pisistratus[24] – falters on the question of what Foucault calls *ethical differentiation*. The latter is the foundation of *parrhēsia* insofar as it constitutes the courage of truth, which is at bottom *always wounding*, because it calls into question the metastabilized processes of transindividuation that it, precisely, destabilizes:

> And if the democratic institutions are unable [according to Plato, then Aristotle] to make room for truth-telling and get *parrhēsia* to function as it should, it is because these democratic institutions lack something. And [...] this something is what could be called 'ethical differentiation'.[25]

The ethical differentiation that distinguishes the one who has the courage to speak the truth – at the risk of wounding and of being wounded in return – is what *dares to oppose* accepted discourse and other forms of rigidified speech [*langues de bois*]. 'Aude, dare, dare to know,[26] and hence dare to oppose the denial of those who do not want to know, beginning with you yourself': such is the moral maxim that everyone should take on wherever it is a question (let's not mince words here) of 'saving the world', that is, of overcoming a danger quite without precedent. And to do so, we must act while caring [*en pansant*]. And to do so, we must 'try to live',[27] that is, to think.

According to Plato, as according to Aristotle, *in democracy*, dominated as it is by an essential demagoguery, *the expression of courage tends always to lose out*. It is this that will lead to the regression of *parrhēsia* itself. And it is an *ethical weakness of democracy*, insofar as it is tied to the impossibility of democracy *giving voice to* ἦθος each time that truth speaks in the city:

if philosophical discourse is not just a moral discourse, it is because it does not confine itself to wanting to form an *ēthos*, to being the pedagogy of a morality, the vehicle of a code. It never poses the question of *ēthos* without at the same time inquiring about the truth and the form of access to the truth which will be able to form this *ēthos*, and [about] the political structures within which this *ēthos* will be able to assert its singularity and difference. Philosophical discourse [...] exists precisely in [...] the necessity [...] of this interplay: never posing the question of *alētheia* without at the same time taking up again, with regard to this truth, the question of *politeia* and the question of *ēthos*. The same goes for *politeia*, and for *ēthos*.[28]

But what does ἦθος mean? In *Protagoras*, ἦθος designates the place of beings endowed with artifices, beings who are mortal, exosomatized, and who must constantly *interpret* this condition, that is, their place, their *locality*[29] within the cosmos, a locality that constantly shifts without ever leaving what, as αἰδώς and δίκη, marks out this place between the gods and the beasts.

If there can be – in the interpretation of this place, of this abode, of this ἦθος – a question of the *courage of truth*, it is because there is the possibility of a *cowardice faced with the truth*: a truth is *always wounding*, because it *always* consists in *pointing out and fighting against some kind of cowardice*, a cowardice that is always also provoked by the *pharmakon* when it becomes toxic, and which always leads to what Nietzsche and Deleuze call stupidity [*bêtise*].[30]

121. Pharmacology of democracy

As for democracy – given that it is *solely through the noetization of the demos*, in law and in fact, and as far as it is possible, that the neganthropic future of *Neganthropos* can be preserved, that is, remain open, and as the Open[31] – it is intrinsically exposed to the threat of regressing into cowardice (that is, into stupidity). To deny this fact is already to betray it: *given that democracy opens access to the pharmakon to everyone, it is inevitably exposed to the risk of aggravating this pharmakon's toxic potential.* Such is the price of the correlative increase of its curative potential, that is, its noetic potential – its neganthropic potential.

Socrates and Plato denounce the toxicity of the *pharmakon* that is literal tertiary retention when, in the hands of the Sophists, it becomes a means by which the ambitious may acquire power. And they're right to do so. It is in order to respond to this that Protagoras makes Prometheus, Epimetheus and Hermes the touchstones of these questions, who *together* suggest (each of these going hand in hand – just as ἀλήθεια, πολιτεία and ἦθος cannot work without working together) that it is the default of origin, and the ὕβρις or *delinquere* in which it consists, that is the challenge of democracy as of any form of πολιτεία. In this way, Protagoras defends democracy. And he's right to do so.

For Protagoras, only democracy draws out all the consequences of the exosomatic situation that is the noetic condition – a situation whose toxicity must be fought against noetically, and a situation that inevitably leads to regression and to the denial of this regression. The toxicity of the *pharmakon* and the problem it poses to democracy more than to any other regime are what the deniers refuse to see, just as they refuse to see its role and its ambiguity, and their responsibility in this regard. In so doing, they ultimately reproduce all the philosophical clichés stemming from Platonism, fighting against Protagoras the democrat at the very moment they believe they are fighting for democracy – and they themselves thereby encourage regressive temptations and denials.

This is, for example, what Jacques Rancière does in *Hatred of Democracy*. In this work, Rancière rightly contests the arguments of Michel Crozier, Samuel Huntington and Joji Watanuki when, in *The Crisis of Democracy*,[32] they argue that (in Rancière's words): 'good democratic government is one capable of controlling the evil quite simply called democratic life. [...] What provokes the crisis of democratic government is nothing other than the intensity of democratic life.'[33] In fact, this type of argument consists in striking down the minority in any real democracy, a minority who must therefore submit to 'expert' tutors in governance. This is what has ruined European democracy, for example, where Greece's humiliating submission to such experts, who are in truth representatives of financial interests, was obviously used as a way of intimidating those democrats who opposed the conditions set by Greek and European creditors, creditors who were then bailed out by the 2008 'mutualization of losses'.

Combatting this anti-democratic argument leads Rancière to argue more generally that the denunciation of the regression borne by democracy as that which always threatens it is itself an expression of the 'hatred of democracy', and that, for example, Gustave Le Bon's analyses of 'crowd psychology' are those of an inveterate reactionary and notorious enemy of the Paris Commune.[34] In this way, it is also Freud who, having put Le Bon's analyses at the heart of his own theory of the ego, comes to be classed among the 'reactionaries'.[35]

That a hatred of democracy exists, and that it is at this moment particularly virulent, and even highly threatening, is not in doubt. But that democracy should be defended by denying its shortcomings is more than doubtful: it is unacceptable, in addition to being completely counterproductive – and firstly because *these shortcomings are also and firstly those of noesis*, and even of truth, as much as of ἦθος, insofar as noesis and truth 'are' *only intermittently*, and finally of the psyche itself, insofar as it is split and polarized by that indefinite dyad analysed by Simondon in *L'Individuation psychique et collective*.[36]

To refuse to consider these organological and pharmacological conditions of democracy, insofar as the latter grants legal rights to every psychic apparatus whomsoever he or she may be (provided, however, that they have passed through the *skholeion*[37]), is to preclude understanding how *it is possible to produce, in an industrial fashion, regressive behaviours that affect, not only the 'people' conceived as the the whole of the poor* (as Aristotle proposes to define the *dēmos*), as those who possess nothing,[38] but *non-inhuman beings in general, inasmuch as they regularly sink into shame and injustice, to the point of generating a 'new form of barbarism'*.

Like any regime of πολιτεία, democracy is exposed to what is originally the result of exosomatization and its shortcomings. Being *founded on noetic sharing* [*partage noétique*] (of which the 'distribution of the sensible' [*partage du sensible*] is one dimension[39]) as a *condition of democracy itself*, the trial and ordeal of these shortcomings is bound to be more radical. To deny this is to refuse to think.

Cynthia Fleury, whose enduring work also investigates courage, shows that, in the current, reputedly democratic world, *it is the rule of law that is threatened by disindividuation*, which obviously amounts to a regression.[40] This means that it is the condition of democracy as well as of the republic, and ultimately of politics, which seems to dissolve with the dissolution of democracy itself insofar as it is no longer able to undergo the test of its intermittent, and therefore regressive, condition – which, consequently, it denies in a quite pathetic and pathological way. Conjoined to so many others already mentioned here, this denial casts upon it a dangerous discredit that combines with the denial and lies of the state we have already discussed.

This is now the case as never before, because disruption outstrips and overtakes law by imposing upon it a thousand states of fact through the creation of structural and chronic legal vacuums that concretize the Nietzschean desert. And after the culture industries come to be engaged in the production of industrial populism[41] – which Lienemann and Quilès deny, joined in this in a certain way by Rancière – the 'data economy' comes to short-circuit individual protentions, replacing them with automatic protentions.

Democracy as defended by Protagoras is an experience of and experiment with the *pharmakon* – which, in the wake of tragic thought, puts this question at the heart of the life of the *polis*. And this means that there is a pharmacology of democracy, and a positive pharmacology of democracy as a *therapeia* for the *politeia* that is required by exosomatization when it becomes that of the letter, where this literal exosomatization is what enables the establishing of public law. To refuse to see these questions is *to be lazy* – intellectually as well as politically. It is to not think [*penser*] – so as to not care [*panser*].

122. Ethical quagmires, suffocations, theoretical vacuums

Intellectual laziness – which has become colossal, seizing hold of every public debate and enlisting its own professional con artists [*embobineurs*], sometimes called 'intellectuals' – took a particularly visible, verbose and tacky form during the crisis triggered by what, in the wake of the law on 'marriage equality' [*mariage pour tous*], has become the *ethical quagmire* surrounding surrogacy [*gestation pour autrui*].

Insoluble and intractable in such conditions of unpreparedness, the immense questions opened up by this childbirth technology, which emerged from organological tinkering rather than from a scientific dream, have become the subject of countless denials with regard to the possibility of changing the reproductive conditions of non-inhuman beings, changes that signal yet more colossal possibilities of radically transforming the conditions of life on earth, an overview of which seems nowhere to be attempted, apart from in transhumanist discourse and at the 'singularity university'.

Here more than ever, Mark Hunyadi has cause to denounce the atomization of 'ethical themes' that allows them to be adopted by what is ultimately a market, without any significant debate ever being held about an overall ethical perspective concerning what is being played out in what amounts to a new stage of exosomatization – where the process of selection (in which, like any organogenesis, it consists) is totally subject to market criteria.

Here as in everything that stems from the rapid advances in the technologies of life, from procreative technologies to synthetic biology, it is clearly a *suffocating disruption* that catches transindividuation off guard, so that any noetizing work with respect to this new possibility of adoption inevitably comes up short. With extreme brutality, this *new possibility* introduces into neganthropology the *question of its différance in relation to negentropy, where the latter relates to life in general.* Where endosomatic life evolves through a so-called natural selection process, exosomatic life must produce artificial selection criteria: those referred to as knowledge (of how to live, do, conceptualize and spiritualize), so that they guarantee a neganthropic differentiation.

From the suffocation of thinking produced by surrogacy [*gestation pour autrui*], it follows that biotechnological disruption, of which it is one of the most visible instances, seems to prohibit any critical work other than denunciation, rather than a thought that ruminates care-fully [*pensée qui panse*]. Opposed to this pseudo-critique is an adherence and legitimation (by 'intellectuals') that is just as flat, weak and ill-considered, and shockingly subject to the 'tyranny of lifestyles', which seems to have made 'like/don't like' their mode of thoughtlessness.

As for the meaning of biotechnological disruption, in a context where exosomatization (stemming as it does from post-Darwinian artificial selection) is becoming a major business of tomorrow, the market has been able to impose its criteria upon exosomatic evolution, defining the non-inhuman not just in the field of conception or the specification of artificial organs, but also and especially in relation to the reproduction of life in general, in agriculture as in the culture of living tissues, not to mention the endosomatization of neurotechnological devices – of which 'medicine 3.0' is an early phase. As for the *meaning* of this new age of exosomatization, this has simply received no attention whatsoever.

The meaning of the biotechnological stage of exosomatization, of which surrogacy is but one of the consequences (dramatizing, via an irresponsible statement by the French government on this subject, the debate on 'marriage equality'), is also something that is *denied* – precisely through the 'tyranny of lifestyles' that accompanies the legitimating discourse of the professional deniers of all kinds.

It is by occupying the *theoretical vacuum* generated by the suffocating aspects of biotechnological disruption, as well as the *legal vacuum* that is its inevitable *correlate* (for law itself needs criteria of veridiction that it lacks precisely because of the absence of sufficient theoretical elaborations), that transhumanism articulates a discourse that gives the appearance of being coherent, and of being the only viable position.

Scientifically inconsistent and economically disastrous, since it is based on a denial of entropy,[42] this discourse is highly dangerous: *just like the 'jihadist offer'*,[43] it amounts to the offer proposed by 'the Valley' (as the disruptors refer to Silicon Valley), which tries to attract noetic souls in search of those pseudo-noeses so favoured by the neo-barbarians.

This state of affairs derives from all those countless denials that define our absence of epoch, which hence finds itself demoralized in almost every sphere, shocked by the brutality of the transformations presently underway, overwhelmed by the effects of disinhibition and paralysed in facing the *wall* towards which all this seems to be rushing headlong, without any prospect seeming to emerge that might resemble a 'dawning awareness' [*prise de conscience*] – which would in any case only resemble it, because it would also need to be a 'dawning unconsciousness' [*prise d'inconscient*].

123. Denial and disavowal

These *denials*, however, are on the whole and as a whole founded on a *more profound disavowal* – it being understood that, for Freud, denial is itself what structures the psyche insofar as it is *sexuated*. Here, denial [*dénégation*] (*Verneinung*) is not the same as disavowal [*déni*] (*Verleugnung*). Denial as *Verneinung* is an *expression* of what is denied, which, being *denied* [*nié*], is also *apprehended through this negation itself*, albeit in the mode of a *repression*:

> Thus the content of a repressed image or idea can make its way into consciousness, on condition that it is *negated*. Negation is a way of taking cognizance of what is repressed; indeed it is already a lifting of the repression, though not, of course, an acceptance of what is repressed.[44]

Verleugnung is of another order: it is the disavowal of castration, which itself presupposes the splitting of the ego, *die Spaltung*.

I maintain, however, that this disavowal is yet more deeply grafted onto what Engels and Marx show in *The German Ideology*, namely, that the organism is constituted organologically, but that *it does not see this exosomatic condition* that conditions both its *soma* and its psyche (and which, today, in the biotechnological disruption, modifies its *germen*).[45]

Engels and Marx show that the idealist vision by which these organs seem to arise from out of the mind [*esprit*], and as its inventions, in fact stems from an *optical illusion* that assumes that this mind would itself be immaterial, that it *precedes* its exosomatic materialization, and that it would be founded on and by ideas that would constitute the 'conditions of possibility' of exosomatization.[46] The denunciation of this illusion might seem to be in opposition to the theory we have maintained here with respect to the noetic dream. But this is not the case: the dream becomes noetic only because it is exosomatized.[47]

The noetic dream is anything but immaterial: it is a psychomotor process that radically changes the trajectory of the sensorimotor loop studied by Jakob von Uexküll,[48] *differing from and deferring* [*diffère*] the looping of this loop, and where noesis is a technesis. Such is noetic *différance*. The noetic dream is possible only for an exosomatized body, and thus conditions – as the hallucination of this body, a body that is *intrinsically fetishistic* and whose instincts have become drives – sexual difference insofar as what it establishes amounts, precisely, to a libidinal economy. This is what Greek mythology describes through the figure of the 'first woman', Pandora, wife of Epimetheus.

The consequences of the *current* evolution of exosomatization, however, are that sexuation itself is:

1. *no longer* a necessary condition for the reproduction of certain genetically modified organisms, which in this way become technical living beings;
2. transformed by the fact that human fertilization can be disconnected from any sexual relationship between parents.

The sexual difference that for Freud was the foundation of disavowal and (as a result) denial thus seems to be faced with the attenuation of its functional imperative, while bringing exosomatization in general into the foreground, both as the condition of noetic sexuation in general, and as the possibility of diminishing the function of sexual difference in the reproduction of life in general.

Exosomatic organogenesis harbours new entropic and negentropic possibilities for non-inhuman life, new possibilities other than those stemming from the organogenesis of endosomatic life. Furthermore, exosomatic organogenesis is de-correlated, *from its earliest moments*, from the bio-logical frameworks of endosomatic organogenesis: no longer being genetically conditioned, exosomatization frees itself from that of which sexual difference remains the condition with respect to endosomatic organogenesis, and as the *ontogenesis* of non-inhuman *individuals*. As for the human *species*, its endosomatic evolution seems to have stabilized during the Neanderthal period.

From all this, we can now conclude that the relegation of sexual difference – as condition of the reproduction of life in general and of parental coupling as the condition of the reproduction and hence the endosomatic organogenesis of the human individual – clearly transforms that situation which Freud believed he could describe while neglecting exosomatic organogenesis, which, as we have seen, he nevertheless *did* care about in *Civilization and Its Discontents*, where he referred to it in terms of 'perfecting [man's] own organs',[49] but without understanding it *as such*.

Because the *selection criteria* that control exosomatic organogenesis are social, and not biological, they raise questions of αἰδώς and δίκη. It is the libidinal economy – insofar as it is the condition of noesis as the process of idealization, projecting consistences that do not exist but that make realizable (and unrealizable) noetic dreams possible – that constitutes the criteriologies characteristic of the ages, eras and epochs of exosomatization. As an economy of the drives, it *produces* collective protentions with respect to the future of what amounts, not just to a human species, but to a non-inhuman kind. These elaborations are made possible by the sublimation processes that are presupposed by spiritual perspectives on the future of noetic life in cosmic becoming.

Purely and simply computational capitalism installs a libidinal diseconomy, amounting to an unbinding of the drives. This unbinding, moreover, liquidates the noetic criteriology, replacing it with market criteria. Therein lies the key fact of denoetization. And herein, too, lies the origin of the malaise that the jihadist offer tries to seduce – and to drive mad – in competition with the transhumanist proposal that tries to package up the disarray resulting from this malaise into promises that foreshadow an oligarchical exosomatization ruling over the global fate of human cattle and claiming to 'go beyond' entropy – something that is made explicit by those who present themselves as 'Extropians'[50] – just as the 'jihadist offer' has the goal of installing the domination of a caliphate on earth.

Here, Freud's own text on the possession of fire – that is, on Prometheus[51] – should be analysed to show how he expresses, in an extremely elaborated and rationalized (in the sense of clinical psychoanalysis) manner, the Freudian disavowal [*déni*] that consists in ignoring the organological condition of any libidinal economy, which is also the condition of noetic sexual difference in libidinal economy: *only* exosomatically constituted (by the jewels of Pandora and by fetishism in general) sexual difference is capable of giving rise to the drives, which are no longer instincts precisely because they can be 'detached' and 'diverted' by the différance in which this economy consists.

The splitting that makes this disavowal possible is, moreover, precisely what stems from the gap opened up by Prometheus and that he brings to 'mortals' as their condition: the same ὕβρις that Roberto Esposito names *delinquere*. Denial, disavowal and splitting are the conditions of regression in general, which is in itself the pharmacological condition: denial, disavowal and regression are aspects of the same split condition, constantly forgotten by an illusion against which we must constantly struggle – failing which all 'remedies' become 'poisons'.

The organological and pharmacological condition in which exosomatization consists, *for better or worse*, is what we experience every time we lie to ourselves – which we do far more often than we imagine, especially in a democracy that has degraded into 'market society': we can lie only after the experience of artifice, which is the lot of noetic life. By denying this condition, we in the West and in the monotheistic world tend to transform these tendencies into entities referred to as Good and Evil. We then seek scapegoats that must be sacrificed on the altar of the Good, and we do evil in the name of the good – hence Daesh, which is a formidable machine for the production of all kinds of scapegoats.

This *Freudian disavowal*, which clearly feeds into the ethical quagmire surrounding surrogacy, *creates a theoretical vacuum* with respect to the current stage of exosomatization. By abandoning this critical subject to transhumanist discourse, which has picked it up and taken it to a much broader level – which is the only way of taking hold of it (in terms of the amplitude of exosomatization) – this disavowal by psychoanalysis renders it *inoperative*, nonworking, which is singularly problematic for a theory that is also, and in an essential way, a therapeutic *practice*.

This is why psychoanalysis now has great difficulty in preserving its noetic and clinical power in the face of behaviouralism, of which transhumanism is a significant radicalization. This state of fact, however, is by no means an inevitability.

124. Not wanting to know: despair

All of us, or almost all, are now more or less caught up in objects that constantly solicit us, to such an extent that we no longer pay attention to ourselves, nor to what, within us, requires reflection: we no longer have the time to do so, nor the time to dream. Without respite, we are piloted, if not remotely controlled. As a result, it becomes very difficult to identify our own practices of denial, that is, it becomes very difficult to think. For to think is also and above all, in some way, to overcome a form of denial into which we have settled.

Denial, which is a way of lying to oneself, is systemically provoked by the fact that 'market societies', in becoming massively addictogenic, erect impediments to thinking by installing a structural infantilization of consumers, whose expensive toys (such as cars and all manner of appliances and devices) lose any transitional virtue – any consistence. As a consequence, they unleash a massively regressive tendency that fails to provide access to 'child's play' [*enfance de l'art*]. All of us thus descend into an 'adolescent' society, that is, a society that is grossly afflicted by immaturity.

It is in this structural state of denial that dreams and promises tend to become synonymous with lies – and, in the first place, with that lying to oneself that is denial in general. As a result, self-esteem decreases, just as aggression massively and proportionately increases – from 'France from below' [*France d'en bas*] to the pinnacle of government. As I have tried to show in *Taking Care of Youth and the Generations*, it is thus the stages of life, stages that are constituted over the course of the formation of the child's psychic apparatus on the way to becoming adult, that are disintegrated. Such is the absence of epoch.

Children confuse the world with their own representations of the world, and this is why, in the world of childhood, children seem to be constantly living through waking dreams that are not yet clearly distinguished from nocturnal dreams. That which makes childhood a kind of *permanent* rather than *intermittent transitional dream* becomes, in the *infantilized adult*, the regression of someone who is permanently lying to themselves, engaged in all those denials through which the abyssal gap that is the absence of epoch widens into a gulf.

When the exosomatization resulting from noetic dreams produces a wave of new types of artificial organs, thereby upsetting the technical system, which is then transformed as a whole – this being what Bertrand Gille describes as a change of technical system – this amounts to the occurrence of the first stage of the doubly epokhal redoubling. The 'understanding that there-being has of its being' is then thrown into question by becoming, which destroys the previously constituted circuits of transindividuation, causing all forms of knowledge (of how to live, do and conceptualize) to enter into crisis. This is what, until now, would lead to the critical stage in which knowledge is transformed, thereby engendering a new epoch.

Today, this is what is no longer true. Disruption is the radicalized ordeal of the second *epokhal* stage of the redoubling, now devoid of any prospect: in the expectation of a new era, negatively foreshadowed in and as the default of epoch and as the prelude to a transition to a panic mode, adults, in denial and totally disoriented, regress to an infantile age while depriving adolescence of its noetic dreams. Yet it is only through the *realization* of such noetic dreams that, in fact, one *can* become adult, but these dreams are what the disillusioned older generation no longer knows how to transmit, in this way becoming themselves powerless.

To pass from childhood to adulthood by going through adolescence is to transform transitional spaces into spaces of idealization granting access to the disciplines to which these are related: it is to learn to *make the difference between childish fantasy and the lying that is specific to the infantile adult*. The transition through which this difference is made (a difference that is a différance) is the *experience of oneself becoming an adult*, where adults themselves learn to *dream otherwise than children dream*, precisely by preserving these adult transitional spaces that are, for Winnicott, works of the mind and spirit.

These works 'work' only by exceeding the opposition of the real and the fictional: their reality is not that of an existence but of a consistence (an ideality) that, for this reason, *can inscribe into the entropy of becoming [devenir] the bifurcation of a future [avenir]*. Such is the power of knowledge, of which the managers of public powerlessness no longer have any inkling.

In a decadent age, of which massive denial is obviously symptomatic, denialists [*dénégateurs*] proliferate, practising their denial upon the most serious and urgent questions. In the context of the Entropocene, which is also the ordeal of the *irreversibility of time that passes and is lost*, we find that the more serious and the more urgent things get, and the more everything turns into a machine that multiplies the reasons for panic, and the closer all this does indeed approach the stage of panic, the more it also becomes the subject of countless strategies of denial, more or less coarse or elaborate, from 'France from below' to the 'pinnacle of government', all of which can do nothing but aggravate the panic to come.

Among the great deniers are, in North America, the Tea Party, whose emblematic historical leader is Sarah Palin, who, to glorify the 'American way of life', stated that she loves 'that smell of [fossil fuel] emissions'. In the absence of a discourse on the state of the world truly capable of providing an understanding of the possibility of a way out – and this is something for which we are all responsible, including Bernie Sanders – and despite the warnings of Pope Francis, Donald Trump may well end up leading the American federal government into a most frightening state of geopolitical denial.[52] Inshallah.

All these discourses exploit the anguish secreted by the reality of this decadence that *we all*, however, more or less deny:

- on the one hand, because it seems to us to be unthinkable, and so untreatable [*impansable*];
- on the other hand, because we are constrained by the structures of our psychic apparatus.

We saw at the beginning of this chapter that our psyche, which is a weave of retentions and protentions, is structured by the archi-protention of the end, itself presupposing the archi-protention of life after this end. In the disruptive context in which no positive collective protention allows us to project ourselves beyond our own end, which seems to condemn all noetic life to having suffered and lived only in vain, we may – in order to appease the negative collective protention that results from this situation and that hollows out the abyss of the absence of epoch within us, as our despair – stuff ourselves with anxiolytics or antidepressants (personally I recommend Laroxyl) so as not to go crazy.

These expedients, these medicines – these *pharmaka* – have the serious disadvantage, however, that they *erase* the gravity and the urgency of a situation for which *we* are *responsible*. Most of the time, these expedients keep us in a state of irresponsibility, locked in that *child's buggy* that Kant mentions as a way of describing the situation of adults who have been kept in a state of immaturity,[53] which an entire industry dedicated to automobility seems to have turned into an immense functional market: by integrating the mobility function that serves the total mobilization of producers and consumers with the infantilization function via the craving for ever faster and more expensive toys.

> Having first infatuated their domesticated animals, and carefully prevented the docile creatures from daring to take a single step beyond the child's buggy[54] in which [their guardians, *tuteurs*] have imprisoned them, they next show them the danger which threatens them if they try to walk unaided.[55]

Because the situation seems unthinkable and incurable [*impansable*], because our psychic apparatus is in structural denial, and because our guardians (who have become private powers, and not just those puppets, themselves infantilized, who believe themselves to be governing us) take advantage of this situation, we, or some of us, *or a part of us*, may eventually (without our being able to see or analyse it) join up, in an underhanded way, with the great ultra-reactionary discourses of denial, if not find ourselves in complete agreement with them.

This generally happens unconsciously, but also, sometimes, dogmatically – from those on the 'radical left' who deny the ecological challenge, such as Alain Badiou, for whom ecology amounts to the 'new opium of the people',[56] to the so-called 'republican' parties, and via Jacques Rancière's denial of the shortcomings of democracy. *All of them agree on the fact that they do not want to know* that the scale of decay is now that of the biosphere, to the point that it will, quite soon, be irreversible. 'Today we are closer to the catastrophe than to the alarm, and this is why it is high time we composed a health of misfortune. Even though it may have the arrogant appearance of a miracle.'[57]

125. Denial and protention: the bifurcation to come

Far more profoundly than those petty accommodations with the organized denial of the worst, there is, in the philosophical discourse of the twentieth century, particularly in France, a kind of protection of what has constituted the very basis of 'metaphysics' since Plato, analogous to Freud's disavowal [*déni*] of ὕβρις as it was thought by the tragic Greeks outside of the morality of guilt, and as the pharmacological condition of mortals – *oi thanatoi*, those who, though they are not gods, yet wield divine fire.

In *Birth of the Clinic* (1963), published after *History of Madness* (1961), Foucault passes over synthetic chemistry, that is, industrial medicine:[58] he passes over the role of the *pharmakon* in the history of modern medicine – which is rather striking, and quite difficult to understand. This impasse is clearly the symptom of a metaphysical blindness still at work in Foucauldianism.

It could be shown that a similar residue in Derrida constitutes a blind spot, when he resists *thinking the supplement in terms of its historicity*, in terms of its being *nothing other than this very historicity*. The historicity of the supplement is the law of the default of origin, and it constitutes noetic différance *as such* [*en tant que telle*], or 'as such' [*comme telle*] – this is what Derrida disavows [*dénie*] under the pretext that, so as to avoid falling into the metaphysical opposition between man and animal, he is in dialogue with Heidegger's 'as such', and, through that, with his reference to *Geist*, spirit.[59]

The *denial of historicity insofar as it constitutes a bifurcation in life qua différance* amounts to a *madness* that believes it is possible to project a 'quasi-transcendental' character onto the *pharmakon* – which would thus be conceivable in terms of a *pure logic* of the supplement, of which one could in addition do the history, but where this never ends up being done. Yet undertaking the *history* of the supplement, including as *natural history*, can consist only in drawing out the separate *regimes* of différance – in thinking with the regimes of individuation – which Derrida himself dispenses with, thereby obstructing in advance, by *obliterating it*, any possibility of thinking a bifurcation to come.

To obliterate:

> prefix *ob* (object) and suffix from *litterae* (letter) [...], [means] literally [...] 'to erase the letters'. The verb [*oblitérer*], has been likened to *oblitus*,

'forgotten' and 'forgetful', past participle of *oblivisci* (to forget), especially in the sense of 'to make forget, to erase the memory'.[60]

In regimes of différance, there are kingdoms, ages, eras (both geological and theological[61]) and epochs. Unless these are drawn out – and, in particular, unless negentropic différance is differentiated from that neganthropology of which exosomatization is the pharmacological advent – the possibility of a bifurcation yet to come cannot be anticipated otherwise than in the mode of a submission to fate: *Geschick*.

Even though (or because) he makes a claim for a transcendental empiricism, one finds in Deleuze the same step back and the same repetition of a metaphysical gesture, for example when, in his commentary on Foucault, he elaborates his discourse on the 'diagram'. This 'metaphysical ὕβρις' can also be found in the way that Simondon neutralizes the medium [*support*] in his consideration of information. Hence with Simondon, too, it is necessary to take a step beyond.

In the epoch of digital disruption and synthetic biology, it is no longer conceivable that we could get caught up in these flights of noetic denial that, at bottom, make a compromise with metaphysics inasmuch as the latter disavows the innumerable organological and pharmacological stakes – that is, the economic and political stakes – of exosomatization, which are first reflected upon by Herder, Engels and Marx contra German idealism. For this materialist heritage to bear fruit, however, we must reconsider it starting from the questions arising from phenomenology, which displace the heritage of idealism. And we must do so on the basis of the Nietzschean question of nihilism, of which the Anthropocene is the fulfilment.

Because we cannot escape the denial that is inscribed within existential horizons as the primordial forgetting of that knowledge of being-towards-the-end giving direction to all protention, this forgetting, which becomes, in the thought of the second Heidegger, the 'forgetting of being' as *Geschick*, amounts to an *automatic obliteration* supported by tertiary retentions, which, at the same time, themselves both effect and occlude being-towards-the-end – and do so precisely as exosomatization.

It is exosomatization that opens being to the world of ek-sistence and its noetic différance, for this being that is 'questioning' only because it is constantly put back into question by the realization of its dreams – a putting back into question that is the 'time of being' and of the dreams of those who are *thereby* thrown into question by what Heidegger calls 'ontological difference'.

126. Calculation and meditation

It is Heidegger who shows that denial is a structural constraint of noesis qua mode of being of Dasein as a 'being who questions', where being *is* time, but where this encloses the latter in what *Being and Time* describes as the 'ontic' plane (qua knowledge of what exists), closing it off from the 'ontological' plane (qua knowledge of what consists).

Constituted by its retentions and protentions, where the temporal ekstasis of the future and of futurity controls its temporizations and temporalizations (its différances), Dasein accedes to this 'ontological' plane only intermittently, through what *Being and Time* calls its resoluteness (*Entschlossenheit*), and when it turns to what, for being, remains to come.

What Heidegger claims to deconstruct under the name of 'calculative thinking'[62] is not reducible to the results of this *Abbau*[63] whose conclusion is that 'science does not think'.[64] It is *only* by *passing through* calculation and determination (as exosomatized understanding) that the experience of the 'indeterminate' in which 'resoluteness' consists is possible. This is what I tried to establish in *Technics and Time, 1*.[65] The fact remains that determination, which, as calculation, controls 'preoccupation' (*Besorgen*, busyness), is indeed a form of the denial of being-towards-the-end (*Sein zum Ende*, which is also *Sein zum Tod*).

And here, we cannot but be struck by the following sentence – taken from one of the most detestable of all the texts in the corpus of this thinker: 'the approaching tide of technological revolution in the atomic age could so captivate, bewitch, dazzle, and beguile man that calculative thinking may someday come to be accepted and practiced *as the only* way of thinking'.[66] It is clear that this is the point at which we have now arrived. 'Then there might go hand in hand with the greatest ingenuity in calculative planning and inventing indifference toward meditative thinking, total thoughtlessness.'[67] To combat this state of fact, however, we must understand that what, in this very text (*Discourse on Thinking*), Heidegger calls 'meditative thinking' is *constituted* by the *pharmakon*, which is what Heidegger will deny from beginning to end, and this is particularly clear in this discourse – although, and I will return to this in *La Société automatique 2*, the lecture on 'The Turn', along with other lectures devoted to *Gestell* and *Ereignis*, do somewhat complicate this situation.

Twentieth-century philosophy was dominated by Heidegger. This major thinker also happened to be one of the most reactionary thinkers imaginable. Not only that: it is clear that he was anti-Semitic. His affiliation with Nazism and his anti-Semitism are particularly obvious in *Discourse on Thinking*, where, *opposing* 'calculative thinking' to 'meditative thinking', he asserts that only *rooted* thinking can meditate – whereas uprooting, of which the Jews are clearly the figure in the Christian West, appears as the Evil in the extremely trivial sense that philosophy inherits from monotheism.

Nevertheless, we cannot think the twentieth century without passing through Heidegger, and we cannot read Heidegger without going back over Husserl's path. Both of them set into *opposition* that which tertiary retention *composes* – and, in so doing, they fail to give tertiary retention the consideration it is due. At the end of their lives, however, both of them seem to bifurcate. 'The Origin of Geometry' makes tertiary retention a condition of possibility of apodictic thinking, while *Identity and Difference*, 'The Turn' and 'Time and Being' make *Gestell* into being itself: 'But being itself essences as the essence of technology. The essence of technology is *Gestell*.'[68]

Today, the hatred of uprooting – and of migrants, immigrants, wanderers – is returning to the forefront, after three decades of unbridled globalization founded on the death drive, which capitalism sets free by unbinding the drives, and as the unleashing of the process of disinhibition. This state of affairs is just part of a generalized demoralization that is also a universal practice of denial in the face of what nevertheless presents itself as an unavoidable terminus – a 'shift', to use the term of the article in *Nature* we have mentioned on several occasions.

127. Reading Heidegger in the twenty-first century

We *must* read Heidegger, and this imperative amounts to a *paradox*. We must read him because he is the first philosopher who put denial at the heart of noesis as well as of epokhality – while showing that one cannot be thought without the other. And we must read him because he himself, by taking to the extreme the metaphysical ὕβρις that begins in Plato with the rejection of the tragic thinking of the *pharmakon*, practises philosophical denial like none before. It is because he radicalizes this paradox that Heidegger is the victim of a form of madness that translates into an adherence to Nazism and an anti-Semitic obsession.

Today, denial consists in refusing to consider the denoetization that Heidegger does indeed discuss in *Discourse on Thinking*, and that has become a massive and striking reality. By deconstructing Heidegger's *Abbau*, Derrida made it clear that it is not possible to oppose calculation and meditation. But he did not himself think the new arrangements between retentions and protentions by which calculation and meditation – that is, understanding and reason – can and must operate the bifurcation that is effected as a new era that is not the destining of being, its *Geschick*, but the inscription of the future that makes (the) différance in becoming.

In the age of disruption, which gives rise to chaotic and febrile situations typical of the generalized denial that characterizes the *Besorgen* of the absence of epoch, where retentional and protentional horizons disintegrate, nothing is further from non-inhuman beings than those consistences in relation to which the 'ontological difference', as Heidegger conceives it, is an attempt to name the extra-ordinariness arising from the ordinary and towards which 'meditative thought' is turned.

Given this, we must say that generally speaking, when we are confronted with an unforeseen situation, we may find ourselves without the ability to analyse, to critique and ultimately to introduce a *bifurcation* into a *state of fact* – a state of law being what produces a bifurcation starting from a state of fact, which thereby becomes lawfully and performatively regulated. We refuse to see the situation in which we are, more or less, involved, and this refusal, which we call disavowal [*déni*], is much more profound than a negation.

Disavowal does not rise to consciousness, whereas negation and denial manifest that which is denied. The contemporary philosophy that today follows in the wake of twentieth-century French philosophy, which does not escape this fate, must overcome Heidegger by redefining the analysis of the 'forgetting of being' as the unfurling of the epochs that lead to the absence of epoch in the disavowal of the organological and pharmacological problems and questions arising from exosomatization.

We deny in order to maintain a process of transindividuation that we believe to be right, which belongs to the same epoch as other similar processes, with which it forms an epoch and which themselves believe they are right, and which we also fight against, even though we also share with them the presuppositions of the epoch. All this constitutes 'the understanding that there-being has of its being'. As far as the philosophy of the twentieth century is concerned, this can last only so long as the absence of epoch remains hidden beneath the feet of this 'understanding'.

It is this of which so-called 'French theory' is the specific experience: it tries to think the absence of epoch, but recoils before the change of era – and it does so because, ultimately, it never truly thinks the *pharmakon*. Through this denial, what we want to preserve at all costs is a *very deeply buried* 'average understanding', one deeply entrenched in the epoch.

For and through this average understanding, we are at bottom willing to give credit to our worst adversaries over what is essential – which then comes to seem secondary or accessory. On the contrary, however, this is what must be analysed as that which enables us not to want to know what is happening, especially when a shock in exosomatization occurs of such magnitude that it amounts to a bifurcation in the natural history of exosomatization itself: this is the meaning of the transhumanist symptom.

The primordial dimension of disavowal is archi-protention as being-towards-death in the test and ordeal through which every epoch constitutes itself as the capacity to overcome the despair that is inevitably provoked by what presents itself as, precisely, inevitable. After the thermodynamic age, however, and well before the atomic age, it is the archi-protention of life beyond my own death that, by turning itself into an archi-protention of the end of life itself, which is also to say of death itself, creates a *panic* that philosophy has yet to begin to think.

It is as this ability to maintain hope beyond the insurmountable fact of death that there-being shares an 'ordinary, vague' understanding of its being with its generation. Through this is formed the horizon of those positive collective protentions that consist throughout that epoch, and which in so doing constitute it – a horizon that for Florian is missing. And it is as courage in the face of the end of life itself that philosophy can and must begin to think again, and firstly to think calculation beyond calculation – as this power to bifurcate that is the 'function of reason'.[69]

Conclusion: Let's Make a Dream

128. 'Universal folly', noetic dreams and τέχνη

Well after the 'great confinement' of 1656, the Marquise de Sablé writes in her *Maximes* that 'the greatest wisdom of man consists in knowing his madness'.[1] The exclusion of madness, consigning it to the exterior of reason, was obviously not something accomplished at a single stroke. In 'Sagesse et folie dans l'oeuvre des moralistes' (published in 1978 but containing no reference to Foucault's *History of Madness*), Margot Kruse shows that, under the influence of Baltasar Gracián's *The Art of Worldly Wisdom*,[2] the moralists of the eighteenth century – in particular François de La Rochefoucauld, Nicolas de Chamfort and Madeleine de Sablé – put the question of a 'universal folly' at the heart of their studies of morality and character: 'Even more often than the idea of the perpetually reversible relation between wisdom and madness [that we find in Montaigne a century earlier], we find in the moralists of the seventeenth century the idea of "universal folly".'[3] This research shows that in the 'classical age', a long process of diluting madness into reason unfolds over more than two centuries.[4] Nine years after the founding of Paris' Hôpital général, La Rochefoucauld writes that the 'most refined folly is begotten by the most refined wisdom'.[5] This in no way invalidates Foucault's thesis according to which 1656 is the year of the institutional turn that inaugurated the 'great confinement' and established a new relationship to madness fifteen years after the *Meditations* of Descartes – heralding a new 'epoch' of madness. On the contrary, it confirms that madness, the definitions of which evolve over time, perpetually haunts the discourses derived from noesis as that with which the latter must constantly compose – whether positively or negatively, intermittently and in many ways, which may specify epochs and, perhaps, specify eras.

We ourselves, latecomers of the twenty-first century, fashion for ourselves, and in the absence of epoch, a new experience of madness – ordinary, extraordinary, reflective, and threatening to turn into a generalized madness, that is, into a panic of unimaginable magnitude. We suffer this ordeal while disruption continues to radicalize the paradoxes of the Anthropocene, which got underway at the end of the classical age as a sudden intensification of disinhibition. Disruption: or what, in *24/7: Late Capitalism and the Ends of Sleep*, Jonathan Crary has described as the *prevention of dreaming*.[6]

In this context, we must rethink the status not just of madness, but of the dream qua origin of noesis. This is a task of the utmost urgency, because we know, today, how the prevention of dreaming can drive us crazy, to the point of becoming criminal, and because only a new noetic dream can yet save us.

If in reading Descartes, and in general, it is necessary to distinguish dream and madness, as Foucault asserts, the articulation between them must also be undertaken with respect to the *pharmakon*, that is, tertiary retention: it is the noetic dream that engenders the *pharmakon*, which, like τέχνη, can drive us crazy. The noetic dream is the condition of exosomatization, which always brings with it the possibility of ὕβρις.

When the ὕβρις provoked by the toxic becoming of the *pharmakon* makes us go crazy, we need a dream with which to cure this madness – which always inclines towards despair, which always turns to despair, from which it can find its energy. Such a dream – *careful, therapeutic* – which bears a resemblance to the creative madness referred to by Erasmus when he reads Saint Paul,[7] is noetic in the sense that it produces hope, and, along with hope, *courage*. In summary:

- What we call madness is what the ancient Greeks always related back to ὕβρις.
- This ὕβρις is the fact of exosomatization, which the Greeks called τέχνη (*tekhnē*).
- After the tragic origin of Western thought, what we call moral philosophy stems from αἰδώς and δίκη, the ordeal of which is embodied by Pandora and her fetishes – and the disavowal of which begins with Plato, who thereby inaugurates metaphysics.
- The experience of ὕβρις arises again today in the ordeal of disruption, as the excessiveness of the absence of epoch, as the becoming without future that provokes a mad panic – strangely echoing the Renaissance as examined by Foucault at the beginning of *History of Madness*.
- The character of the current forms of madness finds its point of departure in the disavowal of the madness in reason, and hence in the disavowal of modern ὕβρις, in and by modern metaphysics.
- Despite all its efforts to shake up this disavowal, twentieth-century philosophy, especially in France, itself remained caught up in this very same disavowal, as did psychoanalysis – thereby feeding into a kind of 'postmodern' metaphysics.
- Heidegger, insofar as he thinks what we are here calling archi-protention,

in addition to thinking *Gestell* and *Ereignis*, remains an indispensable interlocutor. But he himself is in denial. This is reflected in: (1) the opposition he makes between *calculative thinking* and *meditative thinking*; and (2) his neglect of the *archi-protention of life* that we might call *being-for-life*, or again, *being-for-noetic-life*.

129. The folly of the cross, and the dream according to Foucault in 1954

In the pharmacological and organological situation that results from exosomatization, and as the 'wisdom' that 'knows its madness', reason, which designates noesis insofar as it *exceeds* all calculation – but insofar, too, as it *passes through* calculation, with which it distributes the hypomnesic tertiary retentions that spatialize the time of its dreams, which *thereby* become noetic (which Heidegger wants neither to see nor hear, the roots of his anti-Semitism lying in this very fact) – in this situation, then, reason, inasmuch as it is not simply *ratio*, can only ever be intermittent.

In his praise of *stultitia* (a word that means both madness and stupidity – in the sense of the state of stupor), Erasmus, himself referring to Paul's First Epistle to the Corinthians and the 'folly of the cross',[8] distinguishes between

> 'two kinds of dementedness': a calamitous and devastating madness, opposed to a creative and positive madness. [...] At the end of *The Praise of Folly*, positive and Christian *stultitia*, in Saint Paul's sense, is linked to the ancient conception of madness in a positive sense that, according to Plato, distinguishes lovers, prophets and poets. Seneca goes further when he asserts, referring to Aristotle: [...]'No genius has ever existed without a grain of madness.'[9]

Madness is *that through which we may pass*, including as the *madness of calculation*, but it is *where we must not remain*, so that it may become the *remarkable, improbable and neganthropic moment of a noetic intermittence that could never last forever*: that will never be anything more than a différance. A kind of dream [*rêve*]. A reverie [*songe*].

It is in this way that madness, which is not dreaming, is nevertheless *connected to the dream*, which, insofar as it is noetic, is itself connected to exteriorization, and in a way tied to exteriorization, that is, to exosomatization. In exosomatization, the dream is *linked to tertiary retention, both upstream and downstream*. The so-called 'higher' animals, who also dream, and do so because they sleep, nevertheless do not realize their dreams: they do not exosomatize them, they do not express them.

Here it is necessary to go back to the young Foucault and to his dream of 1954, where, reading and interpreting Binswanger, as I have already said,[10] he envisages *an anthropology that would above all be an oneirology*, itself consisting in an interpretation of *Being and Time* on the basis of the clinical psychiatry that lies at the origin of *Daseinsanalyse*. At the beginning of his introduction to 'Dream and Existence', Foucault declares:

> In another work we shall try to situate existential analysis within the development of contemporary reflection on man, and try to show, by observing the inflection of phenomenology toward anthropology, what foundations have been proposed for concrete reflection on man.[11]

What should we expect from such a work with respect to the dream?

The introduction to 'Dream and Existence' undoubtedly shows us the key idea. After stressing that with Freud, who re-evaluates the nocturnal 'oneiric experience' that the modern age had reduced to the 'non-sense of consciousness', the dream is already on the way to becoming *a kind of realization*, Foucault argues that Binswanger makes it possible to think the dream as that which constitutes 'the point of origin from which freedom makes itself world':

> By breaking with the objectivity which fascinates waking consciousness and by reinstating the human subject in its radical freedom, the dream discloses paradoxically the movement of freedom toward the world, the point of origin from which freedom makes itself world. The cosmogony of the dream is the origination itself of existence. This movement of solitude and originative responsibility is no doubt what Heraclitus meant by his famous phrase, *'idios kosmos'*.[12]

Hence the ἴδιος κόσμος lies at the very heart of Binswanger's reflections.[13]

It is a question of going beyond the Freudian function of the dream and its interpretation.[14] Contrary to what this Freudian interpretation suggests (according to Foucault), what is at stake in the dream and in what the dream expresses 'is not the biological equipment of the libidinal instincts; it is the originary movement of freedom, the birth of the world in the very movement of existence'.[15]

Beyond the Foucauldian hypothesis (I note in passing and not without astonishment that Foucault, too, confuses instinct and drive), but in its logical extension, and as what seems to me to be confirmed by the facts, we must now posit that:

- this movement fundamentally belongs to the process of exosomatization, which enables and is enabled by the noetic dream as the power to *bifurcate neganthropologically* – for example, with the onset of expression via parietal art, between the Aurignacian and the Magdalenian;

- the 'inflection of phenomenology toward anthropology'[16] emphasized by Foucault in 1954, which for him, as for Binswanger, clearly passes through Heidegger's existential analytic, must be pursued *beyond anthropology and beyond the existential analytic, against the entropology* that anthropology has become (as Lévi-Strauss himself put it in 1955), and *for the Neganthropocene.*

This is why, as in Miyazaki – interpreter of Valéry, and, through Valéry, interpreter of Pindar – the 'dream is not a modality of the imagination, the dream is the first condition of its possibility'.[17] But here we must add, with Pindar, Valéry, Miyazaki and Simondon,[18] that the first condition of this first condition of possibility of the imagination that is the dream is the *image-object*, that is, μηχανή (*mekhanē*), which is to say, τέχνη.

Which is to say, ὕβρις.

130. The madness of capitalism

If madness is what is unleashed by ὕβρις, purely and simply computational capitalism is the madness of our age in the absence of epoch. This age of absolutely computational capitalism is without epoch because *an epoch is always what, as ἐποχή, suspends the time of mere calculation,* but does so by *passing through* it, going *with* it, *leaping beyond it, over and above it,* and, as I have argued in *Automatic Society, Volume 1,* 'above and beyond the market'.[19] That such a *leap*, which Heidegger does indeed discuss,[20] *passes through* calculation, that it *presupposes* calculation – this is *precisely* what Heidegger *cannot manage to think*.[21]

Calculation is the fate of the understanding that, passing through its exteriorization as prescribed by Descartes' fifteenth and sixteenth rules for the direction of the mind, is bound to find itself exteriorized. The question of reason then arises in other terms, which can only be those of a pharmacology – in which there is not, on one side, authentic thinking, that is, meditative thinking, and, on the other side, calculative thinking, that is, 'uprooted' thinking.[22]

Heidegger does not see that what he calls *Eigentlichkeit* (translated as 'authenticity', 'propriety', 'ownmost-being') stems, not from a native land, but from an *idiocy* that provides the *idiomaticities* that express an *originary default of native language, as well as of the original and the origin,* of whatever kind. He does not see this because this idiocy always carries within it a locality that is a given place in which time is given only as its spatialization, which is also and 'always already' its deterritorialization – its exosomatization.

The default of language (the *shibboleth*[23]) of which this idiomaticity is the consequence, always lived and perceived as the other's default of pronunciation, stems from the argument advanced by Derrida against *Being and Time* under the names of archi-trace and archi-writing.[24] But

Derrida himself did not follow his own meditation on différance all the way out – nor did he do so for his own meditation on calculation, which showed, precisely, that différance must always pass through calculation.

All of Heidegger's political shifts, ambiguities and cowardice derive from the disavowal of the default of origin that prevents thinking locality as what takes place through the arrangement of space and time via tertiary retention – an arrangement of space, time and retention by which and in which bifurcations are produced. Through such a bifurcation, a difference is inscribed against indifferent becoming, thereby making (a) différance, but this also, and as *pharmakon*, erases this difference – like Hermes stealing cattle from his half-brother Apollo.

What remains to be produced on the basis of the Derridian deconstruction of the Heideggerian *Abbau* is an organology that enables a positive pharmacology of the present 'monstrosity' that is our situation in the absence of epoch. Such a positive pharmacology passes through a 'history of the supplement' that must be actualized and conceived *as exosomatization*.

Over the history of the noetic supplement, and since the Renaissance, madness has unfolded in relation to a process of disinhibition that leads to the disintegration of the moral being. This disintegrating disinhibition is possible only because the moral being is originally bipolarized by a field of transindividuation *that is metastable only because it is hybrid*, that is, *ceaselessly composing with the* ὕβρις *that it contains*, and that always *returns* to *haunt* it, *dephasing* it.

It is from this hybrid ground that disadjustments and readjustments are possible, oscillating within what Simondon describes as bipolar metastability. Bipolarity is not merely psychic: it forms between collective individuations and psychic individuals insofar as they are traversed by an indefinite dyad *passing through the collective*. It is in this very way that this being is *moral, that is, social*. The dyad that traverses the moral [*le moral*] insofar as it is indissociably psychic and social is a constant test of temptations that are not simply opposed, but composed.

By *opposing* Good and Evil, monotheism will decompose these compositions. In so doing, it ends up unbinding and unleashing the drives – in becoming the 'spirit of capitalism', and ultimately by doing the precise opposite of the prescription it had offered through the life of Jesus. Transformed into capitalism, and as a process of disinhibition, monotheism now threatens to end in a war of *all against all*, of *everyone against everyone*, including against *themselves*, through an *immense discord* generated by this madness that is despair (like the 'calamitous and devastating' dementedness of which Erasmus speaks).

Nothing is more threatening than the mortiferous energy of despair that we see accumulating everywhere – and not just among those young people deprived of idealization whom the 'jihadist offer' tries to seduce. This mortiferous energy, the energy of desperation, can be 'rectified' [*redressée*] (just as one rectifies an alternating current into a direct current – by *differentiating* polarities) only through the quasi-causal transformation of this hyper-entropy into an unprecedented neganthropic possibility. So-called 'de-radicalization' is in this regard bound to be inconsequential if it does not lead to *de-radicalizing the new barbarians* and all the forms of despair that they themselves represent.

A true 'de-radicalization' can *consist* only in a *new noetic dream*. It is in this context that disruption requires a moral philosophy, as a way of struggling *against* transhumanism and *for* the Neganthropocene – via Bergson and *The Two Sources of Morality and Religion*.[25] In the West, ὕβρις becomes the madness of *logos* in the ordeal of the ἴδιος κόσμος, and then the folly of the cross, and hence this ὕβρις requires a moral philosophy, and, more generally, those forms of knowledge of the moral (non-demoralized) being that are all the therapies and therapeutics of ὕβρις insofar as it always returns, and always returns via *pharmaka* – which alone, however, allow it to be contained.

The true object of Keynes' critique is the unbinding and unleashing of the drives, which leads to moral disintegration, and which stems from the Anthropocene as disinhibition – an Anthropocene whose discourses on 'ethics', such as those described by Mark Hunyadi, are rationalizations that deny and repress. *The morale of the moral being, however, is the most complex, most fragile and most necessary dimension of the doubly epokhal redoubling and the process of transindividuation* – at the heart of which collective protentions are formed and deformed. But it is these collective protentions that, as *moral ends and neganthropic strengths* in *being-for-life*, and as being-for-*noetic*-life, are lacking in Florian.

The moral is what stems from αἰδώς and δίκη. Αἰδώς, as shame, will become the guilt that is opposed to divine justice, with monotheism's interpretation of the facticity and artificiality of exosomatization as original sin. The Good will thus come to be opposed to Evil – which with bourgeois capitalism will result in bourgeois and petit-bourgeois morality, of which Emma Bovary's demoralization and the 'copyism' of the 'two imbeciles'[26] are ordeals characteristic of what, in the collapse of the Austro-Hungarian empire, thereby becoming Kakania, will install the neurosis that will then be discovered on the couch of Sigmund Freud.

Composed, the tendencies that *extend* the moral being as a *dynamic temptation* in bipolarity, of which the feelings for αἰδώς and δίκη are the primordial forms of knowledge, require the existence of a *nomos* without which there can be no *Sittlichkeit*: the morale of the moral being cannot do without the law. Disruption, however, *sets the real outside the law* [*loi*], and does so *by realizing the real beyond any right* [*droit*] – through the creation of legal vacuums, which amount, so we are claiming here, to a de-realization of reality that leads to entropic decomposition.

It is this abyss, which disruption hollows out between αἰδώς and δίκη, that radicalizes contemporary barbarism as totally accomplished nihilism. Along, then, with all the caring professions – along with all the carers, from doctors to teachers and via artists, curators and all those who officiate as well as *politicians of a kind yet to come* – psychiatry must open up a new question of law and of its relationship to the *pharmakon*. In France, this will undoubtedly involve the initiation of a dialogue with Pierre Legendre and Alain Supiot.

There are close links between the drives, lies and the madness of suicidal acts, some of which are also homicidal: these forms of acting out [*passages à l'acte*] are inseparable from the diseconomy and lawlessness afflicting psychic life and collective life, inasmuch as these are themselves dis-integrated by the capitalist diseconomy of disinhibition.

In 'À la recherche d'une autre histoire de la folie', Marcel Gauchet, along with Gladys Swain, offers a critique of Foucault and of everything that accompanied the transformation of the psychiatric institution in the second half of the twentieth century.[27] But in so doing, and by restricting himself to medical questions, Gauchet misses the primary meaning of the Foucauldian approach, which consisted in *politicizing madness*.

The issue of madness in noetic life is, in a general way, what ties ὕβρις to the *pharmakon*, not only in the form of Laroxyl or psychotropics in general,[28] but as exosomatization in all its forms. In the twenty-first century, this question arises in a completely different way, because this century heralds a new era of exosomatization that calls for a new form of noesis – that is, a new form of ὕβρις, and hence a new form of care.

131. One step forward, two steps back: accelerationism and its denial

Heralded everywhere – as the 'quantified self', medicine 3.0, synthetic biology, bionics, nanotechnological 'enhancement', neurotechnology, the discourse on 'singularity',[29] and so on – the new era of exosomatization gives rise to two types of denial:

• the first, quite classically, consists in simply ignoring what has become obvious, so as not to be held back by it;
• the second denounces, in what presents itself as a coming catastrophe, the inherent evil to which technics that has become autonomized technology would amount, and so refuses to understand that the condition of the future is technological.

To say that the condition of the future is technological in no way means that this *condition* is a *solution*: it means, *on the contrary*, that this condition is a *problem* (and not just a question), and that what is required is a 'great politics' of technology, which must become a 'great health', *that is, a transvaluation*.

It is in this context that we should read and compare, on the one hand, the encyclical by Pope Francis, *Laudato Si'*,[30] which clearly amounts to a bifurcation in Christian dogma with respect to man's place in Creation, and, on the other hand, the 'accelerationist manifesto' of Nick Srnicek and Alex Williams.[31] I will not comment here on *Laudato Si'*: this is a project currently underway within Ars Industrialis. The present work, however, is a kind of preparation for a dialogue with the theses put forward in the encyclical, and a way of saluting the courage of Pope Francis – and, in a way, his *parrhēsia*.[32]

Faced with the 'continued paralysis and ineffectual nature of much of what remains of the Left',[33] as well as of the 'new social movements which emerged since the end of the Cold War [and which] have been similarly unable to devise a new political ideological vision',[34] Srnicek and Williams propose a consideration of the question of technology in Marx from the perspective of 'increasing automation' and the 'coming apocalypses' of 'neoliberalism 2.0'.[35]

This 'Manifesto' is in the first place a discourse on speed today, which is also to say on disruption: 'We experience only the increasing speed of a local horizon, a simple brain-dead onrush rather than an acceleration which is also navigational.'[36] A true acceleration would *take up Marx's initial inspiration*:

> [Marx] was not a thinker who resisted modernity, but rather one who sought to analyse and intervene within it, understanding that for all its exploitation and corruption, capitalism remained the most advanced economic system to date. Its gains were not to be reversed, but accelerated beyond the constraints [of] the capitalist value form.[37]

But when they then refer to Lenin,[38] Srnicek and Williams pass over the central question, which is the *philosophical* question of pharmacology, as well as those questions to which it leads, which are the *economic, political, social and ecological* questions that this pharmacology poses through the problems that it raises – and which all fall within general organology.

But in the exosomatic perspective put forward by Marx and Engels, a perspective that lies at the origin of all their work, the pharmacological question fails to emerge. In *The German Ideology*, which is the first formulation of the organological situation of noetic life, pharmacology is understood only in terms of class domination – which means that it is denied in terms of pharmacology.

Marxism, and then Marxism-Leninism, failed to problematize the pharmacology of proletarian productivism, and hence failed to see that *it is not the negative power of the proletariat that effects a 'revolutionary' bifurcation*, but, on the contrary, a *quasi-causal process of de-proletarianization* – that is, an *exceeding of the dialectic* by what then becomes *a therapeutics composed of knowledge* (of how to live, do and conceive).[39]

Srnicek and Williams are right to say that the 'material platforms of production, finance, logistics, and consumption can and will be reprogrammed and reformatted towards post-capitalist ends',[40] even if it is less sure that this will not involve a *new type of capitalism*. (On this point, I am also not in full agreement with the analysis of Michel Bauwens:[41] even if, eventually, capitalism will disappear, the transformation currently underway is unlikely to lead to a straightforward exit from capitalism any time soon. In the short term, the question and the problem are to ensure that we do not disappear along with capitalism.)

Srnicek and Williams are right to want to reprogram the material platforms of what has become the 'data economy' that is leading to full and generalized automation. This is also what the Institut de recherche et d'innovation is working towards through its projects in the service of a negentropic web.[42] A redeployment of digital pharmacology in the service of long-term post-capitalist goals, and undoubtedly composing with capitalism in the short and medium term, requires new forms of knowledge capable of providing new prescriptions – of how to live, do and conceptualize.[43]

Such *prescriptive capabilities*, as new forms of knowledge and of sharing knowledge, have everything to do with 'capabilities' in the sense in which Amartya Sen uses this term. They make it possible to overcome the 'tyranny of lifestyles' – which promises to be *even more tyrannical* with the transhumanist marketing we can soon expect. Such prescriptive capabilities, which are obviously at stake in the commons in the sense of Elinor Ostrom[44] and Benjamin Coriat,[45] must collectively individuate the *new culture* that *true* digital culture[46] will be, rather than the current 'tyranny of digital lifestyles'.

This culture will be that of a new moral being who, de-proletarianized, that is, once again capable of noetic dreaming, will expect nothing from 'technological solutionism'. The moral being of the digital culture to come will be a practitioner, not a consumer. Sharing with other practitioners common and singular capabilities of producing bifurcations, that is, *pressing the time saved by entropic automation into the service of neganthropological dis-automatization*, this moral being will cultivate new forms of knowledge founded on a contributory economy whose canonical value (value of values) will be negentropy[47] – which is the specific feature of the commons economy.

From a completely different perspective, but by relating the political and economic questions they raise to those involving the *place of dreaming*, Srnicek and Williams emphasize that the accelerationist project 'must [...] include recovering the dreams which transfixed many from the middle of the nineteenth century until the dawn of the neoliberal era, of the quest of *homo sapiens* towards expansion beyond the limitations of the earth and our immediate bodily forms'.[48] Here, again, the conquest of space is seen as a prospect for the development of humanity and

> it is only a postcapitalist society, made possible by an accelerationist politics, which will ever be capable of delivering on the promissory note of the mid-twentieth century's space programmes, to shift beyond a world of minimal technical upgrades towards all-encompassing change.[49]

This *spatial politics* would thus rehabilitate the 'mastery' abandoned by the discourse of 'postmodernity'.

This return by the accelerationists to a discourse of mastery (which critical theory and the deconstruction of metaphysics saw as an impasse to be overcome – and I have argued throughout the preceding that this impasse amounts to the Anthropocene itself) is not without inspiration from the same concerns that lie behind transhumanist 'storytelling'. Hence Srnicek and Williams declare that their perspective opens towards 'a time of collective self-mastery, and the properly alien future that entails and enables. Towards a completion of the Enlightenment project of self-criticism and self-mastery, rather than its elimination.'[50] So long as it fails to see or refer to the shadows borne by the Enlightenment, so long as it does not take up these shadows as a theme, this return to mastery is bound to repeat the denial of ὕβρις, and, with this disavowal, the accelerationist manifesto takes *two steps backwards after its first forwards step*.

The disavowal of ὕβρις is also the disavowal of desire: it is desire that is everywhere at stake in these questions – insofar as it is, par excellence, the process of noesis as accomplishment of neganthropic capability, from dreaming to sublimation[51] and via idealization.

132. Psychotic capitalism

What is a noetic dream that is left unrealized, that doesn't come true? What becomes of it? Its non-realization proves that it was *not yet truly noetic* (this is the whole meaning of phenomenology as the exteriorization of Spirit in Hegel). Hence the frequent result is that a dream turns into a nightmare. In disruption as the *de-realization of dreams concretized through an 'age of devices' that turns into a nightmare*, we live through a *generalized denoetization* where dreams are *concretized without being realized*.

Insufficiently noetic dreams de-realize themselves to the extent that they produce *dis*-individuating and trans-*dividuating* dead ends that exhaust and annihilate the potential for individuation – that is, they exhaust and annihilate powers of the future: promises. It is the apparatus and devices of this de-realization of dreams that form the infrastructure of accomplished nihilism. It could be developed only because no pharmacological analysis has enabled the formation of this apparatus to be counteracted by *another politics of technology*. The various forms of denial – methodically maintained by professionals – have prevented the emergence of such a politics, by preventing them from dreaming, that is, from thinking.

This *disappointing* state of affairs is today widely observable as a proliferation of negative expectations and protentions, whether by citizens or by financial markets, insurers and investors – a situation of which speculators always know how to take advantage. In this context, depression, which is both moral and economic, becomes 'psychotic', *clinically* insane, so to speak, and, increasingly often, it leads to acting out – including as this criminal economic act that is speculation about the worst. Having de-realized the real, capitalism itself becomes psychotic. It is no longer just that it has lost its spirit:[52] it has lost its reason.

In *Acting Out*,[53] I recounted that, having myself acted out, I made an experiment out of my new retentional and protentional modality that was incarceration by trying to *understand how and why I had managed to stop loving the world* (which also happened to Claude Lévi-Strauss[54]) to the point of risking my life and ultimately finding myself locked away from the world. In 2003, twenty-five years after my incarceration, I concluded *Acting Out* by stating that, between the moment in 1978 when I entered prison and the moment when I would write my little book, the world had become much worse than before: it was *in the course of being befouled [immonde]*.

Today, it is clear that the situation is again vastly more serious than it was in 2003: the world *has become foul [le monde est devenu immonde]*. This worsening of the situation is the *crossing of a limit* – one of whose names is disruption. This crossing consists above all in the fact of denoetization insofar as it is not simply the domination of stupidity: denoetization involves the question of a madness that *goes beyond the bounds* of νόησις (*noēsis*) and in so doing de-realizes the real. This has been *made possible* by functional stupidity, that is, by generalized proletarianization, which leads to that stage of rupture within which we are trying to live, but *this is a stage that goes beyond systemic stupidity*: it goes beyond these bounds and to the very *extremities* of the Anthropocene.

In this extremity, 'extremisms' proliferate – and first of all the extreme mediocrity of multiple forms of impotence, powerlessness and incapability. The various forms of collective regression disavowed by Jacques Rancière – from the far right to so-called jihadism and via the lies, indiscretions and speculations of the restless souls who govern this Titanic that the 'Earth Ark'[55] has become – all this is directly produced by the process of denoetization.[56] The result is an immense feeling of disorientation on the verge of turning into global panic.

How to renoetize? Is it possible? To pose these questions consists in affirming that it *is* possible, which means: it is *still possible* to be *put into question*, and *therefore* to *noetize*. But to pose these questions also consists in envisaging that *it might no longer be possible* – were it otherwise, we would not even have raised the question: it would only have 'seemed' to have been posed.

133. The conversion to come

The question that now imposes itself, the question that truly arises in the age of disruption, is the question that *can* envisage its own *impossibility* – which is still the question of ὕβρις, but ὕβρις *brought to its final extremity*. Only such a question, imposing itself as the putting into question of the very possibility of questioning, and capable of envisaging its *own impossibility*, only such a question could produce a bifurcation capable of *realizing* this putting into question *as a question, and not as the end of all questioning*.

Here I use the verb *réaliser* also in the sense it has in English: *to realize*. Hence to realize the putting in question also means to discover and confront the countless organological and pharmacological problems posed by the putting into question of the possibility of questioning provoked by the fact that exosomatization reaches this new stage. In this stage, the question is that of a *neganthropology that would be capable of producing the impossible*, namely, the Neganthropocene. To produce the impossible is to create a kind of miracle.

The Neganthropocene is a noetic dream that 'in all likelihood' *has no chance of being realized*. It is *for this* reason that it *is* a dream: a true dream is always what presents itself as something that *cannot* be realized. It is in *this* way that it is oneiric. On those occasions when it *is* realized, however, it is because it has become capable of becoming a *desire* – and a *shared* desire. This desire is that of what, in the 'Letter on "Humanism"', Heidegger called *Möglichkeit*.[57]

In the highly specific context of the ordinary madness provoked by the absence of epoch, the question of *renoetization through the reconstitution of desire* – for which *reason* is first of all *motive*, that is, *motor*, and, therefore, the *protentional dynamic* as a whole inasmuch as it is the source of all quasi-causality qua *imagination* founded in dreaming – this question of renoetization through the reconstruction of desire must be investigated with the involvement of the world of psychiatry and the various forms of therapy for mental, psychological, cognitive and sapiential suffering. And it must thus confront *the state of emergency that is disruption*.

To confront this state of emergency that is denoetization qua disruption, and in order to inscribe within it the possibility of an impossibility that for this reason we are calling a bifurcation – which can only be the desire for such a bifurcation – also presupposes, beyond the therapeutic scene properly speaking, an encounter and discussion with scientific disciplines, with carers in general and with those from the spiritual world – ranging from yoga to the Vatican and from the 'faithful' to theologians – as well as with citizens and with *historial politicians yet to come*.

In the *entropic state of emergency* that is the concrete reality of the Anthropocene, the struggle against denoetization must understand noesis first and foremost in terms of the neganthropological faculty that is the 'function of reason'. This requires a *redefinition of the concept of negentropy* (or of negative entropy, or anti-entropy) *with respect to exosomatization* as a reality that is both organological and pharmacological.

Neganthropy is not just negentropy: exosomatization modifies the terms of the question in the sense that, between negentropy, as characteristic of the organogenesis of life, and exosomatic organogenesis, which would be an organogenesis that is no longer just that of life, a new conception of organogenetic bifurcation is required, which I am here calling neganthropology – and which I oppose to transhumanist 'extropianism'.

Insofar as it concerns all forms of knowledge, the whole set of which constitutes capabilities in Sen's sense, and because denoetization is first and foremost the proletarianization of knowledge in all its forms (of living, doing and conceiving) – this annihilation of knowledge being its devaluation, that is, the fulfilment of nihilism – renoetization necessarily involves *a vast and highly improbable process of conversion*.

What our century would contain of the 'religious' – most of the time in the form of fantasy rather than of vocation, a fantasy resulting from a fundamental frustration exploited by the 'jihadist offer' – ultimately stems from the *necessity of a conversion*. Here we should read the legend of Saint Julian, which, like many others, tells the story of the *conversion of a dealer of death* – hunter, then warrior, then monk.[58]

Saint Julian the Hospitaller,[59] as well as the samurai discussed by Hidetaka Ishida,[60] and the Roman legionaries,[61] individually and collectively invented new *arts of living*: new *forms of peace*, through *noetic transformations of the conflict* to which ὕβρις always amounts, and where this ὕβρις lies within every noesis, being the source of the *noetic phase-shifts* that are transindividuated into eras and epochs as the *reality* of noesis (this is what Gladys Swain highlights in Hegel's relation to madness[62]) and through processes of conversion.

Foucault's final period is completely devoted to the question of techniques of the self, as conversions of forms of noetic life. Noesis is first and foremost a kind of conversion, and an apprenticeship in new forms of life, *tekhnē tou biou*, and these apprenticeships, these lessons, lie at the origin of all forms of knowledge: conversion, as the moment of transition from existence to consistence, is the very dynamic of the life of the mind and spirit in all its forms – whether religious or otherwise.

We ourselves, we who belong to the twenty-first century, we are not warriors like the samurai who invented Zen culture, or like the legionaries who developed the culture of *otium* in the Roman Empire, or like the gathering of the heads of Homeric Greece that develops into the peaceful ἀγορά of the πόλις as the site of λόγος: we are in the midst of a global economic war – in which an oligarchy of the lords of economic war sit on boards of directors, and the masses of producers and consumers are its troops.

We can save the world from disruptive and entropic collapse only by negotiating economic peace treaties founded on an *economy of reconstruction* in the service of a *new noetic era* – cultivating new knowledge both as life-knowledge [*savoir-vivre*], work-knowledge [*savoir-faire*] and conceptual and spiritual knowledge. It is not a question of knowing if we should or should not bifurcate: we are *going* to bifurcate, beyond a shadow of a doubt.

The only real question is to know if it is possible for such a bifurcation to occur as the conversion of our way of life, or whether we will simply undergo the shift announced in 'Approaching a State Shift in Earth's Biosphere' – a question that no one dares to formulate, or knows how to formulate. The conversion to come is that from becoming [*devenir*] to future [*avenir*]. This is what conversion has always meant – and it is always involved in every truly noetic act, in every new occurrence of truth, which always occurs by echoing, whether near or far, a shaking-up [*ébranlement*] produced via ὕβρις – a shaking-up that the Bible calls *original sin*, which lies at the origin of *knowledge* [*connaissance*],[63] and which I describe here as noetic exosomatization – of which προμήθεια (*promētheia*) and ἐπιμηθεια (*epimētheia*) are the two inseparable faces[64] through which the *pharmakon* is constituted, and to which the default of origin has amounted since the very beginning.

134. The onrush into the computational

Derrida relativizes Foucault's opposition between, on one side, Descartes, and, on the other side, Montaigne and Pascal. In so doing, he argues that madness *constitutes* the Cartesian *cogito*: not simply reason, but what in Descartes amounts to the very name of thought. The *constitution* of the *cogito* is God. It is God as infinity – as infinite power.

After the death of God, what possibility remains for madness to yet be, in thinking, that which would *com-pose analysis and synthesis, infinitizing the end as its différance?* Disruption is the dramatization of this question, which raises a *thousand new organological and pharmacological problems.* First among these problems is the status of this noetic organ that is Turing's 'universal machine' as a *noetic dream of noesis as exosomatization.*

We can discover no formulation of these questions and these problems in the work of the philosophers of whom we are the heirs. They have just barely transmitted to us this question that they have not themselves been able to formulate: the philosophical legacy of the twentieth century has transmitted only conditional elements of questions that come to be formulated in the disruptive context of the twenty-first century, questions that these earlier forms of thought were unaware of, and that they failed to see coming.

The 'conditional elements' furnished by twentieth-century philosophy, oftentimes alongside psychoanalytic and psychiatric therapy, must be taken into account, not by rehashing them, but by interpreting them.

The infinite resurfaces in the metaphysics of our time with computationalist cognitivism: it presents itself discreetly, as the infinite tape of the Turing machine[65] – which is infinite in the way that the memory of God is infinite in Leibniz. This is what cognitivism continually denies. In those 'Turing machines' that are computers, of course, the tape is not infinite. The tape is infinite only in principle [*en droit*], that is, mathematically: in the noetic dream, but not in its realization.

The passage from the dream to its realization is, from the beginning of the Anthropocene, the transition from science to technology. The tape is infinite only ideally. The industrial concretization of Turing's noetic dream – as the retentional basis of calculation in totally computational capitalism – is its finitization. Turing, whose computationalist metaphysics will be utilized in both the 1936 article on computing machines[66] and the 1950 article on the intelligence test,[67] will himself put in question this neutralization of the finitude of the memory support by turning to biology, as Jean Lassègue has shown.[68]

In *Automatic Society, Volume 1,* I endeavoured to show that the finitization of Turing's noetic dream by computationalist metaphysics is reflected in the implementation of an algorithmics that exosomatizes the analytical functions of the understanding by separating them from the synthetic function of reason. Algorithmic exosomatization then necessarily becomes a pharmacological problem: one which consists in *prescribing a fecund – that is, truly noetic – arrangement between the finite memories of the new exosomatized computational organ and the neganthropic cerebral organs, via social organizations.* As a *pharmakon,* however, this artificial organ makes it possible to short-circuit these social organizations, setting up the tragedy of disruption proclaimed by Chris Anderson's 'end of theory'.[69] But this would then amount only to the first stage of a doubly epokhal redoubling, which could and should lead to a second stage: not just to a new epoch, but to a new era, which we are dreaming of here as the Neganthropocene.

The pharmacological problem posed by the algorithmic organology in which the exosomatized understanding consists – which is also and *firstly* a political, economic and ecological problem – is what computationalist and libertarian metaphysics denies. This denial constitutes the framework of the theoretical legitimation of the current madness of capital. Herbert A. Simon, both an economist and a computationalist cognitivist, is a perfect example of this link between cognitivism and a capitalist economy that has lost its reason.

The madness of capitalist and cognitivist computationalism consists in believing and in making others believe that the computing machine could be infinitized, which is impossible otherwise than in principle [*en droit*], which is to say that it is impossible other than for an abstract machine. If we also want to infinitize it in fact, that is, concretely, then this can lead only to what Hegel called bad infinity – as a factual impossibility of finishing, for example of finishing the series of integers, and, therefore, as the *impossibility of bifurcating* – which is also to say of *deciding*, which leads to a *headlong onrush into the computational*.

We are living through the *finitization of the infinite, that is, the attenuation of desire*: such is contemporary ὕβρις. Capitalism has finitized the infinity of Christ's power by realizing it on earth. In so doing, the dream of Christ has turned into a nightmare – in being concretized through calculation, and as the de-realization of this dream. Descartes will be finitized in the same way, through the rationalization by which the *Aufklärung* will enter into decline, constituting disciplinary techniques and then psychotechnics in place and instead of any autonomy, and engendering a 'new form of barbarism'.

Founded as they are on the calculability of the audience market, and on an economy of attention that destroys this very attention, the culture industries are now being replaced in the age of disruption by the 'data economy', which can only intensify barbarism qua finitization of this infinite – on the basis of Noam Chomsky's neurocentric gesture, replacing Cartesian ideas with neuronal 'wiring', thus erasing the question of idiomaticity that fundamentally arises from exosomatization as ex-pression, and liquidating the Saussurian dynamic of diachronic and synchronic tendencies as that through which idioms are metastabilized.

135. Creating a miracle: despair, salvation, fidelity

Lost in disruption, wondering how it is possible to avoid going mad, proclaiming that noesis is, nevertheless, the quasi-cause of madness, that is, of ὕβρις, like Simon, Nicolas and Camille at the beginning of the final scene of *Same Old Song*, I hereby confess that I am very depressed, and that I am sometimes *overwhelmed*, literally *laid low*, by what seems to me to be, as perhaps it does to Florian, *evidence of the end*.

I am often overwhelmed because it seems to be *absolutely irrational* to *believe* that a positive bifurcation could arise from out of the chaotic period into which we are rushing at the high speeds imposed by disruption. It is *totally improbable*. And this is a motive for despair.

In the ordeal of *absolute despair* that results from such considerations, one conclusion seems unavoidable: only a miracle could overcome the absence of epoch into which nihilism has led us. Heidegger put it another way: in a 1966 interview he gave to *Der Spiegel*, he declared that 'only a god can still save us'.[70] Having incited irony, including mine, and sometimes hatred, today this statement resonates in my own ears with renewed vigour. What I believe, however, is that the question is less that of a *god* than of a *miracle*. This is my belief because I also believe that the miracle, on the one hand, and dust, on the other hand, are the expression of the way that monotheistic thought prior to thermodynamics was able to conceive what is not conceivable – that which lies at the origin of life. This origin was long thought on the basis of what we have for a long time called God, but, since the twentieth century, it has been thought on the basis of what has been referred to variously as negative entropy, negentropy or anti-entropy.

To *continue to struggle*, faced with the *evidence* of the *absolutely desperate* nature of the situation – that is, to stop denying the state of emergency into which the Anthropocene has led us, taken to the extreme by disruption – it is necessary to *believe in the possibility of a miracle*, and, more precisely, to believe in the miracle of what I am here calling the Neganthropocene.

Let us repeat it once again: such a miracle is a dream. This dream must be *immensely noetic*, capable of projecting the pharmacological situation in advance by turning it into its point of departure – by positing the irreducibility of this pharmacological situation from the outset, which is to say the impossibility of eliminating entropic consequences, and, correspondingly, the constant need to *neganthropologically* take care of the *pharmakon*.

In the past, the name given to unconditional belief in the possibility of the miraculous was *faith*, itself founded on the form of desire that Christianity called ἀγάπη (*agapē*). In the epoch of the absence of epoch to which the death of God has led, such an *affirmation* of the possibility of what can only appear impossible, which is the *improbable as such*, that is, *neganthropy* with respect to entropic becoming, *can no longer* constitute a faith in Providence, nor can it be the expectation of a divinity. This, however, raises questions about a new form of belief and a new relationship to fidelity and infidelity.[71]

It is in this sense that we should, not reject Heidegger's statement in *Der Spiegel*, but subject it to critique, that is, take it seriously,[72] analyse it in terms of its *fundamental inadequacies* – in this case, the *exclusion of questions of entropy and negentropy from Heideggerian thinking*. Here, *it is no longer a question of the ontological difference between being and beings, but of the neganthropic (and exosomatic) différance between future [avenir] and becoming [devenir].*

It is not a new god who alone could still save us (even temporarily, through a différance of and within entropy that can but remain the fate of the cosmos – this fate being, in local terms, the cooling of the solar system[73]), but the new belief required for the transvaluation of all values, which presupposes that we 'transvalue' Nietzsche himself.

Miraculous narratives are *parables of the neganthropic condition* inasmuch as it is always exceeding itself, and such that it can never amount to a simple 'humanism'. Because it is hybrid, and because the ὕβρις it contains always exceeds it, what we call 'man' – the non-inhuman being that we should henceforth name *Neganthropos* – is *both in excess and in default of itself*: it is *never itself.*

It is this *excess*, inasmuch as it 'transcends' the default that it is, that takes the name of spirits, gods, God, History or *Gestell*, and this is what jihadism as well as transhumanism and neo-barbarism try to recuperate at the moment when this excess shows itself to also be the default as default of origin.

Excess and default are what Heideggerian existentialism, in spite of all that it brings to such a perspective, is ultimately incapable of conceiving. This is the objection to Heidegger that Derridian deconstruction is attempting to make. The consequence of this inability is that the place (*Ort*) of the ontological difference (that is, of 'meditative thinking') becomes the 'native land', 'autochthony'.[74] The excess and the default from which neganthropological différance stems, however, are at the same time *what localizes the default as idiomatic difference* and *what de-localizes (exceeds) it as what is necessary beyond the locality* where it is constituted, drawing it out towards the non-place of what, in becoming, *remains* to come, and as the quasi-causal taking place of a promise. For centuries, this excess was experienced as that transcendence whose name was God.

Heidegger could not manage to think this. And yet these are the stakes involved in what he terms *Ereignis*. Derridian deconstruction is itself insufficient to overcome the way in which Heidegger fails to think ὕβρις (as excess and as default, as having a hybrid and 'factical' character), because it fundamentally downplays noetic différance, and hence downplays the fact of exosomatic organogenesis as the condition of its own possibility (and its own impossibility).

This retreat of deconstruction faced with its own consequences equally amounts to an inability to think the fact of proletarianization – which means that it is also an inability *to think capital*, and ultimately the *supplementary* history of calculation, as well as *calculation as the impossible fate of the supplement that must be*, not just 'sublated' [*sursumé*] or 'raised' [*relevé*], but *therapeutically neganthropized*, which also means, *transvalued.*

136. Denial and the obsolescence of man, according to Günther Anders

It is the question of a modest miracle that lies on the horizon of the words of René Char previously quoted[75] – and this horizon forms the question of locality: 'Today we are closer to the catastrophe than to the alarm, and this is why it is high time we composed a health of misfortune. Even though it may have the arrogant appearance of a miracle.'[76] To the hypothesis of just such a miracle without transcendence, without arrogance (despite its appearance), in some way an ordinary miracle – which would pass through the 'health of misfortune', which can only be a worthy form of courage, and of the courage of truth, which is also to say of *parrhēsia*, and, like Canguilhem and Vernant, Char too had the courage to fight – to this hypothesis, Günther Anders might object with what he presented as a 'Molussian dictum': 'Courage? A lack of imagination.'[77]

If I understand it correctly, this statement suggests that courage would be a form of illusion and denial. And I believe that Anders, like Heidegger, whom he claims to be opposing here, on the one hand ignores the pharmacological question resulting from exosomatization, and on the other hand does not investigate the neganthropic possibility that lies within anthropy, because he says nothing about the questions opened up by the theories of entropy and negentropy.

Nevertheless, Anders is in all likelihood the first to have unwaveringly questioned the situation in which we find ourselves today, by reflecting on the consequences of the atomic age, which he himself understood as the age of the bomb – from an angle completely different from that of Heidegger. To live 'under the sign of the bomb' is to find oneself thrown back onto a *terra incognita*. And on this 'unknown terrain', this is firstly to *deny* the unknown, that is, the meaning of the bomb as a new 'invisible' object:

> While it should be constantly present in the glare of its threat and its fascination, it remains [...] hidden at the very heart of our negligence. The great affair of our age is to act as if we do not see it, as if we do not hear it, to continue living as if it did not exist.[78]

> It is systematically kept incognito [...] the ears into which one tries to speak become deaf as soon as this subject is mentioned.[79]

At stake with the bomb is the obsolescence of man. For with the bomb, and more generally with the 'second industrial revolution' and its scientific technology, human beings have become the 'lords of Apocalypse' and are 'themselves the Infinite' qua infinite power *of destruction*: 'If there is something in the consciousness of humanity today that is absolute or infinite, it is no longer the power of God [...]. It is our own power [...]: the power to annihilate, to reduce to nothing.'[80] Anders shows that this infinite power exhausts desire – which was still that of Faust – and brings forth new 'Titans'.[81]

All the questions raised by Anders, which in the essay 'On Promethean Shame' anticipate transhumanist ideology, arise again and proliferate in our absence of epoch – where the bomb remains a major threat, but where the question of new Titans is infinitely more specific and complex, and where denial is vastly increasing, as Anders himself noted more than twenty years after *The Obsolescence of Man* was first published: 'Now atomic power stations obstruct our view of nuclear war and have made our "apocalypse-blindness" more blind than ever before.'[82] As for the bomb, it poses for Anders the question of a decision – which, through a process of transindividuation, should eventually lead to a collective renunciation.

I believe, however, that Anders, like Heidegger, fails to ask the overriding question. Anders insists that

> the desire to revolt against the machines, to reject this Titanic condition we have acquired (or that we have had imposed upon us), [is] a highly dubious, extremely dangerous desire, for [...] it [...] strengthens the position of those who effectively hold total power in their hands.[83]

After insisting on this fact, after emphasizing that it is futile to oppose Titanic becoming and the infinitization of Promethean power, he nevertheless has nothing else to propose except such opposition – and in this way his reasoning seems highly contradictory.

This is so because, like Heidegger, Anders ignores the issues of entropy and negentropy, and exosomatization and its consequences, which are here characterized as neganthropological, organological and pharmacological. He begins his analysis of the atomic situation by condemning Heidegger's discourse on this score, but he does not succeed in fundamentally distinguishing himself from Heidegger's position.

137. Conversion as the taking place of locality

The atomic question, inasmuch as it stems from a technology that seems to replicate, on the local scale, the cosmic process in its totality as local thermodynamic combustions and transformations, requires that we situate noesis in the cosmos and as locality within the cosmos.

In *noetic locality*, a neganthropic différance is produced through exosomatization, which locally *defers* not just the law of entropy, but also the law of anthropy, namely, the toxicity of the pharmacological condition, so that locality is organized and ordered within universal becoming but *against the current*.

This cosmic dimension was taken on in 1936 by Eugène Minkowski, who, in the wake of Binswanger, inscribed it within the psyche. After having placed attention at the heart of his reflections, Minkowski emphasized that 'we are accustomed to considering attention as an individual faculty, varying from individual to individual [...]. Attention, however, can be envisaged from a completely different angle.'[84] Attention is 'one of the salient features of the general fabric [*contexture*] of life',[85] whereby

> phenomena connected with the self surpass it [...] towards the concrete environment [*ambiance*], but, at the same time, do so in the form of a vast arc that can encompass both the self and this environment, revealing to us, above and beyond it, the general fabric of the cosmos.[86]

And, in this context, it is striking to read Montaigne's description of Socrates' 'fuller and wider' imagination as one that 'embraced the universe as his city'.[87]

Hence Minkowski raised the question of a *locality within the universe*, which was no doubt already at stake in the Aristotelian thought concerning place, which we find evoked by Italo Calvino at the end of his *Invisible Cities*, as the question of *admiration and the admirable, which is also to say of the miracle* (in the sense that Char, too, tries to let us hear):

> The inferno of the living is not something that will be; if there is one, it is what is already here, the inferno where we live every day, that we form by being together. There are two ways to escape suffering it. The first is easy for many: accept the inferno and become such a part of it that you can no longer see it. The second is risky and demands constant vigilance and apprehension: seek and learn to recognize who and what, in the midst of the inferno, are not inferno, then make them endure, give them space.[88]

To make room [*faire place*] is to give a place (for) [*donner lieu*].[89] To take place [*avoir lieu*] is always, for a noetic soul, what gives a place for meaning [*sens*] (for significance[90]) that *arrives* with the non-sense (in the insignificance) that exosomatization is perpetually *also* producing – a nonsense that, when it approaches the insane or the senseless [*insensé*], is called the Devil. The question of sense is tied to that of the taking place that *gives* a place for, which has room for, which is a place, a site, *Ort*.

But the question of place, of taking place, that is, of what, as *event*, happens, arrives, and, more generally, the question of locality in a sense that is not just that of meaning [*sens*], and that has to do with space, spatialization, and hence in some way with exosomatization, which is also to say with geography, with geology, with urbanity, with territoriality – this question has today been totally transformed by the relationship between entropy and negentropy.

All negentropy is *local*, and comes at the cost of an increase in the rate of entropy outside this locality. Hence arises the question of an *economy of localities*, and of what *constitutes* localities as their organology

– which in this case always involves territorialized processes of exosomatization, that is, of the production of tertiary retentions, which always generate processes of deterritorialization due to the detachability of these organs, which are also objects of exchange, including as verbal organs, and which in this sense constitute an economy. It is precisely at this point that we must read Georgescu-Roegen.

In the Anthropocene, that is, in the epoch of the absence of epoch to which the death of God has led, the *affirmation* of the possibility of what is bound to seem impossible – and which is the *improbable as such*, that is, negentropy as such, but which we must think and specify beyond negentropy and as neganthropy – can no longer amount to faith in Providence, nor therefore in divinity. Miraculous narratives are parables of the neganthropological condition inasmuch as it always exceeds itself, and which for this reason can never amount to a simple 'humanism'.

How could such a conversion be produced from out of the absence of epoch? *By creating worlds in the befouled unworld [immonde]*, by *again giving room (for)*, by *proliferating acts of taking place in a thousand places making (the) différance*. This is what we are told by René Char and Italo Calvino, giving birth in sites of urbanity to *miraculous relations of mutual admiration* where the hell or the inferno recedes before what, 'in the midst of the inferno, are not inferno, then make them endure, give them space'. This place, which is also a χώρα (*khōra*), is the place of neganthropological différance – whose advent in *Gestell* must occur as a leap into a new era.

138. Dragons and serpents

In the cell where I accumulated tertiary retentions, which were of diverse forms but all literal (there were no images, no objects, nothing but letters), between which I created links (which I annotated), I caused a world to emerge from within the carceral desert that is an absence of world. This proceeded through the spatial exteriorization of the temporal fluxes and flows of my primary, secondary and tertiary retentions, from which protentions arose that even today constitute the *motives* by which and for which I strive to let my concepts take shape [*faire corps*].

The therapeutic power of this *meletē* was due to the *spatialization of my time* through the tertiarization of my reading, then the reading of my writings, that is, of my previous tertiarizations. This spatialization opened up my *analytical* access – critical, discerning, discriminating, mobilizing the conceptual powers of the understanding – to the *synthesis* in which my readings and writings in general consisted.

Hence I acquired a very practical notion of différance as thought by Derrida, and highly 'différant', if I may say so, from that of the orthodox 'Derridians'. Similarly, I understand Heidegger's ontological difference very differently than do the Heideggerians – including Granel.

From this différance, I had an irreducibly idiomatic experience in the sense that, reading a great deal of Mallarmé, I projected into the linguistic idiom an irreducible mark of locality. This is the condition of all singularity, which is also to say of any significance of the non-insignificant (the insignificant being itself the condition of the idiom and of the ἴδιος in general).

The *trials and ordeals of my différance* led me to indeed posit that readers, that is, psychic individuals ex-pressing the significance of the non insignificant in the process of noesis, can signi fy (make signs) only because *they already harbour within themselves the synchronic and diachronic tendencies that condition all transindividuation* – in the form of rules that constitute their idiolect, and that are inherently local.

This is why, during these years, I questioned what I called *local-ity*, that is, the irreducible belonging of any noetic différance to a place, to a giving place (for), including and even always on the basis of a non-place – the default of origin, *delinquere*. What was happening to me amounted to the formation of what I decided to call an *idiotext*, itself always included within *other* idiotexts, an innumerable, indefinable number of other idiotexts. I maintained that what happened to me in this way was an accidental localization of an irreducible local-ity of which noetic différance was the processual and idiomatic test, trial and ordeal.

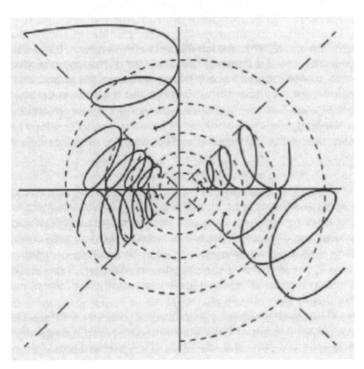

I tested out the irreducible character of the idiomaticity of all noesis, and the fact that an idiom is what, giving place (for), is caught in and by what, by this very fact, takes place. Exosomatization is both what gives place (for) and what wrests away from place: it is the non-place (the default) wherein everything happens *somewhere*, that is, partially, and, provided that a mutual admiration can thereby emerge, as *that which is necessary*. I decided to call this emergence 'virtue' – as that which can turn necessity into virtue.

Given that exosomatic organogenesis is organological and not organic, the artificial organs this forms are producers of entropy as well as negentropy inasmuch as they provoke an epokhality that always *begins* and *ends* in a fundamental disorder – that of the *hubris* starting from which noesis transindividuates new epochs and new eras. Eras and their epochs are the temporal and spatial processes of différance over the course of which the arrow of time, insofar as it amounts to both an *anthropic dissemination* and a *neganthropic economy*, becomes that *historical* irreversibility founded on the infidelity of milieus, and, through them, of places.

Over recent years, and in the framework of the seminars and summer academies of pharmakon.fr, I began to re-elaborate this theorematic body on the basis of neganthropological différance, in particular when I attempted to interpret, through a reading of Maurice Godelier's *The Metamorphoses of Kinship*,[91] Lévi-Strauss' 'entropology' and the consequences stemming from anthropology's rejection of Leroi-Gourhan.

Here, confronted with the challenge of the conversion required by the advent of a new era, and, as was the case during my imprisonment, in order to *struggle against the madness* it can provoke, and by turning this possibility of madness into the source and resource of noesis, I came to revisit the story that began in a cell in Building A of Saint-Michel Prison, and that ended in a cell of Building F of the Muret detention centre. I had to leave the latter in February 1983: this was the moment when I had to convert once again back to the normal life of men and women free to come and go in what had become befouled and worldless [*immonde*].

For more than fifteen years, I wrote a good portion of my books by recording into a dictaphone the thoughts that came to me as I drove on Mondays and Tuesdays along the northward-bound motorway on the way to seeing my students at the Université de Compiègne. Then I began to make recordings while cycling in the beautiful hills of the Bourbonnais, in the south of the department of Cher, on the border with Allier, close to the Creuse river. This desert locality has always made me dream, and sometimes made me delirious.

It was there that with my wife Caroline I created the pharmakon.fr course dedicated to Plato, then the summer academy held each August at the Épineuil mill. When I first began to use a digital dictaphone, Caroline would transcribe the audio files by typing them up using a keyboard. Then she began to use Dragon, a software that enables semi-automatic transcription. In writing these confessions that took me back to my time at Saint-Michel prison, I was suddenly struck by the name of this software, in which I had never heard anything except a brand name.[92]

Saint Michael fights the dragon, who is a figure of the diabolical, dialogical, demonic and diachronic serpent that, as Philippe-Alain Michaud[93] has pointed out, also symbolizes lightning for the Hopi Indians of Mexico and for Aby Warburg who visited them – of which Warburg rendered an account in his celebrated 'Lecture on Serpent Ritual',[94] delivered, after he lost his reason, at the Bellevue Clinic in Zurich, at the invitation of Ludwig Binswanger, who was director of that psychiatric facility.

The dragon is a serpent – like the Aztec feathered serpent, Quetzalcoatl, and like all the monsters that haunt mythology all around the world, and therefore also in Greece and Judea. The serpent who tempts Eve is, in duplicating the mark of Asclepius and Hermes, a therapist and hermeneuticist of the *pharmakon*. The serpent that haunted Warburg lay behind so many of the revenances of which his Atlas was composed – from the Laocoön to the rattlesnake whose head the Mexican Indians placed in their mouths.

The serpent symbolizes lightning – the bolt of lightning – and hence electricity, and therefore divine, celestial, Olympian fire, which in human hands becomes the *pharmakon*. In disruption, thinking would be perpetually outstripped and overtaken precisely because optoelectronic digital tertiary retention can move *twice as fast as lightning*. Overtaken by what has become quicker than the fire of Zeus, noesis would arrive always too late, much too late, and would find its dreams turned *immediately* into nightmares. Such would be denoetization in the befouled unworld of the absence of epoch.

How can we ensure the *possibility of questioning* – of becoming the quasi-cause of what thus puts us into question – which is the possibility by which the noetic dream begins? How to ensure that noetic dreaming is not outstripped and overtaken? How can it be renoetized in order that it could *become the quasi-cause of denoetization itself*? It is possible only on the condition that noesis can move *faster than the algorithms* that are themselves twice as fast as Zeus, a fact that equates, not just to the death of God, but to the death of the gods of Olympus, and, along with them, of Heidegger's god who alone could still 'save us'.

To go faster than what goes twice as fast as lightning is, however, possible – and it is not just possible: it is *the only possibility*, if the possible is what is fundamentally different from the probable, and as the *Möglichkeit* through which a desire is realized. *This possibility is precisely that of the bifurcation, which moves infinitely faster than every trajectory pursued in becoming*, since, as its quasi-cause, it reverses this becoming within which it *opens up* the sole motive for hope: the future as improbable possibility.[95]

What then is desire? This is the question posed to us by Charles Perrault's 'The Fairies'. It is *serpents*, *snakes* that fall from the mouth of the wicked sister, a shameless liar – and, 'for good measure', they are mixed with toads: desire is not envy, or covetousness, or ambition, which are only the deadening decomposition of what, in desire, affirms itself as *being-for-life*. Desire is already courage, which is admirable only because it admires – which makes it courageous, just as Pascal is disturbing because he is disturbed.

> 'Why is Pascal disturbing? Because he is dis...dis...' 'Turbed', put in M. de Charlus.[96]

As I finish this book, I find myself reading Pierre Jacquemain's column in *Le Monde*, after his resignation from Myriam El Khomri's Ministry of Labour over their disagreement concerning the law on the right to work. We must read this courageous text, not just because it is *exemplary*, and exemplary precisely in the fact that it is *courageous*, but also because it speaks of nothing other than what forms the flesh and blood of this book:

> In order to do politics, we must dream.[97]

A Conversation about Christianity

with Alain Jugnon, Jean-Luc Nancy and Bernard Stiegler

23 April 2008, Paris

The following discussion was organized at the initiative of Alain Jugnon for an issue of the journal Contre-Attaque *that was to be entitled 'Why We Are Not Christians'.*

I at first declined to participate in this issue, on the grounds that I did not agree with the question as formulated. Alain Jugnon then proposed that I write an article to defend this point of view, which I accepted.

A week later, he invited me to discuss it with Jean-Luc Nancy, which I agreed to do, as did Jean-Luc Nancy.

Alain Jugnon: The volume in which this discussion will appear is entitled 'Why We Are Not Christians'. I have addressed this question to you, Bernard Stiegler, and to you, Jean-Luc Nancy, this question...

Bernard Stiegler: ...and I wrote to you that, even if, in fact, I am not a Christian, I did not want to respond to *this* question. Indeed, I fear that this negation ('we are *not* Christians') is in reality a kind of *denial* that flees from the need to think another, more delicate question. Today, one talks about 'cultural Christians' to refer to and perhaps to conceal and above all to dispense with having to think the fact that, today, the whole world is Christian because of the market: along with its commodities, the market exports Christianity and its Latinity everywhere. Christianity is not just a faith. It is an economic, historical, political and even techno-logical reality. This is what the globalization of *Gestell*, too, means, and it passes through what Sylvain Auroux describes as a process of grammatization – which the Jesuit missionaries made sure to carry out during colonization.

So I responded that I did not want to respond to this question, not because it is not a question that arises, but because, when a question is asked in philosophy, when we posit a question as being philosophical, we suppose that this question imposes itself in our time as what would, in the language of Heidegger, amount to what we could call, for example, its 'epokhality', overdetermining other questions in terms of priority. But I think that your question is *too old* to be philosophical today.

Some people would no doubt respond that they are not Christians by claiming that this non-belonging is the starting point of their way of living – and for a long time this was undoubtedly the case for me. Ten years ago [the discussion having taken place in 2008], I would have responded to your question without hesitation, and I would have argued with you, I mean, I would have told you *why* I am not a Christian. But today, this question is for me no longer a question. I think that today, it is no longer here that the real question lies (if it ever was an issue at any time throughout the twentieth century, for, in reality, it's a nineteenth-century question, which Sarkozy, for example, would have us believe arises again in the twenty-first century, this being one deceptive strategy among many, and I'm afraid that you may fall into the trap set by this lure): the challenge today is not to emancipate ourselves from the question of God, from the presence of God; this is, *unfortunately*, not the issue, if I may say so. Because it is much more *trivial* than that. The big question today is that of marketing. This is obviously not entirely unrelated. But the relationship between them requires us to completely redefine the terms of the question, and the relations between these terms. The question does interest me *from this point of view*, but what *at the same time* does interest me is to say something about why this is not how I would have posed it.

Jean-Luc Nancy: Putting things into a little bit of order is complicated. I would say some things that are much the same as what Bernard has said, and then, at a certain point, things that are not at all the same. In the question: why are we not Christians, who is this 'we'? You, Bernard, you have rather responded as if this question were addressed to you. At the same time, when you say 'cultural Christian', I was unaware of the existence of this expression. If it exists, therefore, then it immediately marks out another 'we': 'we' who? 'We' Westerners? Indeed, I agree that Christianity has been exported everywhere, and, moreover, that there is a delicate point here: when we refer to Christianity today, we are, for reasons we understand, immediately connected to the three monotheisms, and even Buddhism lies on the immediate horizon. So an adjacent question immediately arises: how to conceive and ask about Christianity in relation to, at least, Judaism and Islam? And here I would say, in fact, that it really is Christianity that is exported and that has shaped Western and now global civilization, and that this is undoubtedly partly tied to what we call capitalism. Furthermore, in what Derrida called 'globalatinization' [*mondialatinisation*], the word 'Latin' fundamentally covered Christianity. One of the things that makes this complicated is that I always feel the need to be scrupulous both about respecting others and about the theoretical as well as practical necessity of doing justice to the other great branches of monotheism, and to succeed in resolving what distinguishes Christianity within monotheism. I also find this very intricate and I have a lot of difficulty making this distinction, because I would tend to say that Judaism and Islam (even Buddhism) are accentuations, ways to bring out valences of something that is in Christianity anyway: for Islam, the infinite distance of an absolutely incomparable, unnameable God, one absolutely not human, and by taking into account everything that the Quran says against Christianity, against incarnation, against the Trinity in particular, all this Christianity can very well take into account. After that, I'm always slightly embarrassed because I feel as if I'm on the triumphalist side: Christianity, it's the best religion! It is not at all clear that it is the best religion, but it certainly has the strongest theoretical, or let's say metaphysical, elaboration. That I believe. I'm not saying that this elaboration contains anything better than Judaism or Islam, because I recognize that there are very strong things in the Talmud, just as there are very strong things in Ibn Arabi. But all this, for it to work, must at least be put through, I was going to say through the mill, by the Christian engine. But on the other hand, I myself, I would respond to this question, to return to it: how can we say we are not Christians, while we are *nolens volens*? But what does that mean? If I try to reply: it means that, regardless of any profession of faith, any ecclesiastical affiliation, any church whatsoever, we are Christians ... but what makes us Christians? It is Pascal's phrase: 'Man infinitely surpasses man.'[1] Voilà, that's it. Perhaps even that the shift from the ancient world to the modern world, if we say the modern world began in the twelfth century (but I believe that indeed we must), that it is what effects this transition, and that until then, and even in the earliest Christianity that had not yet in the least developed its modern possibility, that is, as long as it had not yet entered into a relationship with history (until the year 1000, the Christians had awaited the end of the world), that before that, man had not infinitely surpassed man. In his human determination, which essentially assumed mortality, man was in a relationship with immortals, but immortals were themselves caught within the whole circumscription of the universe, and, in a way, completely within a finitude, within a good finitude, well-circumscribed, a circumscription of circumscriptions, sphere of spheres: there was no infinite surpassing. And, suddenly, there was hesitation between good infinity and bad infinity, opening onto an indeterminate. If Christianity has a kernel, it lies there. And in that regard, we are all Christians. We are all exposed to this call, demand or expectation, or to this anguish of having to deal with a bad infinity. The worst form of evil is undoubtedly given by the indefinite and indeterminate extension of both technology and commodities, that is, the proliferation of ends, to the infinity of every end, where this in turn suggests new ends. At the same time, and unlike you, Bernard, I would say that this is where our problem lies: what are we doing, today, with this infinite exposition? – of which we no longer know how to do anything in metaphysical or religious terms. This is why the whole return of the religious is not on this register, it is only a return to assurances, to securities, not at all to an infinity; we must still lend meaning, have meaning, find meaning, values, references, and then add a spoonful of love on top, whereas I think that by saying that Christianity is the religion of love we are also saying something about this infinity.

Bernard Stiegler: We agree much more than you think. If I say, 'my question is not to know whether I am Christian or not', it is precisely because the question is for me infinite. This question is more alive and more problematic now than it has ever been. Christianity is indeed the religion of love. The question that arises today – and which means that a question like the one you asked, Alain Jugnon, *can* be asked today (for even if I do not recognize myself in the question you have asked, I do indeed recognize a 'we' who is not Christian) – is the question of libidinal economy.

Today, the *question* – inasmuch as it is a philosophical question – is no longer, for 'us', to be or not to be Christians, and I include myself in this philosophical 'we' who wanted, *on the one hand*, to get rid of the Church, that is, a certain way that Christianity had of dominating public and noetic space, the space of the mind [*esprit*], spirituality, and wanted, *on the other hand*, to get rid of Christianity.

Today, the *question* is to *rethink* and *reconstruct* desire. Today, the *question* is the *destructibility* of desire, and, through desire, the destructibility of the unconscious, the super-ego and the id, that is, ultimately, *of the intergenerational relationship*.

Now, Christianity was a great intergenerational machine, as was, beyond that, all monotheism. But Christianity was so specifically as a discourse on the son, which is not the case for the others, and, beyond that, on the father. Moreover, I believe that it is by surpassing and displacing the place of the father that Christianity constitutes itself as a test or ordeal of love.

I further maintain – after Pasolini and his wonderful opening scene in *The Gospel According to St. Matthew* (1964) – that Jesus became Christ by accident and through love. By accident, because he is only a natural son, adopted by his father, Joseph, who is not his father, and adopted out of love, because his father has adopted him for the love of Mary, as the son of Mary, as the child of Mary of whom he must take care. Christianity is this love story. It must have been said, however, in the town of Nazareth, that this child was not Joseph's. So Joseph had to do all he could to make this child believe that he was Joseph's child, so that he would not doubt his paternity, so that he could believe in his paternity; unfortunately, malicious tongues prevented this from working. Jesus then became the son of God. This is what I believe: it is my way both of not being Christian and of thinking that the question of not-being-Christian is not a philosophical question. For, in reality, *the philosophical question is that of adoption* (and this is what marketing does know how to think, whereas philosophers seriously ignore it). If we had time, I would also give some commentary here on Thomas Mann and Sigmund Freud with respect to the adoption of Moses the Egyptian by the Jews – an adoption that constitutes Judaism – and I would point out how and why Islam, in claiming that filiation is by milk and not by blood, is also a religion of adoption.

The issue we face today is *no longer at all* that of struggling against some pre-emptive right claimed by the Church over intergenerational relations. If this was the case for a long time, and if the struggle against

patriarchy, in which I more or less consciously and voluntarily partici-
pated when I was an adolescent, still belonged to this order of questions,
which pass through the question of the place of the Church and religion
in French society, and more generally modern society, this is *no longer at
all* the issue. The question is the *exclusion of intergenerational relations*:
it is the *destruction* of intergenerational relations, inasmuch as they are
short-circuited by the relational technologies of marketing and control
societies generally. Here, I would even be ready to *defend the Church*
against the threat to it represented by these short circuits. The Church
and Christianity form a system of care. And there are things much worse
than these systems, which can themselves obviously become awfully
perverse: there is *carelessness*.

Religion, whether Christian or otherwise, is what knows how to
organize practices of care as practices of a 'we': techniques of care are in
general techniques of the 'we'. Such care articulates a psychic self with
a collective self. There are, of course, practices of the care of the self
and others that are not religious, such as the techniques of the self on
which Foucault focused at the end of his life. Religion is a major case,
and Christianity is what organizes techniques of the self and others as
a certain relation to the infinite, and according to a modality that was
a dominant social system up until the nineteenth century. But generally
speaking, a practice of care is what makes it possible to surpass the
psychic in the collective and vice versa.

The problem is not at all to defend myself, today, from a religiosity
or a Christianity that would threaten me: I do not feel threatened in any
way by Christianity, or indeed by any religion. I have more of a feeling
of responsibility to protect Christianity – and all forms of spirituality,
which I consider to be extremely threatened, and in particular by the
so-called 'return of the religious', which returns only in the way that Parc
Astérix[2] brings back the Gauls, and as this strategy of lures and decep-
tions that I mentioned at the outset.

Jean-Luc, you raise the question of practice in this text entitled 'The
Judeo-Christian':[3] you discuss the Epistle of James [in French, 'Jacques'
– *trans.*] where the question of practice arises, which you refer to a faith,
and you say that faith cannot be reduced to belief, in a discussion with
another Jacques – Jacques Derrida... I would like to discuss with you the
question of trust [*confiance*]: you say that faith is not reducible to belief,
Derrida says the same thing by saying that justice is not reducible to
belief. He says: there is no belief or faith without a justice that precedes
or succeeds them. What you call faith, he calls absolute or messianic
justice. However that may be, as for *fidelity*, *confiance* and faith have
the same root, and fiduciary, too, shares this root. I have argued that
Protestantism is the programme that transforms belief into trust, into
calculability, into administrable belief. Belief must become calculable in
the name of trust.

Jean-Luc Nancy: Here things get very complicated, because there occurs, perhaps, a kind of exchange of meanings, of the most profound values that we attach to both faith and to belief. You make faith slide towards trust [*confiance*] in the sense of calculability, hence at bottom from trust towards reliability [*fiabilité*], and towards fiduciary trust [*fiduciarité*]. If I understand you correctly, this is not so much on the side of systems of care... What I want to understand by faith is that which is based on the other, on what is indeed an absolute alterity, a non-calculable alterity, where I always take 'belief' as weak knowledge, yet on the side of knowledge. Belief, for me, is when I say: 'I believe it will be a nice day tomorrow.' This is what I said to Cluny at a conference. In everyday life, in religious *doxa*, when someone is said to believe in God, whoever the God may be, it is taken to be an operation of weak knowledge. And at the moment, this weak knowledge strongly leads straight to full metaphysics in the Nietzschean-Heideggerian sense of the term: that is to say, we make hypotheses about a being, in reality about a supreme being. Here, Kant's destruction of ontological proof is, for me, always important: I say to myself that we may need one day to take up a commentary once again on the critique of the ontological proof in Kant, because it is worth doing; we don't have this in mind enough; what Kant does is basically show that philosophy can do nothing other than (and it must do it, it is the greatness of thinking) receive the contingency of the world, and hence to end up with a search for a necessary cause involves a contradiction. All this is swept away from philosophy first of all, but the blow strikes all religious thought. For me, 'faith' comes from the Hebrew word *amen*, which means to affirm one's trust, that trust is placed precisely where there is no assurance.

Now, I'll just add something else, which has nothing to do with it, but so as not to forget it.

Myself, I do not go to Mass at all, not at all. The last time I went was two years ago, to accompany a friend's daughter in Strasbourg, who complained that she'd never seen it; moreover she had something to complain about, because her father, who is African, Burkinabe, is an ethnologist, who knows the religion of his people very well; so I took this girl to the cathedral; I was not at all happy to be in the Strasbourg cathedral on a Sunday morning, in this huge nave, with forty people, it was cold, the heating system wasn't working properly, the people who were there did not know how to sing the few songs that had to be sung. I say all this to indicate that I still have a keen sensitivity for Catholic worship, for the Gregorian chant, for all sorts of things that, for twenty years, Philippe had laughed at...[4]

Here I might agree with you, but with a slight difference. I don't want to advocate a return to Latin Mass, I don't know what... to religious practice, but when I see what actual religious practice has become, I say to myself, this is sad. And I don't know where to place, within my mental

space, as well as in our social and cultural space, I don't know where
to place what I have had such a taste for, but as an aesthetic taste, that
is, for church singing and all that... Almost every year, I get a strange
feeling around the time of Holy Week, I recall that for me, as a child, it
was something very serious... On Ash Wednesday, a cross of ashes was
drawn on the forehead... I don't know what to do with this.

Well I, unlike you, I have been marked some years ago as a 'Catho', as
a former Catholic, even as a current Catholic, or crypto-Catho... After
that, I'm embarrassed to say that I'm no longer on the inside...

Alain Jugnon: There is also the expectation of a second volume of your
book on the deconstruction of Christianity: *Dis-Enclosure*...

Jean-Luc Nancy: The issue for this second volume is, for me, a highly
complicated one, because from one day to the next I no longer know
exactly how to put my finger on it... It shifts all the time... In fact, I have
a title: *Adoration*.[5] I really believe that I'm going to stick with this title.
As a word that could indeed gather together a little of what can happen
in worship and then also of the relationship to the infinite.

Let me just add this: I have the feeling that what we are talking
about, you and I, is what Rousseau called 'civil religion'.[6] In *The Social
Contract*, 'civil religion' relates to the foundation of democracy, of our
democracy, that is, to the achieving, the flourishing and the ontological
and metaphysical binding together of all theories of the contract: it is
at the same time the production of man himself. The contract is what
makes an intelligent being and a man, says Rousseau, and the whole
machinery of the contract is the machinery of a system that would truly
be good only for a people of gods; as we are not a people of gods, this
requires civil religion; civil religion, according to Rousseau, is what must
make the heart of the citizen sensitive to the system of government. And
at this, it failed. The obsolescence of the French Republic, which without
doubt inherited this, is the obsolescence of civil religion.

I have come to the conclusion that the break I mentioned between
the ancient world and the modern world is the rupture of civil religion;
when we talk about Athenian democracy, we end up forgetting that
it was a system of civil religion; Athens is a great civil religion. The
proof is that philosophy itself began with Socrates, that is, with a death
sentence handed down for a breach of civil religion. Rome was the
most successful civil religion, but the fact that the emperor was deified
shows that Roman civility, and the republic itself as sacred, begins to
fail, and it is not by chance that after this Christianity will take over.
Christianity rightly does not bring with it another civil religion – it brings
the separation of Caesar and God. It introduces the possibility of the
gaping of the infinite that threatens every terrestrial polis. What follows,
then, will be the state, sovereignty and, for Rousseau, democracy. But

in coming into operation, civil religion was made impossible. Well this is where we are.

Bernard Stiegler: I would like to follow up on what you were just saying in terms of the question of the relationship between faith and trust. And firstly to clarify that like you, I think that faith obviously cannot be reduced to trust and calculability. On the other hand, what Weber described as disenchantment, rationalization and secularization is indeed the transformation of the power of faith into fiduciary instrumentality, and thus into systems of trust. That this leads to discredit is precisely what I argue in *Disbelief and Discredit*.[7] This process of disenchantment is, however, possible, and it is possible only because Christian and monotheistic noesis – commonly referred to as 'spirituality' – is tied to the letter, depends on the letter, is à la lettre; you use the expression, 'God à la lettre'. And I think that here, an essential issue is to reinvest the *pharmacological* question of the letter. The letter is a *pharmakon*. It is because it is a *pharmakon* that faith *can* be turned into calculable trust. This is what is at stake in Weber's analysis of the Protestant origin of capitalism.

Posing this question, I want to go back to what you said about Kant, ontological proof and infinity. For centuries, and in its 'onto-theological' destiny, philosophy was structured by the question of the proof of the existence of God – even if this is not exactly a Greek question, in which case it is not a question that lies at the origin of philosophy. If philosophy can rid itself of this question – which is what leads me to think that the question of knowing why we are not Christians is not a philosophical question – it is because this question reaches philosophy only at a certain moment in its history (even if this moment comes quite quickly) and is not already present in its 'initial conditions'. As for us, we have passed through the death of God. You stressed that Christianity is an experience of mortality in the most radical sense of the term, which for me is where we diverge: I have always conceived of Christianity as belonging to an onto-theology that, precisely, breaks with the tragic – with the tragic age of the Greeks, where there were mortals, immortals and those that 'perish', as Heidegger said. I have always thought that the forgetting of the tragic, that is, of an irreducibly mortal condition of mortals, of a *mortality without remission*, and that begins with Plato himself, this moment, which is to say this moment as the *very origin of philosophy*, was the catastrophe that led to the opposition of soul and body, and hence to metaphysics insofar as it essentially consists in a system of oppositions founded on this primordial opposition – which thus makes it possible, from the outset, for philosophy to become monotheistic and Christian, and therefore to pose the question of the proof of the existence of God.

Now, you come along to break this, or rather to make it more complex, which interests me, which in a way I'm delighted about... I have always thought that Christianity fundamentally involved an immortalization of the soul – and that Christian infinitization came at this price. You, on the other hand, mortalize the Christian soul, if I can put it like that. But I'm not sure that I'm quite convinced. And in particular, I still conceive Christianity as being essentially founded on the *opposition* of body and soul, and therefore of life and death, of the eternal and the temporal, and so on. Moreover, it is only in this way that I could respond positively to Alain's question: if I *can* say that I am not a Christian, it is because I vehemently reject this opposition, and in particular *inasmuch as, along with it, it necessarily implies guilt*. But rather than say 'why *I am not* Christian', I prefer to state 'why *I am* "Nietzschean"' – to the extent that such a statement is not entirely ridiculous and out-of-place.

Be that as it may, what we call the death of God is the proof of the *non-existence* of God. God does not exist: 'God is dead' means that God does not exist. But for several years now, I have said – and I say – that the question is not existence. The question is 'consistence'.[8] Which is to say, the infinite. The problematic field within which, Alain, you can pose this question-affirmation, 'why we are not Christians', is constituted by the fact that others say: 'why we are Christians'. If you have some need to say why we are not Christians, it is because others say, we *are* Christians. These questions, however, are too old. God being dead, the question that remains for us is to know what to do with the question of the infinite – without which there is no desire. Let's leave these old dead questions to the necrophages, and put up with the life of what returns with the ghost of God, who evidently still haunts us all, as spectrality, as spirit, and with what returns, this returning, in all infinity, of some object of desire of which this infinity is the consistence, as long as it is indeed a matter of *desire* and of the *power to infinitize* in which, precisely, it *consists*.

Jean-Luc Nancy: Jacques Derrida, vis-à-vis faith, had the same mistrust [*méfiance*] that he had vis-à-vis his community; it was highly empirical and highly emotional, he just couldn't stand them, that's it; he couldn't stand to put a kippah on his head, and yet, at the same time, he had his tallit and he touched it every evening... It's quite a complicated business...

Bernard Stiegler: Here, I think there is something we have to defend, and, in what we have to defend, the division is not between Christian and non-Christian...

Alain Jugnon: But don't we have to defend it today despite Christianity itself, as it has been historically and where it stands now?...

Bernard Stiegler: My current work, what I am doing right now, is rethinking systems of care. In this regard, I am trying to enter into a dialogue with Michel Foucault. The overarching question of systems of care has been greatly underestimated, and Foucault showed how this question is raised in *Alcibiades*, and how it is forgotten: how Plato will replace the question of *epimelēsthai* with that of *epistēmē* (and I believe that, in the end, Heidegger, against all appearances, repeats such a substitution – which amounts in him to a retreat, *his* retreat – in *Being and Time*). Now, religions are systems of care – where this also involves taking care of the pharmacological character of faith, which is to say of the letter, of *hypomnēmata*. A system of care is a therapeutics that tries to '*deal with*' [*faire avec*] the *pharmakon*, which is to say to 'handle' its indissolubly poisonous and beneficial character. What Weber says is that when the printed letter is deployed for textual study, whether of the Bible or the Gospels, this also corresponds to the deploying of bookkeeping for accountants, and it is what will develop a new relationship to writing, which will introduce a new pharmacology – one in which *logos* has become *ratio*. Faith thereby leads to the development of its opposite: calculation. Here, we should talk about systems of care in the age of what Foucault called biopolitics, and about the transition from disciplinary society to control society, which is, I believe, a society that ultimately destroys care – which is also to say, desire.

Jean-Luc Nancy: I want to ask you about two things: I want to come back to God. And then I want to ask you about this business of systems of care. Firstly, do you differentiate, in the system of care, care for 'social security'? That's my question. And then, beyond that, and from a completely different angle, when you say that you agree on the subject of the infinite, it is true, but there is another aspect of the question for me, and that's the name of God. Behind the issue, 'why we are not Christians', there is also the question, 'why are we not theists?', 'why do we not believe in God?' Perhaps, for me, the most vivid question is: can we or can we not simply scratch the name of God from the vocabulary? Can we be satisfied with talking about the infinite, with using a concept? This is a discussion I had with Jean-Christophe Bailly (now there's someone who stands resolutely without God), who, in his book, *The Animal Side*,[9] praised the Open. I said to him that it is very dangerous to say 'the Open'. Especially with a capital letter. He told me that the capital is only to substantify a concept. The name of God has the advantage of being a substantified concept, turned into a proper noun, and hence does not say anything, but has the properties of a proper noun. The history of Christianity does this too, and it does it between two others who have names, unpronounceable names, but one is called Yahweh and the other is called Allah. In the middle, there is the one who has either the name of a man, Jesus, or else is called God. But this is also a Greek affair: it is Plato who first uses *theos* in the singular. The Jesuit translators of *Theaetetus* translated the passage where Plato says that we must flee towards *theos*, they translated it as 'we must flee to God'.[10] Plato did not at all invent God as proper noun, but he made something possible in philosophy, he made possible the singularization of God.

Putting all this together, what is left with this name of God? When I have an occasion to talk about this, I say something that is very very weak. I say: God remains in the language when everyone says 'my God'. Even Alain Jugnon will on occasion say 'my God', I bet... But maybe he watches out for it ... he might say 'for God's sake' [*nom de Dieu*]...

Alain Jugnon: That's what I was about to say: I do say '*nom de Dieu*', never '*mon Dieu*'...

Jean-Luc Nancy: '*Nom de dieu*' is much more loaded than '*mon Dieu*'! It's a blasphemy, an expletive... No but does it make sense or not to pick out this little linguistic incident in order to ask if it is just a cultural remnant or if it is not a function of language, which says, as we say, '*mon amour*'...?

Bernard Stiegler: I would like to come back to the question of social security... A little earlier, I said that these questions have recently led me to attempt a dialogue with Foucault on the question of biopower. I am trying to show now, against Foucault, that this biopower is never simply a biopower: it is a psychopower, and even a noopower (precisely because care is not 'social security', even if it requires this type of care *too*), while contemporary capitalism tries to eliminate the noetic character of this noopower. In this work, I have been very interested in the figure of Jules Ferry and in the whole Enlightenment tradition, Ferry being engaged in a great struggle against the Christians – which at that time was very necessary – and intending to build a new system of care: a system of care founded by a majority, in Condorcet's sense as well as Kant's, a democratic majority, a majority that must be able to vote (as Jules Ferry says, one is not born French, one becomes French by going to school), but also a maturity in Kant's sense – as becoming adult. And this is how I understand the question of care – an adult is someone who is capable of taking care of a minor, in a relation of intergenerational responsibility, an obligation bound by adoption or by kinship – and this is where I think we have to revisit Christianity, Judaism, Islam and monotheism in general as constituting other systems of care (which I believe are also systems of adoption). As for us, we are enduring the ordeal of the *absence* of care, of the *carelessness* produced by the crisis of desire in the capitalism of psychopower.

To come back to your question about therapeutics, we cannot take care of the bodies of individuals without taking care of their souls. In what Foucault describes as biopolitics, which is supposed to take care of producers insofar as they constitute a resource and an economic, proletarian and labour force function, it seems to me that Foucault does not see, in the history of capitalism, the consequences of the internal contradiction in this system of care that Marx called the tendency of the rate of profit to fall, and which forces capitalism to develop, from the mid-twentieth century, this psychopower that we refer to as marketing, first implemented, it turns out, by Freud's nephew, Edward Bernays. The latter bases his thought of what he called at that time 'public relations' on an instrumentalization of the libidinal economy theorized by his uncle.

One can manipulate the libidinal economy *just as* faith can be manipulated. And here we find pharmacological questions, wherein faith becomes one instance of libidinal economy, which, in being transformed into calculable trust, engenders an absolute infidelity. For Freud's nephew, it is a matter of teaching capital how to capture the desire of consumers, and how to build a system for capturing libidinal and sublimatory investment – it is a matter of channelling the life of the spirit in totality, which inevitably leads to the destruction of all spirituality, and to what I have elsewhere called spiritual poverty [*misère spirituelle*]. To answer your question, all this falls within a general therapeutics. Questions of the religious, or if we prefer the post-religious, are cases of a general therapeutics: *today, we have to pose the question of a general therapeutics that could face up to the industrial development of a general pharmacology*. The question of Christianity, and more generally of the deconstruction of Christianity or of monotheism, is a question of the regional pharmacology of religion. Today, we inherit this pharmacology, and this is why it is so interesting when Derrida analyses what he calls globalatinization. The Christian *pharmakon* leaves indelible traces in the industrial *pharmakon*.

I'll add a word about the infinite. For me, God, what we call God, is the object of desire. It is the object of *all* desires. It is the desire of all desires. And consequently, it is the *object of all attentions*. What is named by the name of God is the object of all attentions: the absolute expectation [*attente*] of an absolute future that contains all desires. Whether we want to give this the name of God or whether we don't, it's all the same to me.

All of this raises the question of sublimation – and there is no system of care without sublimation: whether a religion, a state or a capitalist industrial society, if it does not take care of this infinite desire for the infinite, it is not viable. Nor it is reliable [*fiable*]: it is doomed to death. Today, I have come to the point of saying that I defend capitalism against itself, or Christianity against itself, because we live in a terrible age of Christianity transformed into capitalism. As a result, we have the responsibility to revisit and re-evaluate all of this. It is not a question of saying: I am not Christian. I'm not, actually, myself, Stiegler, but I am ready to work with Christians. My problem is not that of struggling against Christians, it is that of *struggling without God against misbelief* [*mécréance*] – and the lures and dubious beliefs[11] that are its lot. In '*mécréance*', there is no 'faith', but there is '*créance*', which also means debt, and so many other things that go into forming systems of credit, which are *all* systems of care, and which, subject as they are to a capturing of desire that is *exhausting*, today turn into *systems of discredit* and confront us with the trial and ordeal of exhaustion – long after the death of God, who in any case will not be resurrected.

Jean-Luc Nancy: I understand... But then here, we diverge infinitely... Myself, I would not take up all these things in terms of care: why do we have to start from care? From this perspective, Foucault, too, bothers me because of this care for the self, in which, moreover, he very carefully skirts around Christianity. For him, the model instead lies between stoicism and epicureanism. All those practices that do indeed make up a system of care. I say to myself that the issue precisely does not lie there, it is not in care, it is rather in risk, it is in this that Christianity contains resources, that is, in the exposure to the infinite, to God. For me, this is not a matter of taking care: it is a matter of letting it be exposed... For what is capitalism? Fundamentally, it is still the erecting of a general rule, which is that of general equivalence: it is a profound choice of civilization. Modern civilization is the one that has chosen general equivalence. And this goes back to your business about pharmacology, because this choice was a choice of difference of value: the first large-scale commerce, and even the institution of credit and loans, were created within a regime of difference, of values, of forces, of intensities, but what they ultimately generated was the exchange of universal fluidity, of which that marketing to which you referred is, for me, more of a side effect than a threatening cause. The question this raises is: how might another civilization be created on the basis of another fundamental choice, which would be the choice of absolute difference?

Bernard Stiegler: I do not believe that capitalism amounts solely to general equivalence. Capitalism is that, clearly, but it is also and above all investment. And there is no capitalism without investors. Where Marx is very strong is when he says that capitalism will destroy itself because it will be forced to speculate rather than invest. This is where it becomes very interesting to read Weber with Marx, and not against Marx. We must combine the two of them. What does Weber say? He says that there is no capitalism without investment and that investment is noetically charged. The capitalist investor is not just a speculator. He is someone who invests himself in what he does. Investment is essential to the capitalist machine. What a combined reading of Weber and Marx shows is that general equivalence does not work without that which exceeds all equivalence. Capitalism is obliged to have a spirit. This spirit came from Christianity. Capitalism is the spirit of Christianity.

Jean-Luc Nancy: Yes, but it is the bad infinity version of the spirit of capitalism...

Bernard Stiegler: I agree with you, but *this too* is the spirit of Christianity.

Jean-Luc Nancy: Sure, this is basically the ambivalence of Christianity itself. Myself, I kind of want to get back to Christianity and to 'why we are not Christians'... in order to reply that, if we are Christians – because we still are, capitalism or not – then we are always, in one way or another, connected to love. And to the commandment of love. We can do whatever we want... It's a bit like when Kant says that even a villain is still connected to moral law, he knows very well that the law is there. He does not want to respect it, but he is connected to it. In the same way, I would say: despite everything, we are still connected to love, even if we do not know what to do with it. I refer to love very precisely in the way that Freud refers to the Christian commandment of love in *Civilization and Its Discontents*, when he says that the only answer to the depths of human violence is Christian love.[12] Of course, he says that it is not possible. For me, the question becomes: what is implied in the fact that Christianity has, in full knowledge of the facts, put forward this impossible commandment? I say indeed: in full knowledge of the facts [*connaissance de causes*]. So there is a criterion that separates the good from the bad infinity: the bad is what presents universal love as possible, and the good is what knows that it is impossible. And which also knows that, in the love for one person, what is at stake is the impossible. This, for me, is not care.

Bernard Stiegler: Actually ... for me, it is the same thing, care. To take care of my wife is to live with this, and even *thanks* [*grâce*] to it – and it is also to take care of myself. Care is a grace, if not 'grace'. What I call care is what you call practice. One needs practice. But if there is practice, this means that there is a prosthesis – a *pharmakon*, always *between* good infinity and bad infinity. There are no practices without prostheses: there is always a moment when we must trust in the rosary, that is, *habit, assisted by what can become a fetish*. Every morning, I make myself do writing and reading exercises just as others go to morning prayer or to vespers. I know that I must submit to this decree – without God, by God [*sans Dieu, nom de Dieu*]! I know this, but I know it like a believer who no longer believes and who goes to church because he knows that he is absolutely exposed to misbelief [*mécréance*].

Jean-Luc Nancy: Here, no... I cannot even feel that. I'm a little too marked, maybe, by what I see happening in Judaism ... and a bit in Islam. That is, by a kind of system of security. My doctor is Jewish, and, in the twenty-five years that I've spent with him, I've seen him become more and more Jewish. Now that he has retired, he has gone to Israel. When I told him one day that I didn't understand why he was getting back into ritual to such an extent, he said to me: my dear old friend, ritual is reassuring.

Bernard Stiegler: That's precisely it, Jean-Luc, the *pharmakon*. And the bad infinity consists precisely in the belief that we can escape the *pharmakon*, turning the remedy into poison, and what I once referred to as the prostheses of faith as the *condition* of any faith. The bad infinity is the belief that it is *possible* to escape the bad infinity, as you said: it is to believe, in other words, that something exists which in fact only consists. I have tried to describe this as the intermittence of actuality, by commenting on Simonides' famous sentence evoked by Aristotle[13] (and before him by Socrates[14]), and by Heidegger[15] commenting on Aristotle: 'God alone has this privilege.'[16]

As for Foucault's techniques of the self, and more generally for what I call systems of care, they are not at all techniques of reassurance. They are quite the opposite of systems of security, they are techniques of the exposure to risk, to alteration. That they can turn into ritual, mania, habit or obsession is precisely what one cannot escape, and that is exactly what your bad infinity would want: to be able to escape it, for it to be *possible* to escape it, *in total safety [sécurité]*.

The question of care is the question of what Heidegger called *Sorge*. This is precisely because the Catholic Heidegger wanted to think care – solicitude, attention, this is the meaning of what I call care, *cura*, *therapeuma*, *epimelēsthai* – *without pharmacology*, without exposing it to *epimelēia* (a word that comes from *meletē*, which is the question of discipline and rule), without this pharmacology based on the letter, *hypomnēmaton*, and everything that can always be turned into an 'ontic' instrument of calculation, coming to 'determine the indeterminate'. It is for this reason that 'fundamental ontology', which wants to deconstruct the history of being, is ultimately a hyper-metaphysics – it still 'belongs' to 'metaphysics', that is, to 'onto-theology'.

Notes

Epigraphs

1 Bernard Noël, *Monologue du Nous* (Paris: P.O.L., 2015), pp. 9–10.
2 Michel Foucault, *The Courage of Truth (The Government of Self and Others II): Lectures at the Collège de France, 1983–1984*, trans. Graham Burchell (New York: Palgrave Macmillan, 2011), p. 77.
3 Sigmund Freud and Ludwig Binswanger, *The Sigmund Freud–Ludwig Binswanger Correspondence, 1908–1938*, trans. Arnold J. Pomerans (London: Open Gate Press, 2003), p. 212.
4 Pindar, 'Pythian 3', *Pythian Odes*, lines 61–3, available at: https://www.loebclassics.com/view/pindar-pythian_odes/1997/pb_LCL056.257.xml. See ch. 7, n. 17.

1 Disruption: A 'New Form of Barbarism'

1 Theodor W. Adorno and Max Horkheimer, 'The Culture Industry: Enlightenment as Mass Deception', *Dialectic of Enlightenment: Philosophical Fragments*, trans. Edmund Jephcott (Stanford: Stanford University Press, 2002), pp. 94–136.
2 Ibid., p. xiv: 'humanity, instead of entering a truly human state, is sinking into a new kind of barbarism'.
3 I have analysed this text in greater detail in Bernard Stiegler, *Technics and Time, 3: Cinematic Time and the Question of Malaise*, trans. Stephen Barker (Stanford: Stanford University Press, 2011), and in Bernard Stiegler, *States of Shock: Stupidity and Knowledge in the Twenty-First Century*, trans. Daniel Ross (Cambridge: Polity, 2015).
4 According to *Dialectic of Enlightenment*, which in this chapter analyses both the way Hollywood functions and the imminent appearance of television, modernity and the ideas emerging from the Enlightenment transform into an industrial and systemic *Dummheit (bêtise,* stupidity) by engendering these culture industries (also called communication industries), which, together with the Taylorist apparatus of production, form an integrated system for controlling and capturing the gestures of producers and the attention of consumers. Hence is constituted the basis of what, sixty-eight years later, Mats Alvesson and André Spicer would call 'functional stupidity'. The *Aufklärung*-become-rationalization, that is, the absolute reign of calculation, *destroys reason* through tight articulation between the apparatus of production, consumption and design, which leads to the proletarianization of producers and consumers but also of designers and conceptualizers – destroying every form of reason. Today, with 'big data', this affects all intellectual activities. See Bernard Stiegler, *Automatic Society, Volume 1: The Future of Work*, trans. Daniel Ross (Cambridge: Polity, 2016).
5 See the report and order of the FCC, available at: https://transition.fcc.gov/Bureaus/Engineering_Technology/Orders/1997/fcc97115.pdf.

6   *Translator's note*: Patrick Boutot, known professionally as Patrick Sébastien, is a French television host and performer. His programmes consisted of silly tricks and sketches, in which, by the late 1980s, he was involving major French political figures, amounting to the advent in France of what came to be called '*la politique-spectacle*', showbiz politics.

7   See '34 091 habitants des bidonvilles en instance de relogement', *Le Matin* (3 February 2014), available at: http://www.lematin.ma/journal/-/196138.html.

8   'Tuerie de Nanterre', article in the French Wikipédia (no access date).

9   'Les "nous" et les "je": agir ensemble dans la cité'.

10 In the terminology coined by Gilbert Simondon.

11 See the article in the French Wikipédia on 'La Cinq' (accessed July 2015).

12 On this point, see the documentary by Christophe Nick, *Le Temps du cerveau disponible*, available at: https://www.youtube.com/watch?v=amzLnvfaeJM.

13 Bernard Stiegler, 'To Love, to Love Me, to Love Us: From September 11 to April 21', *Acting Out*, trans. David Barison, Daniel Ross and Patrick Crogan (Stanford: Stanford University Press, 2009).

14 Given that I explained this in *Technics and Time, 3*, I will not repeat my account of Adorno and Horkheimer's reasoning with respect to the transcendental imagination. All the arguments I pursue in the present work presuppose shifting to a 'new critique', where the schemas of the so-called transcendental imagination are seen as being, in reality, arrangements of retentions and protentions conditioned by tertiary retention.

15 Stiegler, 'To Love, to Love Me, to Love Us', pp. 81–2.

16 This new policy followed the collapse of the American analogue audiovisual industries. Japan had begun to dominate these industries in the 1970s, and so this was a policy aimed at taking back the lead in hardware, and aimed at doing so during the transition from software to 'dataware'. On this notion, see Christian Fauré, 'Dataware et infrastructure du cloud computing', in Bernard Stiegler, Alain Giffard and Christian Fauré, *Pour en finir avec la mécroissance* (Paris: Flammarion, 2009), and the entry on 'dataware' in Victor Petit, 'Vocabulaire d'Ars Industrialis', in Bernard Stiegler, *Pharmacologie du Front national* (Paris: Flammarion, 2013), pp. 387–88.

17 This is the company that inaugurated 'user-friendly' microcomputing with the Macintosh, released in 1984, a company whose failure was often pronounced, but which ultimately became the global leader in the mobile hardware of the 'smartphone', via the iPod.

18 This included the proposal to create INAVISION, which would have fully integrated the transition from broadcasters to servers, and would have entailed the invention of a completely new way of doing 'television'. Appointed during the Juppé administration, I left INA during the Jospin administration – which persisted in trying to create digital terrestrial television without paying any attention to what ADSL would make possible, namely, the ability to stream digital television along telephone lines. We will see later how far this negligence extended. For my own views on the current state of audiovisual broadcasting in France, see Bernard Stiegler, 'La télévision publique doit à nouveau faire la différence', *Télérama.fr* (24 April 2015), available at: http://television.telerama.fr/television/la-television-publique-doit-cultiver-la-qualite-qui-fera-sa-difference-par-bernard-stiegler-philosophe,125788.php.

19 On 21 September 2015, *Libération* published a dossier showing that the French broadcasting industry and public authorities acknowledged the end of the 'classical model of television' *eighteen years after* this decision of the FCC.

20 Thomas Berns and Antoinette Rouvroy, 'Gouvernementalité algorithmique et perspectives d'émancipation', *Réseaux* 177 (2013), pp. 163–96. See also Antoinette Rouvroy, 'The End(s) of Critique: Data-Behaviourism vs Due-Process', in Mireille Hildebrandt and Katja de Vries (eds), *Privacy, Due Process and the Computational Turn: The Philosophy of Law Meets the Philosophy of Technology* (Abingdon and New York: Routledge, 2013), pp. 143–68.

21 Chris Anderson, 'The End of Theory: The Data Deluge Makes the Scientific Method Obsolete', *Wired* (23 June 2008), available at: http://archive.wired.com/science/discoveries/magazine/16-07/pb_theory.

22 Televisions now come with sensors and cookies as standard. See Michael Price, 'I'm Terrified of My New TV: Why I'm Scared to Turn This Thing On – and You'd Be Too', *Salon* (31 October 2014), available at: http://www.salon.com/2014/10/30/im_terrified_of_my_new_ tv_why_im_scared_to_turn_this_thing_on_and_youd_be_too, and my commentary in *Automatic Society, Volume 1*, p. 16.

23 This is what I have called dividuation and transdividuation, taking my cue from the term 'dividual' coined by Félix Guattari. See Stiegler, *Automatic Society, Volume 1*, pp. 24 and 34.

24 This is also the case for state investment in digital equipment, and therefore for the plan to invest fifteen billion euros in high-speed broadband launched by François Hollande in 2013, which, without a major industrial policy founded on the digital, can amount only to the creation of a network that helps the Big Four and their 'eco-systems' pillage the national economy. See the interview with Bernard Stiegler, 'Entrer dans la troisième époque du Web', *Le Journal du Dimanche* (11 May 2013), available at: http://www.lejdd.fr/Economie/Actualite/La-France-dans-dix-ans-Entrer-dans-la-troisieme-epoque-du-Web-606814.

25 Retentions are, in the language of Husserl, what one retains – and as such they form memories.

26 Protentions are expectations in all their forms. Algorithms enable psychic protentions to be short-circuited, and replaced with automatic protentions. See Stiegler, *Automatic Society, Volume 1*, pp. 111ff.

27 Ibid., p. 140.

28 Disruption provokes chain reactions, and the new leaders of 'disruptive' companies are those who encourage and support these chain reactions for the benefit of their shareholders – but also to the inevitable detriment of the societies in which they occur.

29 On the *pharmakon* and the pharmacological question, see the entry on '*Pharmakon*, pharmacologie', in Petit, 'Vocabulaire d'Ars Industrialis', pp. 421–2. This crucial subject will be recapitulated in what follows. The *pharmakon* refers here to every technique and all technics insofar as they are always both curative and toxic.

30 Florian, in L'Impansable (coll.), *L'Effondrement du temps: Tome 1, Pénétration* (Paris: Le Grand Souffle Editions, 2006), p. 7.

2 The Absence of Epoch

1 The collapse of employment due to automation – heralded everywhere, denied by everyone – will lead to an immense economic cataclysm where, unless a completely new macroeconomic policy is developed (this is the subject of Bernard Stiegler, *Automatic Society, Volume 1: The Future of Work*, trans. Daniel Ross [Cambridge: Polity, 2016]), it will become increasingly unlikely that new generations will be able to find their place. In some French departments, in the current conditions, it is likely that the youth unemployment rate will exceed 50 per cent (and even higher in departments like Seine-Saint-Denis, where unemployment for 16–24-year-olds was in 2011 already 38 per cent) when, in ten years, we see the materialization of the forecasts of Roland Berger, which predicted the loss of three million French jobs by 2025.

2 Including Fukushima, which, since 2011, has continued to spew contaminated water into the Pacific, just as hedge funds continue to rule the roost after the bank bailouts of 2008.

3 Mats Alvesson and André Spicer, 'A Stupidity-Based Theory of Organizations', *Journal of Management Studies* 49 (2012), pp. 1194–220. See also Mats Alvesson and André Spicer, *The Stupidity Paradox: The Power and Pitfalls of Functional Stupidity at Work* (London: Profile, 2016).

4 See http://barbares.thefamily.co.

5 See Bernard Stiegler, *Technics and Time, 1: The Fault of Epimetheus*, trans. Richard Beardsworth and George Collins (Stanford: Stanford University Press, 1998), part II, ch. 1, 'Prometheus's Liver'.
6 See Sophocles, *Antigone*, whose choral ode is discussed in Martin Heidegger, *An Introduction to Metaphysics*, trans. Gregory Fried and Richard Polt (New Haven and London: Yale University Press, 2000).
7 Georges Bataille, *The Accursed Share: An Essay on General Economy, Volume 1: Consumption*, trans. Robert Hurley (New York: Zone Books, 1991).
8 Jacques Derrida, 'Faith and Knowledge: The Sources of "Religion" at the Limits of Reason Alone', trans. Samuel Weber, in *Acts of Religion*, ed. Gil Anidjar (New York and London: Routledge, 2002).
9 This is what, prior to Adorno and Horkheimer, Hölderlin endured as the 'withdrawal of the divine', and which, as Heidegger showed, with Nietzsche became *nihilism inasmuch as it has lost the very notion of the 'default of God'*: 'The time has already become so desolate that it is no longer able to see the default of God as a default. [...] The age for which the ground fails to appear hangs in the abyss. Assuming that a turning point in any way still awaits this desolate time, it can only come one day if the world turns radically around, which now more plainly means if it turns away from the abyss.' Martin Heidegger, 'Why Poets?', *Off the Beaten Track*, trans. Julian Young and Kenneth Haynes (Cambridge: Cambridge University Press, 2002), pp. 200–1.
10 On 'digital natives', see Bernard Stiegler, 'Literate Natives, Analogue Natives and Digital Natives: Between Hermes and Hestia', trans. Daniel Ross, in Divya Dwivedi and Sanil V (eds), *The Public Sphere from Outside the West* (London: Bloomsbury Academic, 2015).
11 Since the nineteenth century, 'teleologism' (or 'finalism') has, in science, become the mark of causal illusion par excellence: science teaches us that there are no final causes – although there are 'attractors', that is, focuses of the potential completion of tendencies harboured by dynamic systems of all kinds – from the vortex of a whirlwind to the universe in totality, as well as all kinds of life on the 'arrow of time'.
   Given that modern humanity, contrary to Aristotelianism, wants to base its behaviour on scientific knowledge, itself defined by the analytical objectification of causes, it cannot appeal to 'purpose' [*finalité*] in order to act.
   This perspective clearly entails a naturalism whose price is the impossibility for humans of giving itself purposes beyond this purely analytical science. This issue has frequently been raised by Alain Supiot. Furthermore, one of the great, enduring epistemological controversies concerns the inclusion – in the dissipation informing the local physical structures theorized by Ilya Prigogine – of *noetic and neganthropological attractions* such as we have been given to think, for example, in the speculative cosmology of Alfred North Whitehead.
   The organological and pharmacological perspective that I have defended for fifteen years argues that: (1) the *pharmakon* that is the artificial *organon* is a neganthropic structure that can equally become anthropic; and that (2) this pharmacology wherein the artificial organ can turn into its opposite precisely defines – and as *neganthropological condition* – *facticity* as such, but also raises, in the very fabric of thought supplied by the sciences, the question of *reason as function and as purpose* [*finalité*], *as function of final causes*.
   The collapse of knowledge caused by general proletarianization is today reaching the stage of the 'end of theory', as Chris Anderson has presented it (and as the fundamental result of intensive computing – see Stiegler, *Automatic Society, Volume 1*, chs 1–2), inevitably generating a *collapse of rational purpose*, that is, of the *credibility of reason* as the ultimate and neganthropological affirmation of *differing and deferring* the end, *différance to the infinite* of the end, *that is, of entropy*. The collapse of knowledge replaced by automatic understanding leads to the sudden and massive increase of entropy, visible at the scale of a single generation, and seen as the fulfilment of the end: this is what Florian is telling us.

12 Stiegler, *Technics and Time, 1*.
13 Martin Heidegger, *Being and Time*, trans. Joan Stambaugh, revised by Dennis J. Schmidt (Albany: State University of New York Press, 2010), §4.
14 This is why what Heidegger called the privilege of Dasein, which is that of being 'the being who questions', must be related, before this 'possibility of questioning', to the *putting in question* that is the techno-logical *epokhē*. See Bernard Stiegler, *What Makes Life Worth Living: On Pharmacology*, trans. Daniel Ross (Cambridge: Polity, 2013), ch. 6; and see p. 281.
15 Bertrand Gille, *The History of Techniques, Volume 1: Techniques and Civilizations*, trans. P. Southgate and T. Williamson (New York: Gordon and Breach, 1986), pp. 54–5.
16 And especially in §16, where Heidegger highlights the effect of an interruption of activity caused by the failure of a tool, through which it appears *as* tool, and, along with it, so does the world appear as world.
17 Jacques Ellul, *The Technological System*, trans. Joachim Neugroschel (New York: Continuum, 1980).
18 Martin Heidegger, 'Positionality', *Bremen and Freiburg Lectures: Insight Into That Which Is and Basic Principles of Thinking*, trans. Andrew J. Mitchell (Bloomington and Indianapolis: Indiana University Press, 2012), pp. 23ff.
19 In the sense of *Being and Time*.
20 Hypomneses are the supports made to store memory that arose at the beginning of grammatization, during the Upper Palaeolithic.
21 But on this point, see especially Stiegler, *Automatic Society, Volume 1*, and Bernard Stiegler, *Symbolic Misery, Volume 1: The Hyper-Industrial Epoch*, trans. Barnaby Norman (Cambridge: Polity, 2014).
22 I have tried to explain why this is so in Bernard Stiegler, *States of Shock: Stupidity and Knowledge in the Twenty-First Century*, trans. Daniel Ross (Cambridge: Polity, 2015), ch. 5, 'Reading and Re-Reading Hegel After Poststructuralism'.
23 *Translator's note*: The author uses *la disruption* to refer both to a particular *phenomenon* and to the particular *period* in which this phenomenon is expressed. Mostly, this translation refers to 'disruption' to refer to the phenomenon, but occasionally we have translated *la disruption* as 'the age of disruption', to indicate this temporal character, where it should always be kept in mind that the author defines this period precisely in terms of its failure to achieve epochal status, and hence as an 'absence of epoch'.
24 This question was undoubtedly also opened by Lyotard, but in such a way that, attempting to think it in terms of postmodernity, I believe that he was ultimately unsuccessful – and I believe that this is something he himself understood.
25 Metastabilization is what maintains a process of individuation (psychic as well as collective) between stability (that is, pure synchrony) and instability (that is, pure diachrony).
26 Social groups participate in social systems that themselves depend on social organizations.
27 On the notion of collective secondary retentions, see Bernard Stiegler, *Uncontrollable Societies of Disaffected Individuals: Disbelief and Discredit, Volume 2*, trans. Daniel Ross (Cambridge: Polity, 2013), pp. 17–19, 94–5, 103–5, 115–17 and 120–5.
28 Heidegger, 'The Turn', *Bremen and Freiburg Lectures*, p. 67, translation modified.
29 This bifurcation is what, occurring in the dynamic system that a process of individuation forms, defers the exhaustion of this system. I posit here that we should understand what Derrida tried to think with the name 'différance' as a negentropic process.
30 *Translator's note*: Stiegler's French translations of *Besorgen* as *affairement* and *pre-occupation* have been followed here, rather than Macquarrie and Robinson's use of 'concern' or Stambaugh's choice of 'taking care'. The latter in particular would risk being misleading in the context of Stiegler's thought, since, as Stambaugh herself

explains, her reason for choosing 'taking care' was 'because it refers more to errands and matters that one takes care of or settles' (Joan Stambaugh, 'Translator's Preface', in Heidegger, *Being and Time*, p. xxv). This finite sense of taking care as 'settling matters' is thus in distinct contrast to Stiegler's much more expansive use of the concepts of care and taking care as, precisely, forms of thought and practice directed towards the infinity of what the author calls 'consistences' (infinite because they do not exist, but consist).

31 *Translator's note*: French employs two terms, *dénégation* and *déni*, to cover the field that in English is mostly occupied by a single word, 'denial'. Whereas *dénégation* refers to the forms of denial that may consist in denying an accusation, or else in denying a fact or a state of affairs (as in the English formulation, 'to be in denial'), *déni* refers to a denial more particularly in the juridical sense when, for example, one speaks of a 'denial of justice' to refer to a miscarriage of justice, or in the sense of a refusal, such as a refusal to discuss something.

In this work, however, the use of these two terms is complicated by the way the author articulates the Heideggerian account of being-towards-death (as a form of knowing existing mostly in the mode of not-knowing and fleeing) with Freudian *Verleugnung*, and consequently with the difficulties faced by Freud's translators in both English and French. James Strachey, for example, sets the convention according to which Freud's use of *Verleugnung* will be translated into English as 'disavowal', whereas *Verneinung* will be translated as 'negation': 'The word *Verleugnung* has in the past often been translated "denial" and the associated verb by "to deny". These are, however, ambiguous words and it has been thought better to choose "to disavow" in order to avoid confusion with the German "*verneinen*", used, for instance, in the paper on "Negation".' James Strachey, translator's note in Sigmund Freud, 'The Infantile Genital Organization (An Interpolation into the Theory of Sexuality)', in Volume 19 of James Strachey (ed. and trans.), *The Standard Edition of the Complete Psychological Works of Sigmund Freud* (London: Hogarth, 1953–74), p. 143. *Verleugnung* refers, in Freud, firstly to his understanding of the way children respond to the discovery of the absence of the penis ('castration') in girls, and is then applied to the way in which the fetishist both recognizes and repudiates this absence, and more generally to this kind of splitting of the ego that operates as a general defence mechanism in the face of external reality. But whereas Strachey decides that 'disavowal' rather than 'denial' is the best way of translating this concept, Laplanche and Pontalis make the case for '*déni*':

> We propose '*déni*' as the best French equivalent of '*Verleugnung*' because it has a number of resonances which the alternative '*dénégation*' does not have:
> a. 'Denial' (*déni*) is often a stronger word. We say 'I deny the validity of your statements.'
> b. As well as referring to a statement which is being disputed, 'denial' is also used to evoke the withholding of goods or rights.
> c. In this last case, the implication is that the prohibition in question is illegitimate: denial of justice, denial of food, etc. – in other words, a withholding of what is due.

Jean Laplanche and Jean-Bertrand Pontalis, *The Language of Psycho-Analysis*, trans. Donald Nicholson-Smith (London: Karnac, 1988), p. 120. As for *Verneinung*, which Strachey translates as 'negation', Laplanche and Pontalis follow suit with *négation* and *dénégation*, but they explain that *Verneinung* does not only mean *negation* 'in the logical and grammatical sense' but also '*denial* in the psychological sense of rejection of a statement which I have made or which has been imputed to me, e.g. "No, I did not say that, I did not think that"' (ibid., p. 262). And hence they note that 'the English reader inevitably loses the ambiguity which derives from the term's meaning both negation and denial' (ibid.).

Were we to assume that Stiegler's use of *dénégation* and *déni* correspond to the Freudian terms, and so decide to follow Strachey, then, we would be compelled to employ two terms, *negation* and *disavowal*, and this would be, in a way, to deny that what is at stake in these terms remains *denial*. For that reason, it is judged preferable to mostly translate *dénégation* (which is the term mostly used by the author here) as denial, hoping that the reader will keep in mind the psychoanalytic sense of negation. On those occasions where Stiegler refers to both concepts serially, or introduces a distinction between them (see especially §123), *déni* will tend to be translated as disavowal, but on other occasions it will more straightforwardly be translated as denial.

32 On this negation [*dénégation*] and its relationship to *Verneinung* in Freud and to disavowal [*déni*] (*Verleugnung*), see §123. *Translator's note*: And see previous note.

33 Yuk Hui, *On the Existence of Digital Objects* (Minneapolis and London: University of Minnesota Press, 2016).

34 See ch. 2, n. 34.

35 *Translator's note*: In German, *Datierbarkeit* and *Zeughaftigkeit*, respectively. Note that Macquarrie and Robinson translate the latter term as 'equipmentality'.

36 See Bernard Stiegler, *Technics and Time, 3: Cinematic Time and the Question of Malaise*, trans. Stephen Barker (Stanford: Stanford University Press, 2011), ch. 3.

37 These retentional systems are themselves composed of hypomnesic tertiary retentions, that is, tertiary retentions that are grammatized and dedicated to recording time in spatial forms.

38 On the process of adoption, see ch. 2, n. 36.

39 On the notion of secondary retention, which I owe to Husserl, and which I have explained many times, see p. 215, and Bernard Stiegler, 'To Love, to Love Me, to Love Us: From September 11 to April 21', *Acting Out*, trans. David Barison, Daniel Ross and Patrick Crogan (Stanford: Stanford University Press, 2009).

40 See Heidegger, *Being and Time*, Division 2, ch. 5.

41 Pierre Schaeffer, *Treatise on Musical Objects: An Essay Across Disciplines*, trans. Christine North and John Dack (Oakland: University of California Press, 2017), p. v.

3 Radicalization and Submission

1 On entropy, negentropy and neganthropology, see Bernard Stiegler, *Automatic Society, Volume 1: The Future of Work*, trans. Daniel Ross (Cambridge: Polity, 2016), 'Introduction' and 'Conclusion'.

2 Genesis 3:19. Quoted from *The Holy Bible: Quatercentenary Edition* (Oxford and New York: Oxford University Press, 2010), no page numbers.

3 See Bernard Stiegler, *Uncontrollable Societies of Disaffected Individuals: Disbelief and Discredit, Volume 2*, trans. Daniel Ross (Cambridge: Polity, 2013), pp. 39–40.

4 Bernard Stiegler, *Taking Care of Youth and the Generations*, trans. Stephen Barker (Stanford: Stanford University Press, 2010), ch. 1.

5 See p. 7, and for more detail see Stiegler, *Automatic Society, Volume 1*.

6 And it is that about which Heidegger's second phase, that of the 'history of being' thought on the basis of the 'epochs of being', has nothing to say. But it is perhaps the question that is raised with the 'final' Heidegger, that of 'The End of Philosophy', 'The Turn' and *Identity and Difference*.

7 And, in a way, these analyses deny that such a projection beyond is always already contained in any projection.

8 *Translator's note*: On 'storytelling', see Christian Salmon, *Storytelling: Bewitching the Modern Mind* (London and New York: Verso, 2010), esp. ch. 3.

9 I am here repeating the terms used in the three volumes of my *Disbelief and Discredit* series.

10 Both as disaffection [*désaffection*] and withdrawal [*désaffectation*]. See Stiegler, *Uncontrollable Societies of Disaffected Individuals*, pp. 86–90.

11 I have introduced this key theme in Stiegler, *Automatic Society, Volume 1*, and I will return to it in more detail in the second volume of that work.

12 See Jean Jouzel, *L'Avenir du climat* (Paris: Fondation Diderot, 2014), and Anthony D. Barnosky et al., 'Approaching a State Shift in Earth's Biosphere', *Nature* 486 (7 June 2012), pp. 52–8.

13 See Bernard Stiegler, 'To Love, to Love Me, to Love Us: From September 11 to April 21', *Acting Out*, trans. David Barison, Daniel Ross and Patrick Crogan (Stanford: Stanford University Press, 2009), pp. 53–4, translation modified: 'A temporal object is a *fabric of retentions and protentions*. These protentional and retentional processes, however, also weave the temporality of consciousness in general, and temporal objects permit the modification of these processes of consciousness in a single blow.'

14 Ibid., p. 53: 'As a general rule, a *temporal object* is an object of time-consciousness, *the flow of which occurs simultaneously with the consciousness of which it is the object – because this consciousness itself flows*. This consciousness is itself *essentially* temporal: it *never ceases* to flow; it has, like all temporal objects, a *beginning* and an *end*, and, between this beginning and end, it is nothing but temporal *flux*. Now, when you who are consciousnesses watch a broadcast or a film, your time-consciousness passes *into* the broadcast or *into* the film, *adhering* to the temporal object that is the object of your consciousness.'

15 Ibid., p. 54, translation modified: 'In the "now" of a melody, in the present moment of a musical object that flows, the note that is present can be a note, rather than merely a sound, only insofar as it retains within itself the *preceding note*, which remains present – a preceding note *still present*, which itself retains the preceding one, and so on. One must not confuse this *primary retention belonging to the present of perception* with *secondary retention* [...]. Since the appearance of the phonogram, which is itself [...] a *tertiary retention* (a prosthesis, for exteriorized memory), the identical repetition of the same temporal object has become possible [...]. *Tertiarized temporal objects* – that is, objects either recorded or converted into a controllable and transmissible signal (such as phonograms, but also films, and radio and television broadcasts) – are materialized time, which overdetermines the relations between primary and secondary retentions in general, *thereby* allowing them, to a certain extent, to be *controlled*.'

16 Ibid., p. 55, translation modified: 'When ten million people watch the same broadcast – the same audiovisual temporal object – they synchronize their flows. Of course, their criteria for selecting retentions vary, and, therefore, they do not perceive the same phenomenon: they don't all think the same thing about what they watch. But if it is true that *secondary retentions* form the *selection criteria in primary retentions*, then the fact that the same people watch the *same* programs every day necessarily leads each "consciousness" into sharing *more and more identical secondary retentions*, and thus to selecting the same primary retentions. They end up being so well synchronized that they have lost their *diachrony*, that is, their singularity, which is to say their freedom, which always means their freedom *to think*.'

17 In Bernard Stiegler, *Symbolic Misery, Volume 1: The Hyper-Industrial Epoch*, trans. Barnaby Norman (Cambridge: Polity, 2014), Stiegler, *Symbolic Misery, Volume 2: The Katastrophē of the Sensible*, trans. Barnaby Norman (Cambridge: Polity, 2015), Stiegler, *The Decadence of Industrial Democracies: Disbelief and Discredit, Volume 1*, trans. Daniel Ross (Cambridge: Polity, 2011), Stiegler, *Uncontrollable Societies of Disaffected Individuals*, Stiegler, *The Lost Spirit of Capitalism: Disbelief and Discredit, Volume 3*, trans. Daniel Ross (Cambridge: Polity, 2014), Stiegler, *La Télécratie contre la démocratie* (Paris: Flammarion, 2006) and Stiegler, *Taking Care of Youth and the Generations*.

18 Frédéric Kaplan, 'Vers le capitalisme linguistique. Quand les mots valent de l'or', *Le Monde diplomatique* (November 2011), available at: http://www.monde-diplo-matique.fr/2011/11/KAPLAN/46925. See also Kaplan, 'Linguistic Capitalism and Algorithmic Mediation', *Representations* 27 (2014), pp. 57–63.

19 *Translator's note*: Here the author is referring to a statement made by Frank Capra in *The Name Above the Title: An Autobiography* (New York: Macmillan, 1971), p. 223: 'Film is a disease. When it infects your bloodstream, it takes over as the Number One hormone; it bosses the enzymes; directs your pineal gland; plays Iago to your psyche. As with heroin, the antidote for film is *more* film. Withdrawal from junk tortures a mainliner's body. But kicking the film habit wracks a filmmaker's soul – his essential nature.' The author discusses Capra's statement in Bernard Stiegler, *The Neganthropocene*, trans. Daniel Ross (London: Open Humanities Press, 2018), ch. 10.

20 See '*Ars Industrialis*: 2005 Manifesto', in Bernard Stiegler, *The Re-Enchantment of the World: The Value of Spirit against Industrial Populism*, trans. Trevor Arthur (London and New York: Bloomsbury Academic, 2014).

21 See '*Ars Industrialis*: 2010 Manifesto', in Stiegler, *The Re-Enchantment of the World*.

22 That viewers believe they are the ones looking when it is really the television that aims at them, as François Chalais said in 1957 ('In the past it was you who made the images, now it is the images who make you', quoted in the 2010 documentary, *Le Temps de cerveau disponible*, directed by Jean-Robert Viallet and written by Christophe Nick), is the idea that was put on stage by Ray Bradbury and then by François Truffaut in *Fahrenheit 451* through the depiction of the television programme 'The Family'.

23 See p. 15.

24 See 'The Family': http://barbares.thefamily.co. It is quite extraordinary that this 'incubator of startups' took the exact name of the powerful television programme depicted in *Fahrenheit 451*.

25 André Comte-Sponville, *Traité du désespoir et de la béatitude. Tome 1, Le mythe d'Icare* (Paris: PUF, 1984), p. 9: 'Our time would be one of despair. The death of God, the decline of churches, the end of ideologies [...]. I see, rather, the work of fatigue. Because they are disappointed, they feel desperate [...]. But if they were truly desperate, they would not be disappointed. Ours is an age not of despair, but of disappointment [*désappointement*]. We live in the age of disappointment [*déception*].' Comte-Sponville was undoubtedly right to describe the morale of what would still have been perceived (in 1984) as an epoch of 'disappointment'. But we must now take the 'measure' (μέτρον) of confronting the excess [*démesure*] (ὕβρις) of the absence of epoch.

26 On this constellation of notions, see Stiegler, 'To Love, to Love Me, to Love Us: From September 11 to April 21', and Victor Petit, 'Vocabulaire d'Ars Industrialis', in Bernard Stiegler, *Pharmacologie du Front national* (Paris: Flammarion, 2013).

27 In using this term here, I am drawing attention to the *punctum* that Roland Barthes, in his phenomenology of photography, himself described in relation to the *studium*. See Roland Barthes, *Camera Lucida: Reflections on Photography*, trans. Richard Howard (London: Flamingo, 1984), p. 77.

28 Florian, in L'Impansable (coll.), *L'Effondrement du temps: Tome 1, Pénétration* (Paris: Le Grand Souffle Editions, 2006), p. 7.

29 On this causal inversion at the basis of ideology, see Stiegler, *Pharmacologie du Front national*. Only a few have spoken out against this infamy – among them Edwy Plenel, in *Pour les musulmans* (Paris: La Découverte, 2014).

30 This possibility is that of the impossible in the sense that any negentropic bifurcation, from the perspective of the system within which it occurs, starts out from the impossible, but not the *impansable*. Thinking is truly thinking only when it thinks the impossible – and on this point I believe I am in complete disagreement with the L'Impansable collective: everything can and must be cared for [*pansé*], and this is what is affirmed by every form of thinking [*pensée*]. To live in the default of origin, that is, in the pharmacological fate that characterizes what I now try to conceive as a neganthropology, is to take as a matter of principle that thinking is always what cares for the impossible without any guarantee of success – Inshallah.

31 Erwin Schrödinger, *What is Life? The Physical Aspect of the Living Cell* (1944), in *What is Life, with Mind and Matter and Autobiographical Sketches* (Cambridge: Cambridge University Press, 1992).

32 *Translator's note*: The author has recently begun to make use of these unusual terms, *pansée* and *panser*, mostly found in old French. The origin of these terms lies, in fact, in the care for, grooming of, and feeding of horses, and by extension comes to be used for care in general, and, in particular, for the care of wounds, in the sense of dressing them in order that they may heal. Similarly to Derrida's notion of différance, they are a modification of the French words for thinking, *pensée* and *penser*. See the detailed discussion in Bernard Stiegler, *The Neganthropocene*, trans. Daniel Ross (London: Open Humanities Press, 2018), ch. 5, n. 171 and ch. 13, §13.

33 In the sense that at one time there existed a philosophical anthropology, that is, an anthropology conceived according to an Idea of man as being himself animated by Ideas of reason.

34 In the sense that there is a positive anthropology – that is, a scientific anthropology.

35 *Translator's note*: See Italo Calvino, *If On a Winter's Night a Traveller*, trans. William Weaver (London: Vintage, 1998), pp. 143–44: 'He is continuing his inspection of the house to which you let him have the keys. There are countless things that you accumulate around you: fans, postcards, perfume bottles, necklaces hung on the walls. But on closer examination every object proves special, somehow unexpected. Your relationship with objects is selective, personal; only the things you feel yours become yours: it is a relationship with the physicality of things, not with an intellectual or affective idea that takes the place of seeing them and touching them. And once they are attached to you, marked by your possession, the objects no longer seem to be there by chance, they assume meaning as elements of a discourse, like a memory composed of signals and emblems.'

36 Gilles Clément, *Toujours la vie invente. Réflexions d'un écologiste humaniste* (Paris: L'Aube, 2008).

37 The age of the new kind of barbarism is indeed also that of a new era of philistinism that includes the 'art market'. See my three lectures given at the University of California in 2011, published as Bernard Stiegler, 'The Proletarianization of Sensibility', 'Kant, Art, and Time' and 'The Quarrel of the Amateurs', trans. Stephen Barker, Arne De Boever and Robert Hughes, *Boundary 2* 44 (2017), pp. 5–52.

38 It is this that becomes the subject of a parable in Gustave Flaubert, 'The Legend of Saint Julian the Hospitaller', *Three Tales*, trans. A.J. Krailsheimer (Oxford and New York: Oxford University Press, 1999).

39 Stiegler, *Uncontrollable Societies of Disaffected Individuals*, ch. 2.

40 See Michel Foucault, *The Government of Self and Others: Lectures at the Collège de France, 1982–1983*, trans. Graham Burchell (New York: Palgrave Macmillan, 2010), and Michel Foucault, *The Courage of Truth (The Government of Self and Others II): Lectures at the Collège de France, 1983–1984*, trans. Graham Burchell (New York: Palgrave Macmillan, 2011).

41 Sigmund Freud, *The Ego and the Id*, in Volume 19 of James Strachey (ed. and trans.), *The Standard Edition of the Complete Psychological Works of Sigmund Freud* (London: Hogarth, 1953–74), p. 30.

42 *Translator's note*: The reference to the 'narcissism of minor differences' derives from Sigmund Freud, 'The Taboo of Virginity (Contributions to the Psychology of Love III)', in Volume 11 of Strachey, *The Standard Edition of the Complete Psychological Works of Sigmund Freud*, p. 199, where he discusses Alfred Ernest Crawley's idea that 'it is precisely the minor differences in people who are otherwise alike that form the basis of feelings of strangeness and hostility between them'. Freud adds: 'It would be tempting to pursue this idea and to derive from this "narcissism of minor differences" the hostility which in every human relation we see fighting successfully against feelings of fellowship and overpowering the commandment that all men should love one another.' Freud returns to the idea in *Civilization and Its Discontents*, in Volume

21 of Strachey, *The Standard Edition of the Complete Psychological Works of Sigmund Freud*, p. 114.
43 See Bernard Stiegler, *States of Shock: Stupidity and Knowledge in the Twenty-First Century*, trans. Daniel Ross (Cambridge: Polity, 2015), part 2.
44 See Maurice Godelier, *The Metamorphoses of Kinship*, trans. Nora Scott (London and New York: Verso, 2011), pp. 35–6, where Godelier shows that each tribe and each idiom (tribes form 'a set of local groups that speak related languages') had differentiated itself in the mythology of the Baruya, 'split off from a common trunk' that 'goes back to the Dreamtime', which is the dream of their 'common origin'.
45 Stiegler, 'How I Became a Philosopher', *Acting Out*, p. 20.
46 In the 'societies of slow history', the 'effection' of this redoubling operates through the slow individuation of metastabilized transindividuations that seem long ago to have stabilized – but this is an illusion. Had they really been stabilized, they could not, for example, accommodate the anthropologists who sometimes come to visit them, nor could they grant hospitality to the spirits who transindividuate within the transgenerationality that is told in myth and celebrated in ritual.
47 *Translator's note*: An idiolect is the 'linguistic system of an individual, differing in some details from that of all other speakers of the same dialect or language'. Entry for 'idiolect' in Lesley Brown (chief ed.), *The New Shorter Oxford English Dictionary*, 2 vols (Oxford: Oxford University Press, 1993), p. 1305.
48 Everything stated here stems from an originary default of origin that I described in Bernard Stiegler, *Technics and Time, 1: The Fault of Epimetheus*, trans. Richard Beardsworth and George Collins (Stanford: Stanford University Press, 1998), through the interpretation of the myth of Prometheus and Epimetheus in *Protagoras*.
49 *Translator's note*: 'C'est l'enfance de l'art' is an idiomatic expression used to say that something is simple or elementary, more or less equivalent to the English 'It's child's play'.
50 Jean-Christophe Bailly, *Le Dépaysement. Voyages en France* (Paris: Seuil, 2011), p. 117.
51 Plato, *Ion* 533d–e, trans. Lane Cooper, in Edith Hamilton and Huntington Cairns (eds), *The Collected Dialogues of Plato* (Princeton: Princeton University Press, 1961), pp. 219–20: 'As I just now said, this gift you have of speaking well on Homer is not an art; it is a power divine, impelling you like the power in the stone Euripides called the magnet, which most call "stone of Heraclea". This stone does not simply attract the iron rings, just by themselves; it also imparts to the rings a force enabling them to do the same thing as the stone itself, that is, to attract another ring, so that sometimes a chain is formed, quite a long one, of iron rings, suspended from one another. For all of them, however, their power depends upon that lodestone. Just so the Muse. She first makes men inspired, and then through these inspired ones others share in the enthusiasm, and a chain is formed, for the epic poets, all the good ones, have their excellence, not from art, but are inspired, possessed, and thus they utter all these admirable poems. So it is also with the good lyric poets.'
52 See pp. 17 and 290–1.
53 *La Grande illusion*, directed by Jean Renoir (1937).
54 And this is why the transitional object, which itself does not exist, is there introduced. See Bernard Stiegler, *What Makes Life Worth Living: On Pharmacology*, trans. Daniel Ross (Cambridge: Polity, 2013).

4 Administration of Savagery, Disruption and Barbarism

1 On the horizon of this radicalization, and more generally of the terrorism typical of the mass media and social network era, lies the West's policy on the Near and Middle East, dominated by the United Kingdom and the United States – from the British administration in Mesopotamia up until the alliances with Ahmad Shah

Massoud and Osama bin Laden, himself long backed by the United States and Saudi Arabia and for whom Massoud had become a competitor, the latter being assassinated by Al-Qaeda on 9 September 2001. Overdetermined by the Israeli-Palestinian conflict, and the policy of the Shah of Iran that would lead to an alliance of communist mujahideen and Shiite mullahs, Anglo-American interventionism would be relaunched at an extreme level with the alliance against Saddam Hussein in the name of alleged weapons of mass destruction – a fable that, in the person of Prime Minister Dominique de Villepin, France would refuse to endorse at the general assembly of the United Nations.

2 A work written between 2002 and 2004.

3 This is Ignace Leverrier's summary, in 'Gestion de la barbarie: stratégie du califat', *L'Obs* (3 August 2015), available at: http://montechristo.blogs.nouvelobs.com/archive/2015/07/23/gestion-de-la-barbarie-strategie-du-califat-566611.html.

4 See Sophie Fay, 'Start-up: ces "barbares" qui veulent débloquer la France', *L'Obs* (20 December 2014), available at: http://tempsreel.nouvelobs.com/economie/20141219.OBS8339/start-up-ces-barbares-qui-veulent-debloquer-la-france.html.

5 At: http://barbares.thefamily.co.

6 Fay, 'Start-up: ces "barbares" qui veulent débloquer la France': 'It all started with a ranking. Published by the Institut Choiseul in February, it deeply annoyed the community of entrepreneurs and some geeks. The liberal think tank kept track of 100 French "economic leaders of tomorrow". In pole position, the "children" of Yannick Bolloré, Bris Rocher, Delphine and Antoine Arnault, Gabriel Naouri. Immediately behind are former advisers of cabinet ministers – mostly *enarques* [ENA graduates – the national school of administration] – parachuted in to head divisions of CAC40 companies. And finally, a host of bankers... No (or almost no) creators of startups, artists, researchers, innovators, software developers, specialists in artificial intelligence or biology, atypical profiles...

'Looking at this, the blood of Antoine Brachet, 36, began to boil. "The France of tomorrow has nothing to do with this", lamented this firebrand, whose day job is at Netvibes – a startup acquired by Dassault Systèmes – and spends his evenings at a think tank, Futurbulences, where with friends he looks ahead.

'Without thinking twice, he created a Facebook page called "Les 100 Barbares", taking as its logo a raised fist and inviting all his friends to take part in a survey: "Vote for your barbarian!" The objective: to identify leaders who are truly capable of changing France. For him, only those who master digital tools would have enough leverage to shake up conservatism and invent solutions that would get the country on the move and society along with it.

'The result – certainly highly unscientific – was stunning. The "barbarian style" ranking was just as elitist as the Institut Choiseul, but much more varied. There one finds entrepreneurs like Frédéric Mazzella, founder of BlaBlaCar, the carpooling site that scares SNCF, or Vincent Ricordeau, founder of the crowdfunding platform KissKissBankBank. Nicholas Colin, author of the Colin-Collin report on the taxation of the digital, figures prominently, along with his associates, Alice Zagury and Oussama Ammar...'

7 Alessandro Baricco, *The Barbarians: An Essay on the Mutation of Culture*, trans. Stephen Sartarelli (New York: Rizzoli Ex Libris, 2014). *Translator's note*: This was originally published in Italian in 2006, but the French translation, like the English, was not published until 2014.

8 See Naomi Klein, *The Shock Doctrine: The Rise of Disaster Capitalism* (New York: Metropolitan Books, 2007), and my commentary in Bernard Stiegler, *States of Shock: Stupidity and Knowledge in the Twenty-First Century*, trans. Daniel Ross (Cambridge: Polity, 2015).

9 Fay, 'Start-up: ces "barbares" qui veulent débloquer la France'.

10 Ibid.: 'Their common ground? The certainty that the world is on the brink of great change and an enthusiasm to forget all the "declinist" prophets in the Zemmour

mode. Why will these self-proclaimed "barbarians" be more successful than the classical elites at getting France moving again? "Because, twenty-five years after the digital revolution began, governments, large corporations and other institutions still have not adapted to it", says Henri Verdier, who with Nicolas Colin wrote *L'Age de la multitude*, published by Armand Colin, and inspired the metaphor of the "barbarians".'

11 Ibid.: 'Those who succeed in this new framework are the radical innovators, entrepreneurs who seize new scientific and technical opportunities to create something unprecedented. Because they come from the outside, they do not care about the usual conventions, since they speak another language.'

12 Bertrand Gille, *The History of Techniques, Volume 1: Techniques and Civilizations*, trans. P. Southgate and T. Williamson (New York: Gordon and Breach, 1986), pp. 54–5.

13 Ibid., pp. 43–4 and 59–66.

14 On incapacitation, see Bernard Stiegler, *Pharmacologie du Front national* (Paris: Flammarion, 2013), §75.

15 See Bernard Stiegler, *Automatic Society, Volume 1: The Future of Work*, trans. Daniel Ross (Cambridge: Polity, 2016), ch. 4.

16 And this is undoubtedly the motive behind the recent mobilization of several scientists around Stephen Hawking, against the new power and new ambitions of artificial intelligence in this context. *Translator's note*: See Stephen Hawking, Stuart Russell, Max Tegmark and Frank Wilczek, 'Transcendence looks at the implications of artificial intelligence – but are we taking AI seriously enough?', *The Independent* (1 May 2014), available at: http://www.independent.co.uk/news/science/ stephen-hawking-transcendence-looks-at-the-implications-of-artificial-intelligence- -but-are-we-taking-ai-seriously-enough-9313474.html. And see also 'Autonomous Weapons: An Open Letter from AI & Robotics Researchers' (28 July 2015), available at: https://futureoflife.org/open-letter-autonomous-weapons, with 3,724 signatories from artificial intelligence or robotics research, and 20,486 other signatories, including Stephen Hawking.

17 Anthony D. Barnosky et al., 'Approaching a State Shift in Earth's Biosphere', *Nature* 486 (7 June 2012), pp. 52–8.

18 Nietzsche was able to think nihilism only within the framework of a physics within which the concept of negative entropy was yet to be incorporated. This is why we must today think Nietzsche *beyond* Nietzsche.

19 That is, at all levels of the proletarianization that strikes the whole world, including and firstly Alan Greenspan. See Stiegler, *Automatic Society, Volume 1*, §1.

20 See Chris Anderson, 'The End of Theory: The Data Deluge Makes the Scientific Method Obsolete', *Wired* (23 June 2008), available at: http://archive.wired.com/ science/discoveries/magazine/16-07/pb_theory, on which I have commented in detail in Stiegler, *Automatic Society, Volume 1*, pp. 32ff.

21 The sharing of these protentions makes possible the difference between facts and law as *différance*: what Derrida named différance is the process of what, in deferring [*différant*] an action, a passage to the act or a deadline, engenders differences. Différance is as such a negentropic economy: it holds in reserve an energy whose dissipation it defers.

22 Fay, 'Start-up: ces "barbares" qui veulent débloquer la France'.

23 See Alain Bonneau, 'Systèmes d'Armes Létales Autonomes (SALA) et Non-Droit à la Vie', *Mediapart* (19 February 2015), available at: https://blogs.mediapart.fr/bonneau-alain/ blog/190215/systemes-darmes-letales-autonomes-sala-et-non-droit-la-vie.

24 See Frantz Marr, 'Sixth Extinction: We are Witnessing the Sixth Mass Extinction of Earth's Species', *The Cubic Lane* (20 June 2015), available at: http://cubiclane. com/2015/06/20/sixth-extinction-we-are-witnessing-the-sixth-mass-extinction-of- earths-species-93463.

25 Plato, *Meno* 81b–82b.

26 According to Sophie Fay, Nicolas Colin stated: 'Make no mistake, people like Mark Zuckerberg [Facebook] or Jeff Bezos [Amazon] think like conquerors, with a global strategy. Don't get left behind.' Following such reasoning, this would count as indisputable evidence that France or Europe must imitate the barbarism of the hegemonic disruptors and through that become capable of fighting this hegemony so as to impose another.

But such a programme could only exacerbate and feed into another barbarism: that of the radicalized reactions against the West, both outside the West and within it. Furthermore, either it is strictly ridiculous – since it is *impossible* to struggle against an adversary *by imitating* them – or it is a matter of struggling *alongside Zuckerberg and Bezos* against every form of public power and political responsibility.

27 In their imagination, Islam is a product of sophisticated marketing, which often comes after this or that toxic expedient, and which stems from a strange relation to the body – of which the epoch is also that of piercing, tattooing, and so on. The experience of taking drugs, too, is in many ways disruptive, which for William Burroughs constituted the essence of capitalism (see *The Naked Lunch*), often provoking such purely phantasmatic 'conversions', as *substitute products* in the suffering of a lack. This is perfectly clear in prisons, a question to which I will return. See p. 71.

28 This concept, coined by Marcel Mauss, has been the subject of works by Ars Industrialis that aim to go beyond the opposition between the international (become globalization) and the national. It was discussed in Stiegler, *States of Shock*, ch. 8, and I will return to it in Stiegler, *La Société automatique 2*.

29 Valéry spoke mainly of the West, including the United States, in its confrontation with other civilizations. Here, we speak of a division within the West, between 'old Europe' on the one hand, which is still the largest market in the world, and which lies at the origin of all these processes, and the United States, which, by suspending continental law, has constituted a new experience of facts – and in particular of technological facts, pressed in an unlimited way into the service of economic development. This state of fact has led to the constitution of a new, admirable Western culture, from which Europe still has much to learn. But a third moment must occur, to confront disruption as the final stage of the Anthropocene, where America, old Europe and the new great industrial civilizations must all *relearn how to admire what has no price*, not by returning to a long outdated past, which is in any case highly toxic, but by building a completely new relationship to the fact of exosomatization, and from a neganthropological perspective that must absolutely reconstitute a horizon of consistences for the collective protentions of a global internation.

30 *Translator's note*: On the translation of *panser*, see the translator's note in ch. 3, n. 32.

5 Outside the Law: Saint-Michel and the Dragon

1 André Leroi-Gourhan, *Gesture and Speech*, trans. Anna Bostock Berger (Cambridge, MA and London: MIT Press, 1993), p. 349.
2 I first tried to show this in Bernard Stiegler, *Symbolic Misery, Volume 1: The Hyper-Industrial Epoch*, trans. Barnaby Norman (Cambridge: Polity, 2014) and Stiegler, *Symbolic Misery, Volume 2: The Katastrophē of the Sensible*, trans. Barnaby Norman (Cambridge: Polity, 2015), then in Stiegler, *Automatic Society, Volume 1: The Future of Work*, trans. Daniel Ross (Cambridge: Polity, 2016).
3 See Stiegler, *Automatic Society, Volume 1*, esp. pp. 60 and 132.
4 Dust (see Genesis 3) being the biblical and archaic name for entropic becoming.
5 See Stiegler, *Automatic Society, Volume 1*, esp. pp. 52, 107 and 149.
6 See also Michel Bauwens, *Sauver le monde* (Paris: Les Liens qui libèrent, 2015).
7 Leroi-Gourhan, *Gesture and Speech*, p. 404, translation modified.
8 Ibid., p. 349.
9 'Qu'est-ce que s'orienter dans le passé?', seminar at the Collège international de philosophie, in the late 1980s.

10 Leroi-Gourhan, *Gesture and Speech*, p. 349. With respect to supraindividual organisms, to what Simondon called the interindividual and to what Gilles Deleuze has to say about related matters, see Bernard Stiegler, *States of Shock: Stupidity and Knowledge in the Twenty-First Century*, trans. Daniel Ross (Cambridge: Polity, 2015), pp. 49ff.

11 See Evgeny Morozov, *To Save Everything, Click Here: The Folly of Technological Solutionism* (New York: Public Affairs, 2014).

12 That is, the libertarians of micro-electronics, which will necessarily change scale as it shifts to the nanometric scale of 'transformational technologies': the discourse of the transhumanists is what paves the way for this change of scale.

13 In Bernard Stiegler, 'L'effondrement techno-logique du temps', *Traverses* 44–45 (1988), pp. 50–7.

14 Anthony D. Barnosky et al., 'Approaching a State Shift in Earth's Biosphere', *Nature* 486 (7 June 2012), pp. 52–8.

15 Roberto Esposito, *Communitas: The Origin and Destiny of Community*, trans. Timothy Campbell (Stanford: Stanford University Press, 2010).

16 See Leroi-Gourhan, *Gesture and Speech*, p. 404.

17 That is, the integration of nanotechnology, biotechnology, information technology and cognitive science.

18 See Allen Buchanan, *Better than Human: The Promise and Perils of Enhancing Ourselves* (New York: Oxford University Press, 2011).

19 Sigmund Freud, *Beyond the Pleasure Principle*, in Volume 18 of James Strachey (ed. and trans.), *The Standard Edition of the Complete Psychological Works of Sigmund Freud* (London: Hogarth, 1953–74).

20 Thinking (noesis) has been defined as a modality of desire ever since Plato's *Symposium*.

21 In the sense indicated by Leroi-Gourhan in *Gesture and Speech*.

22 Here I borrow a term from Vernant. See Jean-Pierre Vernant, *Myth and Thought Among the Greeks*, trans. Janet Lloyd with Jeff Fort (New York: Zone Books, 2006), p. 174, and see also Marcel Detienne and Vernant, *The Cuisine of Sacrifice Among the Greeks*, trans. Paula Wissig (Chicago and London: University of Chicago Press, 1989), p. 81.

23 This genesis of *Of Grammatology*, which was related to me by Derrida himself, is also partially indicated in the preface of the book.

24 The *rachat*, the 'buy back', is what redemption means.

25 This 'nucleus' is what Edmund Husserl, in *Ideas Pertaining to a Pure Phenomenology and to a Phenomenological Philosophy. First Book: General Introduction to a Pure Phenomenology*, trans. F. Kersten (The Hague: Martinus Nijhoff, 1983), will call the 'noema'.

26 Stéphane Mallarmé, *Selected Poetry and Prose*, Mary Ann Caws (New York: New Directions, 1982), p. 76, translation modified. We will see in §92 why and how this absent flower initiated a turn in what, in the silence of my prison noesis, I would begin to experience as language without an interlocution other than what I myself projected into books, during my reading and writing, which I began to perceive as the experience of significance, insignificance and their infinite *différance* – like 'the sea, the sea, ever recommencing'. See Paul Valéry, *Le Cimetière marin* (*The Graveyard by the Sea*), in Hugh P. McGrath and Michael Comenetz, *Valéry's Graveyard: Le Cimetière marin Translated, Described, and Peopled* (New York: Peter Lang, 2013), pp. 4–5, translation modified.

27 There are, therefore, three key moments and works along his course: Edmund Husserl, *Logical Investigations*, 2 vols, trans. J.N. Findlay (London and New York: Routledge & Kegan Paul, 1970), Husserl, *Ideas Pertaining to a Pure Phenomenology* and Husserl, *Cartesian Meditations: An Introduction to Phenomenology*, trans. Dorion Cairns (The Hague: Martinus Nijhoff, 1960), published in 1901, 1913 and 1931, respectively.

28 *Translator's note*: The author is here referring to the lines by Hölderlin in the poem 'In beautiful blue...', which read, 'Full of merit, yet poetically / Humans dwell upon this earth', and which are frequently cited by Heidegger. See, for example, Martin Heidegger, *Hölderlin's Hymns 'Germania' and 'The Rhine'*, trans. William McNeill and Julia Ireland (Bloomington and Indianapolis: Indiana University Press, 2014), pp. 34ff.; *Hölderlin's Hymn 'The Ister'*, trans. William McNeill and Julia Davis (Bloomington and Indianapolis: Indiana University Press, 1996), pp. 137ff.; *Elucidations of Hölderlin's Poetry*, trans. Keith Hoeller (New York: Humanity Books, 2000), p. 113; and, of course, in Heidegger, '...Poetically Man Dwells...', *Poetry, Language, Thought*, trans. Albert Hofstadter (New York: Harper & Row, 1971).

29 With Jacques Derrida, *Of Grammatology*, corrected edition, trans. Gayatri Chakravorty Spivak (Baltimore and London: Johns Hopkins University Press, 1998), but also with Derrida, *Speech and Phenomena*, trans. David B. Allison (Evanston: Northwestern University Press, 1973) and Derrida, *Edmund Husserl's Origin of Geometry: An Introduction*, trans. John P. Leavey Jr. (Lincoln and London: University of Nebraska Press, 1978).

30 On reading, see Maryanne Wolf, *Proust and the Squid: The Story and Science of the Reading Brain* (New York: Harper, 2007). And see Bernard Stiegler, 'Préface', in the French translation of the same work: Wolf, *Proust et le calamar* (Angoulême: Abeille et Castor, 2015).

31 See p. 247.

32 *Translator's note*: On the translation of *panser*, see the translator's note in ch. 3, n. 32.

33 The archangel Michael, slayer of the dragon, that is, the devil, gave his name to this prison house [*maison d'arrêt*] (now disused), that is, this *house of suspension*, and, therefore, of *epokhē*: what I am noting here did not occur to me when I was residing there – I was at that time completely unaware of what this 'archangel' represents, which is mentioned in the Bible, in the Gospels (Revelation) and in the Koran. 'In the Jewish, Christian and Muslim religions, archangels are a category of angels. They constitute one of the nine choirs of angels. In the hierarchy of angels, the *archangels* form the second level, just above the angels themselves (as indicated by the prefix arch-, which means higher). The word *archangel* comes from the Greek ἀρχάγγελος, *arkhangelos*, composed of ἀρχι-, *arkhē*, which means both "commandment" and "commencement" (it is in a way the "head") and ἄγγελος, *angelos*, "messenger".' Entry for 'Archange' in French Wikipédia (accessed 15 August 2015).

34 This can also occur with disease.

35 'Silence' in prison is in reality a constant hubbub – but, thanks to Paul Montès, the Protestant prison chaplain, I had managed to procure earplugs.

36 See §53.

37 See p. 223.

38 This was a real bookstore, run by a true bookseller who was passionate about reading and supported by a municipal library of which my brother later became the director. It was not long before it disappeared from the area, which is almost entirely composed of housing projects, killed off by stationery outlets that also sold newspapers, in turn killed off today by digital networks and hypermarkets.

39 *Translator's note*: In France, high school levels are named according to a descending scale, so the *classe de seconde* refers to the penultimate year of secondary education.

40 Tel Quel (coll.), *Tel Quel. Théorie de l'ensemble* (Paris: Seuil, 1968).

41 Julia Kristeva, *Revolution in Poetic Language*, trans. Margaret Waller (New York: Columbia University Press, 1984).

42 *Translator's note*: A student who has passed the baccalaureate.

43 *Translator's note*: Derrida's 1968 lecture 'La différance' was included in the Tel Quel collection *Théorie de l'ensemble* just mentioned. It was subsequently included in *Marges de la philosophie*, translated into English as Jacques Derrida, *Margins of Philosophy*, trans. Alan Bass (Chicago: University of Chicago Press, 1982).

44 Jacques Derrida, *Dissemination*, trans. Barbara Johnson (Chicago: University of Chicago Press, 1981).

45 Jacques Derrida, *Writing and Difference*, trans. Alan Bass (Chicago: University of Chicago Press, 1978).

46 Initially the incomplete translation of Rudolf Boehm and Alphonse de Waelhens, and then the translations of Emmanuel Martineau and François Vezin.

47 Martin Heidegger, *An Introduction to Metaphysics*, trans. Gregory Fried and Richard Polt (New Haven and London: Yale University Press, 2000).

48 Jean-Luc Marion, *Reduction and Givenness: Investigations of Husserl, Heidegger, and Phenomenology*, trans. Thomas A. Carlson (Evanston: Northwestern University Press, 1998).

49 In Maurice Blanchot, *The Infinite Conversation*, trans. Susan Hanson (Minneapolis and London: University of Minnesota Press, 1993), pp. 264–81.

50 Stiegler, *Symbolic Misery, Volume 2*, pp. 30–1.

51 Blanchot, *The Infinite Conversation*, pp. 265–6.

52 Maurice Blanchot, *The Step Not Beyond*, trans. Lycette Nelson (Albany: State University of New York Press, 1992), pp. 14–15 and 21. *Translator's note*: And see also Blanchot, *The Space of Literature*, trans. Ann Smock (Lincoln: University of Nebraska Press, 1982), p. 229, and Blanchot, *The One Who Was Standing Apart from Me*, trans. Lydia Davis (Barrytown: Station Hill Press, 1993), p. 24.

53 Which is not just the 'exigency of writing' (Blanchot, *The Infinite Conversation*, p. xii). This is why everything that we are questioning here relates to the creation of Ars Industrialis, then IRI, then pharmakon.fr.

54 This work was co-financed by the Ministry of Research and the Ministry of Culture, thanks to the interest of Jean-Pierre Dalbéra, who headed the research and technology mission.

55 Maurice Blanchot, 'The Beast of Lascaux', trans. Leslie Hill, *Oxford Literary Review* 22 (2000), p. 9, translation modified.

56 Rudolf Boehm, 'Pensée et technique. Notes préliminaires pour une question touchant la problématique heideggérienne', *Revue Internationale de Philosophie* 14 (1960), pp. 194–220.

57 Christophe Bonneuil and Jean-Baptiste Fressoz, *The Shock of the Anthropocene: The Earth, History and Us*, trans. David Fernbach (London: Verso, 2016).

58 This theme was introduced in Stiegler, *Automatic Society, Volume 1*.

59 It was a sportsground where the inmates were allowed to go on weekend afternoons, and where I had myself developed a taste for running and athletics. Once I got out, one of my problems would be to maintain these physical exercises of which I had discovered the 'moral' virtue – that is, with respect to the state of my morale. On *morality* and *demoralization*, see Part 3, 'Demoralization'.

60 I had noticed this during two brief periods when I had been allowed out prior to my release: incarceration leads to the loss of basic habits such as checking before crossing the street.

61 Alan Sillitoe, *Saturday Night and Sunday Morning* (London: W.H. Allen, 1958).

62 I understand this term, here, in Foucault's sense, when he analyses *parrhēsia* and the history of what he calls 'regimes of truth'.

63 That is: (1) if there remained enough heart, and therefore courage, to lead this struggle; and (2) if the traces produced by this struggle are not completely and forever erased, as was for Blanchot the very test and ordeal of his thought of the terrifyingly ancient.

64 Quasi-causality is neganthropological causality par excellence, that is, noetic and pharmacological causality, as therapeutic of the exosomatic, organological and therefore pharmacological condition.

65 This association was founded (see p. 24) following the colloquium 'La lutte pour l'organisation du sensible', which I had organised with CNRS while I was the head of IRCAM, and whose theses were at the origin of *Symbolic Misery*.

66 On this point, see Bernard Stiegler, *The Neganthropocene*, trans. Daniel Ross (London: Open Humanities Press, 2018), esp. chs 1–2.

67 *Translator's note*: On *dénégation* and *déni*, see the translator's note in ch. 2, n. 31.

68 Naomi Klein gives an eloquent example by quoting Sarah Palin, who declared that she loves the smell of fossil fuel emissions. Naomi Klein, *This Changes Everything: Capitalism vs. The Climate* (New York: Simon & Schuster, 2014), p. 1.

69 In particular Chapters 8 and 9.

70 See p. 16.

6 Who Am I? Hauntings, Spirits, Delusions

1 Louise Fessard, 'D., détenu pour trafic de stupéfiants et aspirant djihadiste', *Médiapart* (2 May 2015).

2 Ibid.

3 Ibid.

4 Ibid.

5 Entry for 'Malcolm X' in the French Wikipédia (no access date).

6 Malcolm X, letter to 'Brother Raymond', dated 15 February 1950. *Translator's note*: This particular letter went to auction on 7 April 2017, as can be seen here: https://www.the-saleroom.com/en-gb/auction-catalogues/alexanderhistorical/catalogue-id-alexan1-10004/lot-f1ad28b2-5278-4f9d-b550-a73d01455c06.

7 It was his brother Philbert who first introduced Malcolm Little to the Nation of Islam. See Malcolm X, *Autobiography* (New York: Ballantine, 1965), p. 169.

8 Entry for 'Malcolm X' in French Wikipédia. This article alone is sufficient to dispel the myth that, in the United States, racism was mainly limited to the South.

9 Philippe Carles and Jean-Louis Comolli, *Free Jazz/Black Power*, trans. Grégory Pierrot (Jackson: University Press of Mississippi, 2015). In a recent republication of their book, Carles and Comolli write (pp. 188–9):

> to shake history, the history of jazz, the history of the *Blues people*, the history of America struggling against itself, the weight of the ghosts of yesteryear, the weight of ghosts of slaves coming back at night to dance in our heads, the weight of bodies dancing with the ghosts' chains, free jazz shaking the chains of the black body who is in white history, invisible, off screen.
> When we first published this book
> it was 1971,
> Blacks were Panthers, [...]
> Malcolm had been assassinated and
> his prophet was not quite born,
> Julius Hemphill was forming his first band in St Louis
> and Joe McPhee had already recorded his *Nation Time*,
> with Clifford Thornton,
> Spike Lee [director of the film, *Malcolm X*] was in school and the iron curtain was drawn, [...]
> LeRoi Jones was already called Amiri Baraka,
> the Workshop de Lyon was still called Free Jazz Workshop,
> George Jackson had been assassinated in jail and Mumia Abu-Jamal was not yet
> on death row,
> here's for what hasn't changed, here's for what has changed.

10 Spike Lee's film *Malcolm X*, released in 1992, possesses most of the virtues and most of the flaws of good Hollywood movies. It opens with the amateur video that showed the 1991 beating of Rodney King by members of the LAPD, whose later acquittal sparked the LA riots. Derrida commented on the effects of this recording in Jacques

Derrida and Bernard Stiegler, *Echographies of Television: Filmed Interviews*, trans. Jennifer Bajorek (Cambridge: Polity, 2002), pp. 90ff.

11　Listen, for example, to *Let My Children Hear Music*, recorded in 1971 by Charles Mingus.

12　This is a question of the sense given to this word by Peter Sloterdijk, 'Rules for the Human Park: A Response to Heidegger's "Letter on Humanism"', *Not Saved: Essays After Heidegger*, trans. Ian Alexander Moore and Christopher Turner (Cambridge: Polity, 2017).

13　Mahalia Jackson singing 'In the Upper Room' and Sidney Bechet playing 'Move'.

14　Fessard, 'D., détenu pour trafic de stupéfiants et aspirant djihadiste'.

15　Contrary to the young Nietzsche's assertion, *Socrates is the last tragic*: it is Plato who breaks with this age, where the *pharmakon* is the mortal condition emerging from the conflict between the Titans and Olympians.

16　Martin Heidegger, *Being and Time*, trans. Joan Stambaugh, revised by Dennis J. Schmidt (Albany: State University of New York Press, 2010), §17.

17　Ibid., §18.

18　See Michel Foucault, *The Courage of Truth (The Government of Self and Others II): Lectures at the Collège de France 1983–1984*, trans. Graham Burchell (Houndsmills and New York: Palgrave Macmillan, 2011), pp. 113–14.

19　See p. 156.

20　*Translator's note*: The attack on 21 August 2015 was carried out by Ayoub El Khazzani on a Thalys train from Amsterdam to Paris. This attack became the subject of the film *The 15:17 to Paris*, directed by Clint Eastwood (2018).

21　Marie Peltier, 'Terrorisme, complotisme, drame des migrants: et si l'Europe perdait le sens?', *Libération* (2 September 2015), available at: http://www.liberation.fr/debats/2015/09/02/terrorisme-complotisme-drame-des-migrants-et-si-l-europe-perdait-le-sens_1374430.

22　*Translator's note*: Since the French publication of *Dans la disruption*, the author has decided to compose a fourth volume of *Technics and Time*, but this will not be the long-planned volume *Symboles et diaboles*, which will now become the fifth volume (and the others six and seven, respectively).

23　On this *rétention tertiaire numéraire*, see Clarisse Herrenschmidt, *Les Trois Écritures: langue, nombre, code* (Paris: Gallimard, 2007).

24　*Translator's note*: The Mouvement des entreprises de France is a large business and employer association in France, and Laurence Parisot was its president between 2005 and 2013.

25　See Hélène Croizé-Pourcelet, *Slate.fr* (10 April 2012), available at: http://www.slate.fr/story/52591/photos-campagne-1988-generation-mitterrand-seguela: 'It is not a matter of political communication, as would be "France united", at one time the campaign slogan for the Mitterrand candidacy. We have here a "sociological, even psychological" form of publicity, according to the words of Séguéla. "Séguéla is a true publicist, not a politician", explains Antoine Boulay. Moreover, he will later run the campaign of Jacques Chirac and is a friend of Nicolas Sarkozy. "From a political perspective, this campaign poster is not interesting", he continues. According to him, the slogan doesn't mean much: "'Generation Mitterrand' is a slogan without content. I defy you to find behind it any reform proposal or even a social category or targeted age-group." The subtlety of this slogan lies, he says, in the fact that everyone can draw any meaning from it they want, and project into it whatever they want. Very much an "advertising" technique: to use a "new" word but one already inside people's heads.'

26　On this formulation, see Chapter 11.

27　*Translator's note*: In May 2011 Dominique Strauss-Kahn, at that time a likely Socialist candidate for president in the upcoming elections, was photographed getting into a Porsche belonging to one of his advisers. See for example Angelique Chrisafis, 'Porschegate leaves Sarkozy rival with headache of champagne socialism', *Guardian*

(9 May 2011), available at: https://www.theguardian.com/world/2011/may/09/porsche-dominique-strauss-kahn-france. This article, dated 9 May 2011, notes that Strauss-Kahn was at that time far ahead of the incumbent Nicolas Sarkozy in the polls (but also notes that Strauss-Kahn's reputation as a 'seducer' was a potential problem for his candidacy), but, as Stiegler points out, it would be a mere five days later, on 14 March, that Strauss-Kahn's ultimate downfall would occur, with the accusations of sexual assault at the Sofitel Hotel in New York made by Nafissatou Diallo.

28 'Capables et incapables', on the page headed 'Présidentielle J-57: la campagne vue par Bernard Stiegler', *Télérama.fr* (23 February 2012), available at: http://www.telerama.fr/idees/presidentielle-j-57-la-campagne-vue-par-bernard-stiegler,78318.php.

29 These themes are explored in depth in Bernard Stiegler, *Pharmacologie du Front national* (Paris: Flammarion, 2013).

30 Manuel Valls, quoted in 'Départementales: Valls dénonce un "endormissement généralisé" face au "danger" du FN', *Le Monde* (5 March 2015), available at: http://www.lemonde.fr/politique/article/2015/03/05/departementales-valls-denonce-un-endormissement-generalise-face-au-danger-du-fn_4588426_823448.html.

31 Christophe Cornevin, 'Attentat en Isère: Yassin Salhi voulait "frapper les esprits"', *Le Figaro* (29 June 2015), available at: http://www.lefigaro.fr/actualite-france/2015/06/29/01016-20150629ARTFIG00166-yassin-salhi-le-point-sur-les-debuts-de-l-enquete.php.

32 Pascal Ceaux and Eric Pelletier, 'Salhi, Nemmouche, Ghlam... Ces nouveaux fous d'Allah', *L'express* (1 July 2015), available at: http://www.lexpress.fr/actualite/societe/salhi-nemmouche-ghlam-ces-nouveaux-fous-d-allah_1694723.html.

33 Ibid. About the Dijon crime [*passage à l'acte*], *Le Monde* writes: 'The attack "is absolutely not a terrorist act", according to the Dijon prosecutor, who added, as *Le Monde* reported on the morning of Monday, 22 December, that the perpetrator, a man of forty, suffered from an "old and significant psychiatric illness". He made "157 trips into a psychiatric unit between February 2001 and November 2014", according to the magistrate. He has been known to police for committing ordinary crimes dating back to the 1990s.' 'Un déséquilibré renverse treize piétons à Dijon', *Le Monde* (22 December 2014), available at: http://www.lemonde.fr/societe/article/2014/12/21/un-desequilibre-renverse-une-dizaine-de-pietons-adijon_4544483_3224.html#XO4ugp 3djZO3jzdH.99.

34 *Translator's note*: A '*chauffard*' is ordinarily used to characterize someone as a reckless driver, or as a hit-and-run driver, but the term has been here left untranslated, given that these English equivalents do not really cover the deliberate nature of the act in this case.

35 Which is a jar – that of Pandora.

36 All this belongs to what I have referred to as the Antigone complex and negative sublimation. See Bernard Stiegler, *Uncontrollable Societies of Disaffected Individuals: Disbelief and Discredit, Volume 2*, trans. Daniel Ross (Cambridge: Polity, 2013), ch. 2.

37 Peter Sloterdijk, *In the World Interior of Capital*, trans. Wieland Hoban (Cambridge: Polity, 2013).

38 Hannah Arendt, 'The Crisis in Education', trans. Denver Lindley, and 'The Crisis in Culture: Its Social and Its Political Significance', *Between Past and Future* (New York: Viking, 1961).

39 That is, the condition of possibility of all forms of protention.

40 See Aeschylus, *Prometheus Bound*, 252.

41 On life in contemporary Japan, on the difficulty of caring for what is played out in Japan's relation to the admirable tradition of its ancient culture, and on the technological power that it has generated, including as administration, bureaucracy and forms of power, it is fascinating to watch Akira Kurosawa's *Ikiru* (1952) and *I Live in Fear* (1955), to read Tatsuo Hori's *The Wind Rises* (1937) and to watch Hayao Miyazaki's *The Wind Rises* (2013).

42 Sigmund Freud, *Moses and Monotheism: Three Essays*, in Volume 23 of James Strachey (ed. and trans.), *The Standard Edition of the Complete Psychological Works of Sigmund Freud* (London: Hogarth, 1953–74).

43 Aristotle, *Problems*, Book 30, 'Problems Connected with Practical Wisdom, Intelligence, and Wisdom'.

7 Dreams and Nightmares in the Anthropocene

1 Jean de La Fontaine, 'The Milkmaid and the Pot of Milk', trans. Eli Siegel, in Eli Siegel, *Hail, American Development* (New York: Definition Press, 1968), pp. 43–4.

2 Ibid., p. 44.

3 Simone de Beauvoir, 'Existentialism and Popular Wisdom', trans. Marybeth Timmermann, *Philosophical Writings* (Urbana and Chicago: University of Illinois Press, 2004).

4 See pp. 113 and 133–8.

5 Bernard Stiegler, *What Makes Life Worth Living: On Pharmacology*, trans. Daniel Ross (Cambridge: Polity, 2013), p. 2. I hope that the reader will forgive this self-quotation on the grounds it saves having to explain again what was already covered in that earlier work.

6 This is also what is at stake in Hans Vaihinger, as Raphaël Ehrsam shows, commenting on a book by Christophe Bouriau, *Le 'Comme si'. Kant, Vaihinger et le fictionalisme* (Paris: Cerf, 2013). Ehrsam states: 'In the theoretical order, Kant utilizes the expression "as if" in order to characterize the use of certain ideas: the idea of a totality of real objects (the idea of "world"), the idea of an intelligent creator of nature, or the idea of finality in living things. The decisive point is that, even though these ideas have no object that corresponds to them in experience, we can treat them "as if" they have objective reality in order to stimulate certain of our enterprises of knowledge (in physics, biology, etc.). "Every rational being must act as if it were through its maxims always a legislative member in a universal kingdom of ends".' Raphaël Ehrsam, 'La philosophie du comme si', *La vie des idées* (15 January 2014), available at: http://www.laviedesidees.fr/La-philosophie-du-comme-si.html. The quotation from Kant is from Immanuel Kant, *Groundwork for the Metaphysics of Morals*, trans. Allen W. Wood (New Haven and London: Yale University Press, 2002), p. 56.

7 Alexandre Kojève, *Kant* (Paris: Gallimard, 1973), p. 96.

8 Michel Foucault, *History of Madness*, trans. Jonathan Murphy and Jean Khalfa (London and New York: Routledge, 2006).

9 *Translator's note*: The French equivalent of the English phrase 'what makes life worth living' is, as in the title of Stiegler's so-named book, *ce qui fait que la vie vaut la peine d'être vécue*: 'what makes life worth the *pain or effort* of being lived'.

10 Voltaire, 'In Praise of Reason', trans. Adi S. Bharat, *Pusteblume* 8 (2017), available at: http://www.bu.edu/pusteblume/8/issue-1-bharat-translating-voltaire.htm.

11 Eustache Deschamps, quoted in Foucault, *History of Madness*, p. 15.

12 Paul Valéry, *Cahiers Paul Valéry 3. Questions du rêve* (Paris: Gallimard, 1979), p. 42. The italics in the third paragraph are mine.

13 Marc Azéma, *La Préhistoire du cinéma: Origines paleolithiques de la narration graphique et du cinématographe...* (Paris: Errance, 2011), p. 21.

14 Federico Fellini, *The Book of Dreams*, trans. Aaron Maines and David Stanton (New York: Rizzoli, 2008).

15 Gilbert Simondon, *On the Mode of Existence of Technical Objects*, trans. Cecile Malaspina and John Rogove (Minneapolis: Univocal, 2017).

16 See p. 157.

17 'O my soul, cease aspiring to immortal life, / But draw energetically from the sources of *mēkhanē*.' Pindar, *Pythian* 3: 61–3. The original Greek word is μαχανάν, which is an ancient form of μηχανή. *Translator's note*: These lines from Pindar's third Pythian

ode were used by Albert Camus as an epigraph for his work, 'The Myth of Sisyphus', where the English translation renders them as 'O my soul, do not aspire to immortal life, but exhaust the limits of the possible'. See Albert Camus, *The Myth of Sisyphus, and Other Essays*, trans. Justin O'Brien (New York: Knopf, 1955), p. 2. Prior to that, however, and as Stiegler mentions, Paul Valéry had already used the same lines as the epigraph for *Le Cimetière marin*, in which, in the English translation (*The Graveyard by the Sea*), they are translated as follows: 'Do not, dear soul, strive after deathless life, / but use to the utmost the resources in your power.' See Hugh P. McGrath and Michael Comenetz, *Valéry's Graveyard: Le Cimetière marin Translated, Described, and Peopled* (New York: Peter Lang, 2013), p. 5. Lastly, the lines were translated by Conway as: 'Pray for no life immortal, soul of mine, / But draw in full depth on the skills of which / You can be master.' See Pindar, *The Odes and Selected Fragments*, trans. G.S. Conway and Richard Stoneman (London: Everyman, 1997), p. 116.

18 In August 2015, Ludovic Duhem gave an interesting lecture on Icarus during the '(Rencontres) Inattendues de la musique et la philosophie' in Tournai, which enriched my reflection on this figure. I offer him my thanks.

19 *Translator's note*: Clément Ader (1841–1925) was a French engineer and aviation pioneer who constructed several '*avions*', which he attempted to fly prior to the successful efforts of the Wright brothers.

20 Simondon uses this expression in Gilbert Simondon, *L'individuation à la lumière des notions de forme et d'information* (Grenoble: Jérôme Millon, 2013).

21 Paul Valéry, 'La crise de l'esprit', *Variété I et II* (Paris: Gallimard, 1998), p. 32. *Translator's note*: The author refers here not to Valéry's actual essay, 'La crise de l'esprit', but to a 'Note' that was included in the French publication but not the English translation, and that was itself the text of a lecture given by Valéry at the University of Zurich on 15 November 1922. The specific passage to which Stiegler refers is the following (pp. 32–3):

> What, then, is this spirit? How can it be touched, struck, diminished, humiliated by the current state of the world? From whence comes this great pity for the things of the Spirit, this distress, this anguish of men of the Spirit? This is what we must now speak about.
>
> Man is that separate animal, that bizarre living being who opposes himself to all others, who raises himself above all others, by his … *dreams* – by the intensity, the concatenation, the diversity of his *dreams!* by their extraordinary effects, which can ever modify his nature, and not just his nature, but the very nature that surrounds him, which he tries indefatigably to make submit to his dreams.
>
> I mean that man is constantly and necessarily opposed to *that which is* by his concern for *that which is not!* and that he laboriously, or through genius, gives birth to what he must in order to give his dreams the very power and precision of reality, and, on the other hand, in order to impose upon this reality increasing changes that bring it closer to his dreams.

22 *Translator's note*: The phrase '*faire ses besoins*', which translates literally as 'to do (or make) his needs', is equivalent to the English phrase (used mostly for pets), 'to do his business', meaning to defecate.

23 And this is what allows Freud to see, in the child proud of 'doing his business', an early moment in the constitution of the libidinal economy.

24 *Translator's note*: See Bertrand Gille, *The History of Techniques, Volume 1: Techniques and Civilizations*, trans. P. Southgate and T. Williamson (New York: Gordon and Breach, 1986), pp. 43–4. And see Bernard Stiegler, *Technics and Time, 1: The Fault of Epimetheus* (Stanford: Stanford University Press, 1998), pp. 30–44, esp. 36–7: 'Two phases in the process of invention must be distinguished – the phase of adjustment and that of development – and a difference must be introduced (taken from François Perroux) between invention and innovation. Innovation accomplishes

a transformation of the technical system while drawing the consequences for the other systems. In other words, the rules of innovation are wholly different from those of invention. The rules of innovation are those of socialization, as analyzed mainly by economists [...]. Innovation destabilizes established situations: it thereby creates resistance. The socialization in which innovation consists is work on the milieus it crosses through (social, economic, political, etc.). [...] One could say that the logic of innovation is constituted by the rules of adjustment between the technical system and the others. There is for each age a typology of the conditions of innovation that are possibilities of adequacy between the technical system and the other systems.'

25 Gilbert Simondon, *Imagination et invention* (Chatou: Éditions de la Transparence, 2008).

26 Edmund Husserl, 'Foundational Investigations of the Phenomenological Origin of the Spatiality of Nature: The Originary Ark, the Earth, Does Not Move', trans. Fred Kersten, revised by Leonard Lawlor, in Maurice Merleau-Ponty, *Husserl at the Limits of Phenomenology, Including Texts by Edmund Husserl*, edited by Leonard Lawlor and Bettina Bergo (Evanston: Northwestern University Press, 2002).

27 See John Clark and Camille Martin (eds), *Anarchy, Geography, Modernity: Selected Writings of Elisée Reclus* (Oakland: PM Press, 2013).

28 I continue here the analysis undertaken in Bernard Stiegler, *States of Shock: Stupidity and Knowledge in the Twenty-First Century*, trans. Daniel Ross (Cambridge: Polity, 2015).

29 See, for example, the economic assessment available on the Boursorama website (available at: http://www.boursorama.com/forum-adocia-les-defis-risques-du-marche-du-diabete-439203195-1).

30 Georges Canguilhem, *The Normal and the Pathological*, trans. Carolyn R. Fawcett and Robert S. Cohen (New York: Zone Books, 1991), p. 200.

31 In Canguilhem's sense, that is, in the sense of the living being who constantly invents and transforms normality – a transformation of the 'normal' of which exosomatic organogenesis immeasurably accelerates the pace compared with simple organic development.

32 See Stiegler, *What Makes Life Worth Living*, ch. 2.

33 See Maryanne Wolf, *Proust and the Squid: The Story and Science of the Reading Brain* (New York: Harper, 2007). And see Bernard Stiegler, 'Préface', and the concluding dialogue between Maryanne Wolf and myself, included in the French translation of the same work: Wolf, *Proust et le calamar* (Angoulême: Abeille et Castor, 2015).

34 This is how Kant viewed the propositions of Herder.

35 André Leroi-Gourhan, *Gesture and Speech*, trans. Anna Bostock Berger (Cambridge, MA and London: MIT Press, 1993), p. 129.

36 A destruction envisaged by Freud in *Beyond the Pleasure Principle*, in Volume 18 of James Strachey (ed. and trans.), *The Standard Edition of the Complete Psychological Works of Sigmund Freud* (London: Hogarth, 1953–74), pp. 48–9.

37 Jean-Baptiste Fressoz, *L'Apocalypse joyeuse. Une histoire du risque technologique* (Paris: Seuil, 2012).

38 Christophe Bonneuil and Jean-Baptiste Fressoz, *The Shock of the Anthropocene: The Earth, History and Us*, trans. David Fernbach (London and New York: Verso, 2016), p. 289, translation modified.

39 See the journal *Études digitales*, including Bernard Stiegler, 'Dans la disruption. La main, ses doigts, ce qu'ils fabriquent et au-delà', *Études digitales* 1 (2016), pp. 215–30, and the digital-studies.org website.

40 Let us say, for now, that the discourse of Bonneuil and Fressoz ignores the question of transindividuation, does not see the question of the doubly epokhal redoubling and makes it impossible to think disruption. It argues that the processes of transindividuation that have accompanied the evolution of the Anthropocene have always unfolded at the expense of taking account of risk, which is no doubt very true,

and they provide precisely documented facts to support it. Still, there was a public argument and debate that is now virtually gone. And if, according to Bonneuil and Fressoz, these arguments have always led to arbitrations headed in the wrong direction, this is also because the arguments opposed to risk-taking were not correct – and this is therefore a theoretical, epistemological and philosophical question, as well as a scientific one: it is not just a historical or political question.

41 See René Passet, *L'Économique et le vivant* (Paris: Payot, 1979).
42 See p. 87.
43 Joseph Vogl, *The Specter of Capital*, trans. Joachim Redner and Robert Savage (Stanford: Stanford University Press, 2015).
44 On the improbable and the calculus of probabilities, see Bernard Stiegler, *Automatic Society, Volume 1: The Future of Work*, trans. Daniel Ross (Cambridge: Polity, 2016). And on 'big data', which exploits applied mathematical probabilities, as the production of spurious correlations, see Cristian S. Calude and Giuseppe Longo, 'The Deluge of Spurious Correlations', *Foundations of Science* 22 (2017), pp. 595–612.
45 I have also developed this point with respect to posthumanism – see Stiegler, *What Makes Life Worth Living*, p. 104.
46 On the question of adoption, see Bernard Stiegler, *Technics and Time, 3: Cinematic Time and the Question of Malaise*, trans. Stephen Barker (Stanford: Stanford University Press, 2011), and the entry on 'adaptation/adoption' in Victor Petit, 'Vocabulaire d'Ars Industrialis', in Bernard Stiegler, *Pharmacologie du Front national* (Paris: Flammarion, 2013), pp. 371–3.
47 Bernard Stiegler, *Symbolic Misery, Volume 1: The Hyper-Industrial Epoch*, trans. Barnaby Norman (Cambridge: Polity, 2014), ch. 3.
48 See p. 46.
49 On this partial realization, see Stiegler, *Automatic Society, Volume 1*, esp. pp. 229–31.
50 *Translator's note*: Manning's sentence was commuted by President Obama on 17 January 2017, and Manning was released on 17 May.
51 Seneca, 'On the Tranquillity of the Mind', *Dialogues and Essays*, trans. John Davie (Oxford and New York: Oxford University Press, 2007), ch. 17, p. 139, translation modified.
52 Robert Esposito, *Communitas: The Origin and Destiny of Community*, trans. Timothy Campbell (Stanford: Stanford University Press, 2010).
53 Desiderius Erasmus, *The Praise of Folly*, trans. Hoyt Hopewell Hudson (Princeton and Oxford: Princeton University Press, 2015).
54 Michel de Montaigne, 'An Apology for Raymond Sebond', *The Complete Essays*, trans. M.A. Screech (London: Penguin, 2003), p. 548.
55 See Jonathan Crary, *24/7: Late Capitalism and the Ends of Sleep* (London and New York: Verso, 2013), and my commentary in Stiegler, *Automatic Society, Volume 1*, esp. ch. 3.
56 Nevertheless, these are the stakes of *Identity and Difference, Being and Time* and 'The Turning', as I will endeavour to show in Bernard Stiegler, *La Société automatique 2. L'avenir du savoir*, forthcoming.
57 Michel Foucault, *The Courage of Truth (The Government of Self and Others II): Lectures at the Collège de France, 1983–1984*, trans. Graham Burchell (Houndmills: Palgrave Macmillan, 2011), pp. 11–12.
58 Michel Foucault, *On the Government of the Living: Lectures at the Collège de France, 1979–1980*, trans. Graham Burchell (Houndmills: Palgrave Macmillan, 2014).
59 And as a corresponding transformation of the cerebral organs and psychic apparatus. This is the whole meaning of Meyerson's social psychology, which also amounts to a translation of Freud's *Traumdeutung*, and it is in this vein that the works of Vernant, Detienne and Loraux are inscribed.
60 Chris Anderson, 'The End of Theory: The Data Deluge Makes the Scientific Method

Obsolete', *Wired* (23 June 2008), available at: http://archive.wired.com/science/discoveries/magazine/16-07/pb_theory.

61 René Descartes, 'Rule Fifteen', in *Rules for the Direction of the Mind*, in *The Philosophical Writings of Descartes, Volume 1*, trans. John Cottingham, Robert Stoothoff and Dugald Murdoch (Cambridge: Cambridge University Press, 1985).

62 Which is also and necessarily a 'madness' of history, at least in the sense of a blindness. This question lies on the horizon of Derrida's objection to the very possibility of the history of madness undertaken by Foucault.

63 Whatever the justifications and necessities of the objections made by Derrida, I believe that the latter ultimately evinces a little bit of bad faith by refusing to hear the question that Foucault raises, even if the basis on which the latter does so is not completely convincing.

64 See p. 89.

65 For reasons I have already discussed in Bernard Stiegler, *Taking Care of Youth and the Generations*, trans. Stephen Barker (Stanford: Stanford University Press, 2010), ch. 8.

66 *Translator's note*: On *dénégation* and *déni*, see the translator's note in ch. 2, n. 31.

8 Morality and Disinhibition in Modern Times

1 René Descartes, 'Rule Fifteen', in *Rules for the Direction of the Mind*, in *The Philosophical Writings of Descartes, Volume 1*, trans. John Cottingham, Robert Stoothoff and Dugald Murdoch (Cambridge: Cambridge University Press, 1985), p. 65, translation modified.

2 Ibid., translation modified.

3 Descartes, 'Rule Sixteen', in ibid., p. 66.

4 Ibid., p. 67.

5 It is odd that Derrida, to my knowledge, nowhere refers to these *Rules*, and especially odd that he does not do so in his introduction to 'The Origin of Geometry'. Husserl clearly had Rules Fifteen and Sixteen in mind when he gave this lecture, as he did when he analysed the *crisis of the European sciences*.

6 Descartes, 'Rule Sixteen', in *Rules for the Direction of the Mind*, p. 67.

7 Ibid.

8 See Yvon Belaval, *Leibniz, critique de Descartes* (Paris: Gallimard, 1960).

9 Descartes, 'Rule Sixteen', in *Rules for the Direction of the Mind*, p. 67.

10 *Translator's note*: In 1890, for the first time, the United States census utilized tabulating machines, commissioned by the American Census Bureau after the 1880 census. See Alain Desrosières, *The Politics of Large Numbers: A History of Statistical Reasoning*, trans. Camille Naish (Cambridge, MA and London: Harvard University Press, 1998), p. 197. By 1896, these machines were being used for the French census (see ibid., pp. 253–54).

11 In respect to which Erasmus and *The Praise of Folly* will play a major role. See Stefan Zweig, *Erasmus of Rotterdam*, trans. Eden and Cedar Paul (New York: Viking, 1934).

12 See Clarisse Herrenschmidt, *Les Trois Écritures: langue, nombre, code* (Paris: Gallimard, 2007), pp. 226ff.

13 Peter Sloterdijk, *In the World Interior of Capital*, trans. Wieland Hoban (Cambridge: Polity, 2013), p. 54.

14 Peter Sloterdijk, 'Rules for the Human Park: A Response to Heidegger's "Letter on Humanism"', *Not Saved: Essays After Heidegger*, trans. Ian Alexander Moore and Christopher Turner (Cambridge: Polity, 2017).

15 Sloterdijk, *In the World Interior of Capital*, p. 53.

16 See p. 119.

17 See pp. 291–2.

18 On madness and the Reformation, see Michel Foucault, *History of Madness*, trans.

Jonathan Murphy and Jean Khalfa (London and New York: Routledge, 2006), pp. 68–70.

19 I have evoked these transformations in various works, including Bernard Stiegler, *For a New Critique of Political Economy*, trans. Daniel Ross (Cambridge: Polity, 2010), pp. 96–129.

20 Blaise Pascal, *Pensées*, A.J. Krailsheimer (Harmondsworth, Middlesex: Penguin, 1966), p. 148.

21 Foucault removed this preface in the second edition in 1972, but it can still be read in *Dits et écrits 1. Translator's note*: This preface *is* included in the complete English translation of *History of Madness*, where the pensée by Pascal just cited is rendered as: 'Men are so necessarily mad, that not being mad would be being mad through another trick that madness played' (p. xxvii). As the translators of that work, Jonathan Murphy and Jean Khalfa, note (p. 593), Foucault makes great play with the idea of the *tour de folie*, the trick or the turn of madness.

22 Fyodor Dostoyevsky, quoted in Foucault, *History of Madness*, p. xxvii.

23 See p. 120.

24 Foucault, *History of Madness*, p. xxvii.

25 On this point, see 'Aux bords de la folie', a special section in *Esprit* 413 (2015).

26 This is the meaning of Sloterdijk's discussion of 'consultancy': it is that which has the goal of organizing this denial, in close association with marketing and the media.

27 Jacques Derrida, 'Cogito and the History of Madness', *Writing and Difference*, trans. Alan Bass (Chicago: University of Chicago Press, 1978), originally a lecture given in March 1963, then published in 1964 in *Revue de métaphysique et de morale*.

28 Foucault, *History of Madness*, p. xxix.

29 Ibid.

30 Ibid., pp. 12–14.

31 Ibid., p. 14, translation modified.

32 See pp. 89–90.

33 Foucault, *History of Madness*, p. 15, translation modified.

34 The link between Descartes and his epoch is investigated by Gabriel Rockhill in 'Le Droit de la philosophie et les faits de l'histoire. Foucault, Derrida, Descartes', *Le Portique* 5 (2007), available at: https://leportique.revues.org/1473. Through an examination of Foucault's work, Rockhill investigates the relationships between philosophy and the social sciences.

35 Foucault, *History of Madness*, p. 48, translation modified.

36 Ibid., p. 78.

37 This question, which I will not deal with here, involves the reflection I have tried to initiate with respect to the conception of desire in French thought in the second half of the twentieth century, and in particular in Deleuze and Guattari. See especially what I have to say in *Pharmacologie du Front national* (Paris: Flammarion, 2013) concerning Gilles Deleuze and Félix Guattari, *Anti-Oedipus: Capitalism and Schizophrenia*, trans. Robert Hurley, Mark Seem and Helen R. Lane (Minneapolis: University of Minnesota Press, 1983), and the status it accords to Wilhelm Reich, *The Mass Psychology of Fascism*, trans. Vincent R. Carfagno (New York: Farrar, Straus and Giroux, 1970).

38 Foucault, *History of Madness*, p. xxx.

39 Gabriel Rockhill has brilliantly highlighted how this text, and the challenge to it by Derrida, are inscribed within a 'structuralist' context. See Rockhill, 'Le Droit de la philosophie et les faits de l'histoire'.

40 Foucault, *History of Madness*, p. 79, translation modified.

41 Ibid., pp. 59–60, translation modified.

42 Ibid., p. 163, translation modified.

43 This will culminate in Voltaire's 'In Praise of Reason', his reply to Erasmus.

44 Foucault, *History of Madness*, p. 139.

45 Ibid., p. 47, translation modified.

46 Ibid., p. 52, translation modified.
47 Ibid., p. 60, translation modified.
48 Ibid., p. 77.
49 Ibid., p. 517: '*Le Neveu de Rameau* and the whole literary fashion that followed it indicated a reappearance of madness in the domain of language, a language where madness was permitted to speak in the first person, uttering in the midst of the empty verbiage and the insane grammar of its paradoxes something that bore an essential relation to the truth.'
50 Ibid., p. 140.
51 In a sense that must not be reduced to that of psychiatry and its nosology, but precisely in its meaning of ὕβρις, of which clinical madness is a case. See here the replies of Jacques Hochmann in the interview, 'Les contestations de la psychiatrie', *Esprit* 413 (2015), pp. 19–27, part of the section 'Aux bords de la folie', to which we referred previously.
52 I have previously highlighted this in Bernard Stiegler, *Taking Care of Youth and the Generations*, trans. Stephen Barker (Stanford: Stanford University Press, 2010), p. 128.
53 Sloterdijk, *In the World Interior of Capital*, p. 169, my italics. Sloterdijk continues, noting that Dostoyevsky 'instinctively grasped the immeasurable symbolic and programmatic dimensions of the hubristic construction. As the exhibition building bore no name of its own, it seems likely that Dostoyevsky transferred the name "Crystal Palace" to it.'
54 Ibid., p. 54, my italics. *Translator's note*: The phrase translated into English as 'willingness to embrace delusion' is translated in the French edition as '*propension à la folie*'.
55 See pp. 7 and 78, and Bernard Stiegler, *Automatic Society, Volume 1: The Future of Work*, trans. Daniel Ross (Cambridge: Polity, 2016).
56 Sloterdijk, *In the World Interior of Capital*, p. 57.
57 Ibid.
58 This is the case despite the fact that Foucault, in *Discipline and Punish*, initially considers them to be a technology of power. See Laurent Jaffro, 'Foucault et le problème de l'éducation morale', *Le Télémaque* 29 (2006), pp. 111–24. On these questions, which concern *moral philosophy*, see also Stiegler, *Taking Care of Youth and the Generations*.
59 Sloterdijk, *In the World Interior of Capital*, p. 59.
60 Ibid., pp. 61–2.
61 Through those cycles of which Freud provided elements of understanding in *Civilization and Its Discontents*.
62 I have discussed this Hegelian phenomenology of spirit in *States of Shock: Stupidity and Knowledge in the Twenty-First Century*, trans. Daniel Ross (Cambridge: Polity, 2015).
63 Michel Foucault, *The Birth of the Clinic: An Archaeology of Medical Perception*, trans. A.M. Sheridan Smith (New York: Vintage, 1973).
64 See pp. 130–1.
65 Sloterdijk, *In the World Interior of Capital*, p. 72.
66 Bertrand Gille frequently cites Schumpeter.
67 Sloterdijk, *In the World Interior of Capital*, p. 74.
68 Ibid.
69 Ibid.
70 Ibid., p. 75.
71 Ibid.
72 Ulrich Beck, *Risk Society: Towards a New Modernity*, trans. Mark Ritter (London: Sage, 1992).
73 Anthony Giddens, *The Consequences of Modernity* (Cambridge: Polity, 1990).
74 Sloterdijk, *In the World Interior of Capital*, pp. 84–5.

75 Ibid., p. 85.
76 Desrosières, *The Politics of Large Numbers*.
77 See Thomas Berns and Antoinette Rouvroy, 'Gouvernementalité algorithmique et perspectives d'émancipation', *Réseaux* 177 (2013), pp. 163–96.
78 Sloterdijk, *In the World Interior of Capital*, p. 85.
79 Ibid.
80 Ibid.
81 Ibid., p. 86.
82 Jean-Baptiste Fressoz, *L'Apocalypse joyeuse. Une histoire du risque technologique* (Paris: Seuil, 2012), p. 9, my italics.
83 Ibid.
84 Eugène Huzar, quoted in ibid., p. 10.
85 Fressoz, *L'Apocalypse joyeuse*, p. 15.
86 Ibid., p. 16.
87 Ibid., my italics for '*passage à l'acte*' and '*fait accompli*'.
88 Ibid., p. 17.
89 *Translator's note*: In *Technics and Time, 1: The Fault of Epimetheus* (Stanford: Stanford University Press, 1998), Stiegler writes on p. 30: 'Lucien Febvre called attention to the necessity and the lack of an actual history of techniques within general history, to the necessity of a concept founding its method: the history of techniques is "one of those numerous disciplines that are entirely, or almost entirely, to be created".'
90 Fressoz, *L'Apocalypse joyeuse*, p. 19.
91 With Michel Foucault and his introduction (published in English as 'Dream, Imagination, and Existence', trans. Forrest Williams) to Ludwig Binswanger, 'Dream and Existence', trans. Jacob Needleman, in *Dream and Existence*, reprint of a special issue of *Review of Existential Psychology & Psychiatry* 19 (1) (1986).
92 'He is the very type of the universal artist, by turns an architect, sculptor, inventor of mechanical means.' Pierre Grimal, *The Dictionary of Classical Mythology*, trans. A.R. Maxwell-Hyslop (Oxford: Basil Blackwell, 1996), p. 124, translation modified.
93 Jean Jaurès, *L'Humanité* (18 April 1904). *Translator's note*: This statement was made by Jaurès in the first issue of *L'Humanité* as part of the statement of 'our aim', the text of which is available at: https://www.marxists.org/francais/general/jaures/works/1904/04/jaures_19040418.htm. It was discussed by Derrida during a speech to the 'Fête de l'Humanité': see Jacques Derrida, 'My Sunday "Humanities"', *Paper Machine*, trans. Rachel Bowlby (Stanford: Stanford University Press, 2005). And see Bernard Stiegler, *What Makes Life Worth Living: On Pharmacology*, trans. Daniel Ross (Cambridge: Polity, 2013), §50.
94 Christophe Bonneuil and Jean-Baptiste Fressoz, *The Shock of the Anthropocene: The Earth, History and Us*, trans. David Fernbach (London: Verso, 2016), p. 289, translation modified.
95 A point of disagreement is here worthy of mention, concerning the place of Kant's philosophy in this history. Fressoz presents Kant's reflections on inoculation as ultimately retreating from the corporal consequences of the autonomy of reason. I do not at all believe that Kant here retreats: he pursues his position, which consists in distinguishing the powers of the understanding, which is also to say of calculation, from the ends that reason alone can provide.
96 Fressoz, *L'Apocalypse joyeuse*, p. 27.
97 Ibid., p. 31.
98 Guillaume J. de L'Épine et al., *Rapport sur le faite de l'inoculation de la petite verole* (Paris, 1765).
99 Fressoz, *L'Apocalypse joyeuse*, p. 42.
100 We should relate these evolutions of American morality to the decline in the ethics of work in the United States towards the end of the twentieth century, to which Rifkin draws attention in *The European Dream*, a work that is in other respects

quite mediocre. See Jeremy Rifkin, *The European Dream: How Europe's Vision of the Future is Quietly Eclipsing the American Dream* (Cambridge: Polity, 2004), pp. 24–8.

101 Sloterdijk, *In the World Interior of Capital*, p. 87.
102 Ibid., p. 92.
103 Ibid., p. 112.
104 Ibid., p. 113.
105 Ibid., p. 284, ch. 30, n. 2.
106 Ibid., p. 209.
107 I myself introduced this expression, *mal-être*, in Bernard Stiegler, *Technics and Time, 3: Cinematic Time and the Question of Malaise*, trans. Stephen Barker (Stanford: Stanford University Press, 2011), where, unfortunately, it was translated into English as 'malaise', erasing the 'question of being' or the *bad-question of being* that it contains – whose investigation we are attempting to continue here.
108 *Translator's note*: Once again, *'propension à la folie'* refers to a phrase of Sloterdijk's translated into English as 'willingness to embrace delusion' (Sloterdijk, *In the World Interior of Capital*, p. 54).
109 Sloterdijk, *In the World Interior of Capital*, p. 212.
110 Ibid., p. 5.
111 Gilles Deleuze and Félix Guattari, *What is Philosophy?*, trans. Hugh Tomlinson and Graham Burchell (New York: Columbia University Press, 1994), p. 208, quoted in Sloterdijk, *In the World Interior of Capital*, p. 5.
112 Félix Guattari, *The Three Ecologies*, trans. Ian Pindar and Paul Sutton (London and New Brunswick, NJ: Athlone, 2000).

9 Ordinary Madness, Extraordinary Madnesses

1 'Aux bords de la folie', a special section in *Esprit* 413 (2015).
2 Michaël Foessel, 'La folie ordinaire du pouvoir', *Esprit* 413 (2015), pp. 152–64.
3 Marc-Olivier Padis, 'Derrière la folie, les malaises ordinaires', *Esprit* (2015), p. 14.
4 Jean-Paul Delevoye (with Jean-François Bouthors), *Reprenons-nous!* (Paris: Tallandier, 2013).
5 Padis, 'Derrière la folie, les malaises ordinaires', p. 14.
6 Gladys Swain, *Dialogue avec l'insensé: Essais d'histoire de la psychiatrie* (Paris: Gallimard, 1994).
7 *Translator's note*: 'Sectorization' was the name given to a specifically French approach to psychiatric policy-making and organization that was first considered seriously in 1960, and implemented in fits and starts mainly in the 1970s. It aimed to replace confinement models based on the asylum with community-based structures divided according to geographical regions, and was the site of scientific struggles involving the division between psychoanalysis, psychiatry and neurology, and ideological struggles involving the French anti-psychiatry movement.
8 Jacques Hochmann, 'Les contestations de la psychiatrie', *Esprit* 413 (2015), pp. 19–27.
9 Foessel, 'La folie ordinaire du pouvoir', p. 152.
10 Ibid., p. 153.
11 Ibid.
12 Ibid.
13 Ibid.
14 Ibid., p. 154.
15 Ibid., p.155.
16 This is the horizon within which I read Spinoza's *Political Treatise*, and the highly stimulating interpretations of it proposed by Frédéric Lordon in *Imperium. Structures et affects des corps politiques* (Paris: La Fabrique, 2015).
17 Foessel, 'La folie ordinaire du pouvoir', p. 155. The quotation from Pascal is

from Blaise Pascal, *Pensées*, trans. A.J. Krailsheimer (Harmondsworth, Middlesex: Penguin, 1966), p. 39, translation modified.

18 Ludwig Binswanger, 'Dream and Existence', trans. Jacob Needleman, in *Dream and Existence*, reprint of a special issue of *Review of Existential Psychology & Psychiatry* 19 (1) (1986). *Translator's note*: 'Traum und Existenz' was first published in *Neue Schweizer Rundschau* 9 (1930), pp. 673–85 and 766–79.

19 Gilbert Simondon, *Imagination et invention* (Chatou: Éditions de la Transparence, 2008).

20 Foessel, paraphrasing Pascal, in 'La folie ordinaire du pouvoir', p. 156.

21 Michel Foucault, 'My Body, This Paper, This Fire', trans. Geoffrey Bennington and Robert Hurley, in James D. Faubion (ed.), *The Essential Works of Michel Foucault, 1954–1984, Volume 2: Aesthetics, Method, and Epistemology* (London: Penguin, 2000).

22 Foessel, 'La folie ordinaire du pouvoir', p. 156.

23 See Jonathan Crary, *24/7: Late Capitalism and the Ends of Sleep* (London and New York: Verso, 2013), and my commentary in Bernard Stiegler, *Automatic Society, Volume 1: The Future of Work*, trans. Daniel Ross (Cambridge: Polity, 2016).

24 *Translator's note*: An oddity of the French idiom is that '*solution de continuité*' refers to a break in continuity, an interruption, a discontinuity or, indeed, a disruption.

25 Foessel, 'La folie ordinaire du pouvoir', p. 157.

26 *Translator's note*: On the relationship between and politics of power and powerlessness, see also Bernard Stiegler, *The Neganthropocene*, trans. Daniel Ross (London: Open Humanities Press, 2018), ch. 12.

27 Blaise Pascal, 'Three Discourses on the Condition of the Great', in Richard H. Popkin (ed. and trans.), *Pascal Selections* (New York: Scribner/Macmillan, 1989), p. 74, quoted in Foessel, 'La folie ordinaire du pouvoir', p. 160.

28 Foessel, 'La folie ordinaire du pouvoir', p. 161.

29 Ibid.

30 Sigmund Freud, *The Future of an Illusion*, in Volume 21 of James Strachey (ed. and trans.), *The Standard Edition of the Complete Psychological Works of Sigmund Freud* (London: Hogarth, 1953–74).

31 *Translator's note*: This refers to the marches that occurred throughout France on 10–11 January 2015, in response to the Charlie Hebdo attack on 7 January and the subsequent and related siege and murders that took place at a kosher supermarket in Porte de Vincennes on 9 January, and in particular to the vast march that took place in Paris on 11 January.

32 Foessel, 'La folie ordinaire du pouvoir', p. 163. *Translator's note*: The last clause of this quotation refers to events such as those described in an article by Maïa de la Baume and Dan Bilefsky, 'French Police Question Boy, 8, After Remarks on Paris Attacks', *New York Times* (29 January 2015), available at: https://www.nytimes.com/2015/01/30/world/europe/french-police-questions-schoolboy-said-to-defend-charlie-hebdo-attack.html.

33 Foessel, 'La folie ordinaire du pouvoir', p. 164.

34 Jacques Derrida, 'Cogito and the History of Madness', *Writing and Difference*, trans. Alan Bass (London: Routledge and Kegan Paul, 1978), p. 47, translation modified.

35 Ibid., pp. 47–8.

36 Ibid., p. 57.

37 Ibid., p. 48.

38 Ibid.

39 René Descartes, *Meditations on First Philosophy, with Selections from the Objections and Replies*, trans. John Cottingham (Cambridge: Cambridge University Press, 1986), p. 14, quoted in Derrida, 'Cogito and the History of Madness', p. 49. *Translator's note*: In this instance, the translation utilized in the Derrida volume has been followed.

40 Derrida, 'Cogito and the History of Madness', p. 50.
41 Ibid., p. 51.
42 Ibid. *Translator's note*: The word translated as 'insanity' here is *'extravagances'*.
43 Ibid., pp. 52–3, translation modified.
44 Ibid., p. 54, translation modified.
45 Ibid.
46 Ibid., p. 56.
47 Ibid.
48 Ibid., translation modified.
49 Ibid., translation modified.
50 Ibid.
51 Ibid., pp. 56–7, translation modified.
52 Ibid., p. 57, translation modified.
53 See §129.
54 Derrida, 'Cogito and the History of Madness', p. 57, translation modified.
55 Ibid., translation modified.
56 As *to endekhomenon allōs ekhein. Translator's note*: See, for example, Aristotle, *Nicomachean Ethics*, Book 6, 1140a; and see Bernard Stiegler, *Technics and Time, 3: Cinematic Time and the Question of Malaise*, trans. Stephen Barker (Stanford: Stanford University Press 2011), pp. 187–8.
57 I return here to the question that I introduced at the beginning of my work through the analysis of *Being and Time*. See Bernard Stiegler, *Technics and Time, 1: The Fault of Epimetheus*, trans. Richard Beardsworth and George Collins (Stanford: Stanford University Press, 1998).
58 Derrida, 'Cogito and the History of Madness', p. 60, translation modified.
59 Ibid., p. 60.
60 Ibid., pp. 60–1, translation modified.
61 Ibid., p. 61, translation modified.
62 Ibid., translation modified.
63 Ibid., pp. 61–2, translation modified.
64 Ibid., p. 62, translation modified.
65 Ibid., translation modified.
66 Ibid., p. 63.

10 The Dream of Michel Foucault

1 In Michel Foucault, 'My Body, This Paper, This Fire', trans. Geoffrey Bennington and Robert Hurley, in James D. Faubion (ed.), *The Essential Works of Michel Foucault, 1954–1984, Volume 2: Aesthetics, Method, and Epistemology* (London: Penguin, 2000). This was an appendix to the second edition of *Histoire de la folie à l'âge classique* (from which the preface to the first edition was removed).
2 Ibid., p. 395.
3 Jacques Derrida, 'Cogito and the History of Madness', *Writing and Difference*, trans. Alan Bass (Chicago: University of Chicago Press, 1978), p. 51.
4 Foucault, 'My Body, This Paper, This Fire', p. 397.
5 Ibid., translation modified.
6 Ibid., p. 398, my italics.
7 Ibid., my italics.
8 René Descartes, 'Discourse on the Method of Properly Conducting One's Reason and of Seeking the Truth in the Sciences', §6, in *Discourse on Method and the Meditations*, trans. F.E. Sutcliffe (London: Penguin, 1968), p. 78.
9 Jacques Brunschwig, 'Introduction' to *'Règles pour la direction de l'esprit'*, in René Descartes, *Oeuvres philosophiques 1* (Paris: Garnier, 1963), p. 70: 'The *Rules* are anterior to Cartesian metaphysics, but [...] most of their assertions [...] call for and make necessary this metaphysics.'

10 I tried to show in Bernard Stiegler, *Technics and Time, 3: Cinematic Time and the Question of Malaise*, trans. Stephen Barker (Stanford: Stanford University Press, 2011), that with Kant it is not possible to think that ὕβρις that is technoscience, which to a large degree puts into question the great partitions on which his critique is founded, the basis of which lies in the transcendental imagination and the schematism of the understanding.

11 *Translator's note*: The influential *Logique de Port-Royal* of 1662 put forward a reasoned approach to language based not on customs of use but on the notion that language always concerns relationships between the representation of an idea and the representation of a thing. See also Bernard Stiegler, *Taking Care of Youth and the Generations*, trans. Stephen Barker (Stanford: Stanford University Press, 2010), pp. 138–9 and 144, and the translator's note 28 on p. 223.

12 In 1966, Michel Foucault writes in *The Order of Things: An Archaeology of the Human Sciences*, no translator credited (London and New York: Tavistock/Routledge, 1970), p. 73: 'Between the *mathesis* and the *genesis* there extends the region of signs [...]. Hedged in by calculus and geneses, we have the area of the *table*.' Ibid, p. 75: 'The centre of knowledge, in the seventeenth and eighteenth centuries, is the *table*.' And what feeds into the possibility of the table stems from the difference between the simple and the complex, the analytic and the synthetic, which generates the taxonomies. Ibid., p. 72: 'When dealing with the ordering of simple natures, one has recourse to a mathesis, of which the universal method is algebra. When dealing with the ordering of complex natures (representations in general, as they are given in experience), one has to constitute a *taxinomia*, and to do that one has to establish a system of signs.' It is to form such systems that general grammar is constituted.

13 Jacques Derrida, 'Faith and Knowledge: The Sources of "Religion" at the Limits of Reason Alone', trans. Samuel Weber, in *Acts of Religion*, ed. Gil Anidjar (New York and London: Routledge, 2002).

14 In an essential way, 'extended Latin grammar' accompanies and links together:

- globalization, via the exogenous disinhibition described by Sloterdijk;
- the disinhibition of risk, as described by Sloterdijk and by Fressoz;
- the translation of all idioms, on the basis of the linguistic theory emerging from extended Latin grammar, so that what comes to be woven together is a 'homogeneous network of communication centred on Europe. Each new language affected by this increases the power of the network, and increases disequilibrium in relation to other, not yet integrated languages.' Sylvain Auroux, *La Révolution technologique de la grammatisation. Introduction à l'histoire des sciences du langage* (Liège: Mardaga, 1994), p. 72.

It is striking to note how Google, which is in a way the flagship of transhumanism and disruption, amounts to a leap in linguistic technology (anticipated by Auroux in terms of what he called the 'language industries') that reshapes all these questions, while at the same time confirming their profound and ancient dynamics.

15 See Bernard Stiegler, *For a New Critique of Political Economy*, trans. Daniel Ross (Cambridge: Polity, 2010), pp. 73–96 and Stiegler, *States of Shock: Stupidity and Knowledge in the Twenty-First Century*, trans. Daniel Ross (Cambridge: Polity, 2015), ch. 2.

16 *Translator's note*: The author is here referring to a passage in Gottfried Wilhelm Leibniz, *Opuscules et fragments inédits* (Paris: Alcan, 1903), pp. 98–9. This passage from Leibniz is quoted and discussed by Jacques Derrida in 'Psyche: Invention of the Other', trans. Catherine Porter, *Psyche: Inventions of the Other*, vol. 1 (Stanford: Stanford University Press, 2007), p. 41. And see Stiegler, *States of Shock*, pp. 201–2, and Bernard Stiegler, *Automatic Society, Volume 1: The Future of Work*, trans. Daniel Ross (Cambridge: Polity, 2016), p. 222.

17 Françoise Waquet, 'Qu'est-ce que la république des Lettres? Essai de sémantique historique', *Bibliothèque de l'école des chartes* 147 (1989), pp. 473–502, available at: http://www.persee.fr/docAsPDF/bec_0373-6237_1989_num_147_1_450545.pdf.

18 Michel Foucault, *The Courage of Truth (The Government of Self and Others II): Lectures at the Collège de France, 1983–1984*, trans. Graham Burchell (New York: Picador, 2012), p. 14, translation modified.

19 Ibid., pp. 76–7.

20 In the seminar of 15 February 1984, in ibid., pp. 73ff.

21 Georges Dumézil, *The Riddle of Nostradamus: A Critical Dialogue*, trans. Betsy Wing (Baltimore and London: Johns Hopkins University Press, 1999). See Foucault, *The Courage of Truth*, p. 96.

22 In Bernard Stiegler, *What Makes Life Worth Living: On Pharmacology*, trans. Daniel Ross (Cambridge: Polity, 2013), §12.

23 The mother of Asclepius being Coronis, daughter of Phlegyas, as is the case in Pindar, or else Arsinoe, daughter of Leucippus.

24 Pierre Grimal, *The Dictionary of Classical Mythology*, trans. A.R. Maxwell-Hyslop (Oxford: Basil Blackwell, 1996), p. 63.

25 Friedrich Nietzsche, *The Gay Science, With a Prelude in German Rhymes and an Appendix of Songs*, trans. Josefine Nauckhoff (Cambridge and New York: Cambridge University Press, 2001), §340 ('The dying Socrates'), quoted in Foucault, *The Courage of Truth*, p. 99: 'Whether it was death or the poison or piety or malice…'.

26 Alain Frontier, *Lettre à un ami* (3 January 2014), available at: https://www.sitaudis.fr/Incitations/lettre-a-un-ami.php.

27 Ludwig Binswanger, 'Dream and Existence', trans. Jacob Needleman, in *Dream and Existence*, reprint of a special issue of *Review of Existential Psychology & Psychiatry* 19 (1) (1986), p. 102.

28 Ibid.

29 Ibid., p. 103.

30 Michel Foucault, 'Dream, Imagination and Existence', trans. Forrest Williams, in *Dream and Existence*, p. 44.

31 Ibid., pp. 44–5, translation modified.

32 Claude Lévi-Strauss, *Tristes Tropiques*, trans. John Weightman and Doreen Weightman (Harmondsworth: Penguin, 1976), pp. 542–3.

33 Jonathan Crary, *24/7: Late Capitalism and the Ends of Sleep* (London and New York: Verso, 2013).

34 Jacques Derrida, 'My Sunday "Humanities"', *Paper Machine*, trans. Rachel Bowlby (Stanford: Stanford University Press, 2005), originally published in *L'Humanité* on 4 March 1999.

35 Jean Jaurès, 'Our Goal', *L'Humanité* (18 April 1904), available at: https://www.marxists.org/archive/jaures/1904/04/18.htm, trans. Mitch Abidor, translation modified.

36 See §24.

37 John Howkins, *The Creative Economy: How People Make Money From Ideas* (New York: Penguin, 2001). Ars Industrialis opposes to this fundamentally oligarchic discourse (selecting the 'creatives' from out of the masses) the concepts and the project of contributory economy.

38 Immanuel Kant, *Critique of Pure Reason*, trans. Paul Guyer and Allen W. Wood (Cambridge: Cambridge University Press, 1998), p. 397. *Translator's note*: Note that where the German word '*Untunlichkeit*' has been translated into English as 'impracticability', in the French translation it is rendered as '*irréalisable*'.

39 Stiegler, *Automatic Society, Volume 1*, p. 46.

40 We can, however, take this word at its word, and reply that, indeed, as clear-sighted militants have always said, it is not enough to vote, and we must impose pressure upon those who have been elected throughout the period of their mandate.

11 Generation Strauss-Kahn

1 *Translator's note*: No reference is given for this *Washington Post* article, and an online search has failed to find one article that contains both Flanagan's statement about being a powder keg and the observation about the number of mass shootings; however, these two elements can be found in two different articles published on the same day. The first can be found in Joel Achenbach, 'Killer's Ultimate Selfie: Roanoke Horror Becoming the New Norm', *Washington Post* (26 August 2015), available at: https://www.washingtonpost.com/national/health-science/killers-ultimate-selfie-roanoke-horror-becoming-the-new-norm/2015/08/26/5fabd7c0-4c10-11e5-902f-39e9219e574b_story.html. The second can be found in Christopher Ingraham, 'We're Now Averaging More than One Mass Shooting Per Day in 2015', *Washington Post* (26 August 2015), available at: https://www.washingtonpost.com/news/wonk/wp/2015/08/26/were-now-averaging-more-than-one-mass-shooting-per-day-in-2015.

2 Hanna Rosin, 'The Silicon Valley Suicides', *The Atlantic* (December 2015), available at: https://www.theatlantic.com/magazine/archive/2015/12/the-silicon-valley-suicides/413140.

3 Paul Krugman, 'Despair, American Style', *New York Times* (9 November 2015), available at: https://www.nytimes.com/2015/11/09/opinion/despair-american-style.html.

4 Marine Forestier, 'L'épidémie d'héroïne, nouvel enjeu des élections américaines', *Le Monde* (2 October 2015), available at: http://www.lemonde.fr/ameriques/article/2015/10/02/l-epidemie-d-heroine-aux-etats-unis-s-invite-dans-la-campagne-pour-l-election-de-2016_4781695_3222.html.

5 Ibid.

6 Louis Gallois, *Pacte pour la compétitivité de l'industrie française* (5 November 2012), available at: http://www.egee.asso.fr/IMG/pdf/rapport_gallois.pdf.

7 Jean Pisani-Ferry, *Quelle France dans dix ans?* (Rapport de France Stratégie au président de la République, June 2014).

8 Pierre Collin and Nicolas Colin, *Mission d'expertise sur la fiscalité de l'économie numérique* (January 2013), available at: https://www.economie.gouv.fr/files/rapport-fiscalite-du-numerique_2013.pdf.

9 Jean Jouzel, *L'Avenir du climat* (Paris: Fondation Diderot, 2014), pp. 22–3.

10 Fethi Benslama, *Déclaration d'insoumission: à l'usage des musulmans et de ceux qui ne le sont pas* (Paris: Flammarion, 2005).

11 French Wikipédia article on the first Gulf war (no access date): 'On 25 July 1990, Saddam Hussein met the American ambassador, April Glaspie, in Baghdad. The latter, well aware of what was being prepared ("we can only see that you have deployed massive troops in the south"), let him understand that "we [the United States] have no opinion on the Arab-Arab conflicts".' *Translator's note*: The transcript of the 25 July meeting between Saddam Hussein and April Glaspie was obtained by the *Washington Post* and then published by the *New York Times*: 'Confrontation in the Gulf; Excerpts From Iraqi Document on Meeting With U.S. Envoy', *New York Times* (23 September 1990), available at: http://www.nytimes.com/1990/09/23/world/confrontation-in-the-gulf-excerpts-from-iraqi-document-on-meeting-with-us-envoy.html. And see Lloyd Gardner, *The Long Road to Baghdad: A History of U.S. Foreign Policy from the 1970s to the Present* (New York: The New Press, 2008), p. 76, for a description of Glaspie's reaction when confronted with her statements as described in the transcript:

> Glaspie was later cornered by British journalists brandishing the transcript outside the American embassy in Baghdad. 'You knew Saddam was going to invade [Kuwait] but you didn't warn him not to. You didn't tell him America would defend Kuwait. You told him the opposite – that America was not associated with Kuwait.' Another chimed in, 'You encouraged this aggression – his invasion. What were you thinking?' Her reply added still more to the confusion.

'Obviously, I didn't think, and nobody else did, that the Iraqis were going to take all of Kuwait.' The journalists chased after her as she entered her limousine. 'You thought he was just going to take some of it? But, how could you?' The ambassador said no more as she was whisked away.

12 French Wikipédia article on Ahmad Shah Massoud (no access date): 'At that time, American strategy consisted in supporting the most fundamentalist combatants, believing that they would be the most ferocious in the struggle against the Soviet occupation.' Massoud became bin Laden's enemy, and the latter had him assassinated two days prior to 9/11.

13 See p. 35.

14 We have seen that there has also been an increase in the suicide rate in the United States, and this tendency is in fact global. The planet is becoming highly suicidal, and this is what, with Gilles Dostaler, Bernard Maris pointed out six years before his murder. See Gilles Dostaler and Bernard Maris, *Capitalisme et pulsion de mort* (Paris: Albin Michel, 2009), p. 16: 'Has not capitalism, by diverting technics to the benefit of accumulation, opened the floodgates to a death drive buried in the heart of humanity? If the answer is yes, then we may fear that bad times lie ahead for human beings.' We will return to this work in the conclusion to this book. *Translator's note*: On the suicide rate in France, see Muriel Moisy, *Suicide: Knowledge as a Means of Prevention at the National, Local and Association Levels. Summary of the 2nd Report* (Observatoire National du Suicide, 2016), available at: http://drees. solidarites-sante.gouv.fr/IMG/pdf/ons2016_summary.pdf, p. 3.

15 French Wikipédia article on the Fukushima nuclear accident (accessed 2 January 2015): 'Due to the extent of the damage and the extremely difficult conditions of intervention, the complete dismantling of the destroyed facilities will take a long time, and will require close to forty years to complete. On 26 December 2011, Tepco published a roadmap that provides for three phases: for two years or more, spent fuel will be removed from the deactivation pools of Reactor 4, and then, from about ten years on and for twenty or thirty years, the molten fuel will be removed from Reactors 1 and 3, and the contaminated water will be treated. Research and development will be needed to develop technology for investigating, controlling and intervening in a highly radioactive environment.'

16 *Translator's note*: '*Décomplexion*' is a term used to describe the removal of (psychological) 'complexes', meaning a kind of process of disinhibition consisting in coming to accept the legitimacy of one's own prejudices, or in surrendering to a view of life no longer burdened by complex moral considerations (such as, for example, to decide to simply accept one's love of money or riches).

17 Michel Jarraud and Achim Steiner, 'Foreword', *Climate Change 2014: Synthesis Report*, available at: https://www.ipcc.ch/pdf/assessment-report/ar5/syr/SYR_AR5_FINAL_full_wcover.pdf, p. v. *Translator's note*: This and the following quotation have been taken from the English version of the report. They differ slightly from the French version quoted by the author.

18 IPCC, *Climate Change 2014: Synthesis Report*, p. 2.

19 Christophe Cornevin and Jean-Pierre Beuve, 'Maxime, 22 ans: l'itinéraire d'un bourreau français en Syrie', *Le Figaro* (17 November 2014), available at: http://www. lefigaro.fr/actualite-france/2014/11/17/01016-20141117ARTFIG00403-maxime-de-la-normandie-a-l-etat-islamique.php.

20 Carlos Parada, 'La fanatisme religieux comme une drogue dure?', *Libération* (8 December 2015), available at: http://www.liberation.fr/debats/2015/12/08/le-fanatisme-religieux-comme-une-drogue-dure_1419269.

21 *Translator's note*: Dominique Alderweireld, known as 'Dodo la Saumure', was one of the defendants in the trial involving Dominique Strauss-Kahn that was centred on the Carlton Hotel in Lille.

22 See Boris Cyrulnik, *Quand un enfant se donne 'la mort': Attachement et sociétés*

(Paris: Odile Jacob, 2011). The sponsor of the report and Secretary of State for Youth, Jeannette Bougrab, writes in the preface: 'Forty percent of children think of death because they are anxious and unhappy' (p. 9). And as Boris Cyrulnik says in a discussion of his report: 'In France, suicide is the second leading cause of death among youth aged between 15 and 24 (16.6% of deaths from all causes) after road accidents, and in 2008 accounted for 3.8% of deaths from all causes for children aged between 5 and 14' (Boris Cyrulnik, 'Le rapport Boris Cyrulnik: "Quand un enfant se donne la mort"', *Encyclopédie sur la mort*, available at: http://agora.qc.ca/thematiques/mort/documents/le_rapport_boris_cyrulnik_quand_un_enfant_se_donne_la_mort_).

23 I have tried to reflect upon this state of affairs in Bernard Stiegler, *Taking Care of Youth and the Generations*, trans. Stephen Barker (Stanford: Stanford University Press, 2010).

24 Donella H. Meadows, Dennis L. Meadows, Jørgen Randers and William W. Behrens III, *Limits to Growth* (New York: Signet, 1972), a report commissioned in 1970 by the Club of Rome.

25 I began to examine this subject in Bernard Stiegler, *Uncontrollable Societies of Disaffected Individuals: Disbelief and Discredit, Volume 2*, trans. Daniel Ross (Cambridge: Polity, 2013).

26 I have tried to describe this logic in Bernard Stiegler, *Pharmacologie du Front national* (Paris: Flammarion, 2013), on the basis of a reading of Karl Marx and Friedrich Engels, *The German Ideology*, no translator credited (Moscow: Progress Press, 1976).

27 Including the process of forgetting disasters, which Hidetaka Ishida describes in the wake of Fukushima.

28 In the conclusion of this work, however, we will show that we diverge from Dostaler and Maris in terms of the way they read Freud and Keynes in relation to one another – and that this cross-reading leads them into an impasse with respect to economics.

29 This is something that Dostaler and Maris themselves point out: see Dostaler and Maris, *Capitalisme et pulsion de mort*, p. 27.

30 As the judge in the Lille Carlton trial pointed out.

31 Immanuel Kant, *The Metaphysics of Morals*, trans. Mary Gregor (Cambridge: Cambridge University Press, 1991), pp. 255–6.

32 *Translator's note*: Mohamed Mera, armed with a .45 calibre pistol, carried out a series of attacks in Montauban and Toulouse between 11 and 19 March 2012, focused on French soldiers and a Jewish school, killing seven and wounding five, before eventually being killed by police.

33 See Bernard Stiegler, 'Ces abominables tueries peuvent s'expliquer par la dérive de nos sociétés' [These abominable killings can be explained by the drift of our societies], *Le Monde* (29 March 2012), available at: http://www.lemonde.fr/idees/article/2012/03/29/ces-abominables-tueries-peuvent-s-expliquer-par-la-derive-de-nos-societes_1677650_3232.html. In that article, I wrote:

> One hears people say, after the appalling Merah affair, that the monstrous has no explanation. This is what Rabbi Gilles Bernheim argues in *Libération*, and it is what Henri Guaino reiterates on France Culture. Leibniz, on the other hand, argued that the rational conception of the world consists in positing that everything has its reason, that is, its cause – including the most unreasonable, mad and murderous things.
>
> As Goya knew, it is the sleep of reason that engenders monsters, and this is even truer in the contemporary world characterized by the hyper-power of means – the .45 calibre pistol, webcams, the mass media, financial robots – and the powerlessness of ends, that is, their loss, which, leading also to a loss of reason, promotes passages to the act of all kinds, provoked by a constant excitation of the drive to destruction in a world that has itself become inherently and tragically drive-based.

Here, however, it is necessary to add and to repeat that *sueño* also means dream, and that Goya highlights here that reason dreams, and that its dreams can be monstrously realized – such as when Napoleon massacred Spaniards in the name of the French Revolution.

34　Kant, *The Metaphysics of Morals*, p. 256.

35　*Translator's note*: The terms used in the French translation of Kant that here correspond to 'moral being' and 'be improved' are '*être moral*' and '*devenir meilleur*' (to become better).

36　*Translator's note*: From 2012 to 2014, Vincent Peillon was Minister of Education under François Hollande.

37　Unconditional in law, they are conditioned in fact – by the facticity of the *pharmakon* that makes them possible, but which can always turn against them, making them impossible. Law is then what enables the *containment of this condition of impossibility* that the *pharmakon* (that is, tertiary retention) always also constitutes, and which can therefore always be turned into its contrary. On these questions of fact, law and *pharmakon*, see Bernard Stiegler, *Automatic Society, Volume 1: The Future of Work*, trans. Daniel Ross (Cambridge: Polity, 2016), §37.

38　This is the morality that Flaubert endeavours to describe, along with the suffering it brings, in *Madame Bovary*, which first appeared in 1856. And we should recall that in 1857 Flaubert was put on trial for this work by the court that would later condemn Baudelaire (who was not officially rehabilitated until 1949).

39　This point is developed with the concept of 'différante identity' in Bernard Stiegler, *Technics and Time, 2: Disorientation*, trans. Stephen Barker (Stanford: Stanford University Press, 2009).

40　On this question, see p. 227.

41　See Bernard Stiegler, *What Makes Life Worth Living: On Pharmacology*, trans. Daniel Ross (Cambridge: Polity, 2013), §2.

42　We will return to this point in the conclusion to this book.

43　Slavoj Žižek, *The Ticklish Subject: The Absent Centre of Political Ontology* (London and New York: Verso, 1999), p. 61.

44　Immanuel Kant, *Critique of Pure Reason*, trans. Paul Guyer and Allen W. Wood (Cambridge: Cambridge University Press, 1998), pp. 271–7.

45　*Translator's note*: On interpassivity, a term that Žižek borrows from Robert Pfaller, see Slavoj Žižek, *The Plague of Fantasies*, 2nd edition (London and New York: Verso, 2008), pp. 144–7. Žižek takes the concept of interpassivity as a counter to the fashionable concept of the 'interactivity' attributed to digital technology, that is, as a way of drawing attention to the ways in which this 'new electronic media' may foster a situation in which, delegating to the object itself the possibility of 'enjoying the show', the viewer is discharged of the positive and potentially political duty arising out of the super-ego. In relation to interpassivity and madness, see esp. p. 145.

46　*Translator's note*: This is a reference to Jérôme Cahuzac, a cosmetic surgeon who became a minister in the Hollande government, with specific responsibility for investigating tax fraud, until he was forced to resign due to allegations that he had himself been committing tax fraud, and for a long period of time. His resignation occurred on 19 March 2013, and he was eventually sentenced to three years in prison on 8 December 2016. Strauss-Kahn was charged in the 'Carlton hotel' case on 26 July 2013 (and acquitted on 12 June 2015).

47　*Translator's note*: This statement by Rocard is quoted (translating *pulsions* more conventionally as 'impulses' rather than 'drives'), for example, in Kim Willsher, 'Dominique Strauss-Kahn arrives back on French soil', *The Guardian* (4 September 2011), available at: https://www.theguardian.com/world/2011/sep/04/dsk-strauss-kahn-back-in-france.

48　Søren Kierkegaard, *Either/Or*, 2 vols, trans. Howard V. Hong and Edna H. Hong (Princeton: Princeton University Press, 1987).

49　Orphaned in this way, the young generation, which cannot constitute itself without idealization, at times rushes headlong into the worst traps generated by the phantom of this God, who becomes all the more powerful in being dead, just as was the case for the father of the primitive horde killed by his sons, thereby sending them into what Freud believed should be identified with guilt: 'The dead father became stronger than the living one had been.' Sigmund Freud, *Totem and Taboo*, in Volume 13 of James Strachey (ed. and trans.), *The Standard Edition of the Complete Psychological Works of Sigmund Freud* (London: Hogarth, 1953–74), p. 143.

50 Alain Juppé, *L'Avenir de la politique* (Paris: Fondation Diderot, 2014), p. 7. I said to myself, as did many Frenchmen, that *in no case* is it a matter of voting for Hollande. The poverty of Juppé's political proposals, his impressive lack of understanding of the issues tied, on the one hand, to education, and, on the other, to what he still believes can be called the 'digital path' (the digital is *anything but* a 'path'), his total ignorance (that is, his denial) of the stakes of disruption: all this leads me to think that the worst-case scenario is without doubt inevitable – and that, at best, it can only be postponed in an absence of epoch where there is not a minute to lose.

51 This is why Kant can posit that the condition of any judgement of others is firstly respect for this other.

52 *The example is a 'monstrator', a 'deictic' in the sense where deixis refers to what is here and now,* what cannot be de-monstrated being what cannot be established apodictically – that is, independently of any *deixis.*

53 Singularity is brought to its most vivid expressions in what is experienced as extraordinary madness. Here, we must revisit the modern history of moral philosophy and of those moralists who, between the seventeenth and eighteenth centuries, are like replicas of those tremors embodied by Montaigne and Pascal, precisely as regards the relations between wisdom and madness [or folly], and who, like Chamfort or La Rochefoucauld, complicate the sharp opposition between reason and madness that is found, better than anywhere else, in Voltaire. See Margot Kruse, 'Sagesse et folie dans l'oeuvre des moralistes', *Cahiers de l'Association international des études françaises* 30 (1978), pp. 121–37.

54 *Translator's note:* The first text mentioned here was published as Martin Heidegger, *Identität und Differenz* (Pfullingen: Verlag Günther Neske, 1957) and the second as 'Die Kehre', in Martin Heidegger, *Die Technik und die Kehre* (Pfullingen: Verlag Günther Neske, 1962). But whereas *Identity and Difference* is based on lectures and seminars delivered in the year of its publication, 'The Turn' is based on a lecture first given in 1949.

55 Graham Harman, *The Quadruple Object* (Winchester: Zero Books, 2011).

56 See especially Stiegler, *Automatic Society, Volume 1*, pp. 60, 113, 132–3 and 149.

57 Against this blocked horizon, Stephen Hawking and some of his colleagues have tried to alert global public opinion to the probability of the worst, and to the new dangers that have arisen with reticulated artificial intelligence – which is, precisely, the key element of the disruption.

58 On 'ordinary' and 'extra-ordinary' men, see Maurice Godelier, *The Metamorphoses of Kinship*, trans. Nora Scott (London and New York: Verso, 2011), to which we will return.

59 *Translator's note:* In addition to its usual meanings in French, *déchéance* is also the term used in French translations of Heidegger to correspond to *Verfallen.*

60 Michel Foucault, *The Courage of Truth (The Government of Self and Others II): Lectures at the Collège de France, 1983–1984*, trans. Graham Burchell (New York: Palgrave Macmillan, 2011), pp. 27ff.

61 Jean-Claude Englebert sent me a message at the end of 2015 in which he recommended that I see the film *The Big Short* (2015), directed by Adam McKay, which provides a masterful explanation of the genesis of the subprime crises, and, therefore, the lies of Alan Greenspan when he referred to the overly complicated nature of the models used by the financial industry. I understood this invitation as an objection to what I argued in the first volume of *Automatic Society*, namely that Greenspan bore

witness, in his replies during the Congressional hearing in Washington before which he appeared on 23 October 2008, to the fact that he himself was proletarianized by the automation of financial decision-making. Here is the reply I gave to Englebert: 'I haven't seen the film but I believe, as you do, and as it seems the film suggests, that these "overly complicated models" are an alibi for Greenspan. And yet I think that Greenspan can summon up this alibi, and that doing so does not seem totally wrong or deceitful, only because the system is indeed based on a systemic dissimulation, the consequence of which is that risks are taken that can no longer be calculated at all other than by algorithms that make impossible any decision about the precise level of risk they involve. I make these points by referring to the question of risk that lies at the heart of the Anthropocene, as shown by Sloterdijk in *In the World Interior of Capital* and by Fressoz in *L'Apocalypse joyeuse*. In this regard, the proletarianization of Greenspan himself is indeed actual.'

62 It is because every true hope bears reason that there are *non-logical* forms of reason within the wholly reasonable social forms that were destroyed by disinhibition during Western reason's worldwide expansion.

63 'Aux bords de la folie', a special section in *Esprit* 413 (2015).

64 This is what Caroline Fourest did during a French Culture programme with Brice Coutourier, in which they both showed an incredible lack of consideration and respect for Jean-Claude Ameisen, who was trying to respond with admirable probity to the questions he was asked about surrogate motherhood.

65 Stiegler, *Pharmacologie du Front national*, esp. §§42–43.

66 On negative sublimation, see Stiegler, *Uncontrollable Societies of Disaffected Individuals*.

67 I reserve a discussion with Frédéric Lordon on this point for another occasion.

68 This is the time described in Stiegler, *Uncontrollable Societies of Disaffected Individuals*.

69 Jacques Derrida, *Of Grammatology*, trans. Gayatri Chakravorty Spivak, corrected edition (Baltimore and London: Johns Hopkins University Press, 1998), p. 5, translation modified.

70 Gilles Deleuze, 'Postscript on Control Societies', *Negotiations 1972–1990*, trans. Martin Joughin (New York: Columbia University Press, 1995).

71 Jean-Luc Godard, *Histoire(s) du cinéma, tome 1* (Paris: Gallimard, 1998), p. 4.

72 Henry Ey, quoted in Georges Canguilhem, *Knowledge of Life*, trans. Stefanos Geroulanos and Daniela Ginsburg (New York: Fordham University Press, 2008), p. 180, n. 25.

73 Canguilhem, *Knowledge of Life,* p. 133. *Translator's note*: The English translators of *Knowledge of Life* cite *Doctor Faustus* as the source of this quotation, but state that they have been unable to locate it. The solution to this dilemma seems to be that Canguilhem is in fact quoting from two different sources. The first is indeed taken from Thomas Mann, *Doctor Faustus*, trans. H.T. Lowe-Porter (London: Penguin, 1968), p. 229: 'And where madness begins to be malady, there is nobody knows at all.' The second is taken from Thomas Mann, 'Dostoevsky – In Moderation', preface to Fyodor Dostoevsky, *The Short Novels of Dostoevsky* (New York: Dial Press, 1945), pp. xiv–xv:

> The truth is that life has never been able to do without the morbid, and probably no adage is more inane than the one that says that 'only disease can come from the diseased'. Life is not prudish, and it is probably safe to say that life prefers creative, genius-bestowing disease a thousand times over to prosaic health; prefers disease, surmounting obstacles proudly on horseback, boldly leaping from peak to peak, to lounging, pedestrian healthfulness. Life is not finical and never thinks of making a moral distinction between health and infirmity. It seizes the bold product of disease, consumes and digests it, and as soon as it is assimilated, it is health. An entire horde, a generation of open-minded, healthy lads pounces upon

the work of diseased genius, genialized by disease, admires and praises it, raises it to the skies, perpetuates it, transmutes it, and bequeathes it to civilization, which does not live on the home-baked bread of health alone. They all swear by the name of the great invalid, thanks to whose madness they no longer need to be mad. Their healthfulness feeds upon his madness and in them he will become healthy.

In fact, this remarkable text, signed by Mann in California in 1945, contains many passages on Dostoevsky and Nietzsche, and on the relationship between madness and crime, that are highly pertinent to the author's argument.

74 See Laurent Cherlonneix, 'Après Nietzsche et Canguilhem', in Anne Fagot-Largeault, Claude Debru and Michel Morange (eds), *Philosophie et médecine: en hommage à Georges Canguilhem* (Paris: Vrin, 2008).

75 The infidelity of the milieu can become the source of normativity (that is, of the passage from the first to the second moment of the doubly epokhal redoubling) only as this morality and this logic of quasi-causality that Deleuze reactivated on the basis of Stoic moral philosophy, to which Foucault will himself turn at the end of his life in order to think education as moral experience. See Laurent Jaffro, 'Foucault et le problème de l'éducation morale', *Le Télémaque* 29 (2006), pp. 111–24.

12 Thirty-Eight Years Later

1 Sigmund Freud, *Civilization and Its Discontents*, in Volume 21 of James Strachey (ed. and trans.), *The Standard Edition of the Complete Psychological Works of Sigmund Freud* (London: Hogarth, 1953–74), esp. ch. 3.

2 See Bernard Stiegler, *Symbolic Misery, Volume 1: The Hyper-Industrial Epoch*, trans. Barnaby Norman (Cambridge: Polity, 2014), and Stiegler, *Symbolic Misery, Volume 2: The Katastrophē of the Sensible*, trans. Barnaby Norman (Cambridge: Polity, 2015).

3 See Bernard Stiegler, *Pharmacologie du Front national* (Paris: Flammarion, 2013), ch. 15. *Translator's note*: This should be understood in a way that includes the sense of 'capability' discussed by the economist Amartya Sen. See also n. 6 in this chapter.

4 See Jacques Généreux, *La Dissociété* (Paris: Le Seuil, 2006).

5 Heraclitus, *Fragments: A Text and Translation with a Commentary by T.M. Robinson* (Toronto: University of Toronto Press, 1987), p. 19. Robinson translates the fragment as follows: 'If he doesn't expect the unexpected, he will not discover it; for it is difficult to discover and intractable.'

6 It is important to note, here, that 'capability', in Amartya Sen's sense, can be made to serve a reconsideration of knowledge in Canguilhem's sense.

7 See Roberto Esposito, *Communitas: The Origin and Destiny of Community*, trans. Timothy Campbell (Stanford: Stanford University Press, 2010).

8 See p. 311.

9 I have put forward this notion in Bernard Stiegler, *Technics and Time, 3: Cinematic Time and the Question of Malaise*, trans. Stephen Barker (Stanford: Stanford University Press, 2011).

10 *Translator's note*: Jacques Séguéla is a French advertising guru who headed the marketing campaigns of several French presidents on all sides of politics, beginning with the successful campaigns he organized on behalf of François Mitterrand.

11 The promise is what promises the possibility of a negentropy that differs from and defers calculation and that calculation prevents.

12 This is what I argued in Bernard Stiegler, *Automatic Society, Volume 1*, trans. Daniel Ross (Cambridge: Polity, 2016), where I endeavoured to understand, analyse and comment on Jonathan Crary's analysis, which I have tried to take further in this volume (see pp. 160–1 and 287).

13 There are two kinds of games, finite and infinite. See James P. Carse, *Finite and*

Infinite Games: A Vision of Life as Play and Possibility (New York: The Free Press, 1986). *Translator's note*: 'Play' and 'game' are in French designated by a single word, *jeu*.

14 Walter Benjamin, *Rêves* (Paris: Le Promeneur), p. 94. I owe the reading of these dreams to Anne Alombert.

15 See Donald Winnicott, *Playing and Reality* (Abingdon: Routledge, 2005), esp. ch. 7.

16 Charles Perrault, 'The Fairies', *The Complete Fairy Tales*, trans. Christopher Betts (New York: Oxford University Press, 2009), pp. 128–9.

17 This threefold individuation is expounded in the two volumes of Bernard Stiegler, *Symbolic Misery*.

18 See Bernard Stiegler, *The Neganthropocene*, trans. Daniel Ross (London: Open Humanities Press, 2018), ch. 12.

19 And, as we have already seen, Mitterrand called on Séguéla at the moment when he also invited Berlusconi to take over the Le Cinq television network, opening the door to the privatization of TF1.

20 See Hélène Croizé-Pourcelet, '1988, l'invention de la marque Mitterrand', *Slate. fr* (10 April 2012), available at: http://www.slate.fr/story/52591/photos-campagne-1988-generation-mitterrand-seguela: '"With 'Generation Mitterrand', I created the Mitterrand brand", explains advertiser Jacques Séguéla. [...] "Séguéla is a true advertiser, not a politician", explains Antoine Boulay. [...] "This campaign poster is not interesting from a political perspective [...]. 'Generation Mitterrand' is a slogan without content, I defy you to find any hint of a reform proposal or even a social category, a target age." The subtlety of this slogan, according to Boulay, lies in the fact that anyone can draw any meaning they want from it and project onto it whatever they want.'

21 *Translator's note*: TEN was a consulting firm created by the left-leaning Claude Neuschwander in 1976, specializing in economic development and regional planning.

22 *Translator's note*: The École nationale d'administration (ENA) is the major institution involved in the training of French officials and bureaucrats, including many French politicians.

23 See 'Ars *Industrialis*: 2005 Manifesto', in Bernard Stiegler, *The Re-Enchantment of the World: The Value of Spirit against Industrial Populism*, trans. Trevor Arthur (London and New York: Bloomsbury Academic, 2014).

24 Paul Valéry, *Regards sur le monde actuel et autres essais* (Paris: Gallimard, 1945). *Translator's note*: See in particular Paul Valéry, 'Freedom of the Mind', in Jackson Matthews (ed.), *The Collected Works of Paul Valéry, Volume 10: History and Politics*, trans. Denise Folliot and Jackson Matthews (New York: Bollingen, 1962), p. 190: 'All these values, rising and falling, constitute the great stock market of human affairs. On that market, *mind* is "weak" – it is nearly always falling.'

25 Yves Thréard, 'DSK: le propriétaire de la Porsche identifié', *Le Figaro* (5 May 2011), available at: http://www.lefigaro.fr/politique/2011/05/05/01002-20110505ARTFIG00741-dominique-strauss-kahn-assure-la-pub-de-porsche.php.

26 See 'Présidentielle J-57: la campagne vue par Bernard Stiegler', *Télérama* (23 February 2012), available at: http://www.telerama.fr/idees/presidentielle-j-57-la-campagne-vue-par-bernard-stiegler,78318.php.

27 See Robert Maggiori, 'Où sont les politiques?', *Libération* (19 March 2015), available at: http://www.liberation.fr/societe/2015/03/19/ou-sont-les-politiques_1224250.

28 *Translator's note*: Laroxyl is a brand-name of the SNRI Amitripyline.

29 I will return to this question of the 'act' of reading in Bernard Stiegler, *Mystagogies 1. De l'art et de la littérature*, forthcoming, Inshallah.

30 See pp. 59–60.

31 *Translator's note*: This discipline, little known in the Anglophone world, is largely associated with the work of Jean Ricardou. In an interview, Ricardou states that textics 'observes that fiction and theory, even though they might poorly accept their reciprocal contamination, are nevertheless, in a certain light, écrits of the same

nature'. Jean Ricardou, in 'An Interview with Jean Ricardou: "How to Reduce Fallacious Representative Innocence, Word by Word"', *Studies in 20th and 21st Century Literature* 15 (1991), p. 281.

32 Here, Marcel Gauchet's critique of anti-psychiatry in his introduction to the work of Gladys Swain seems to me to be at once totally justified and far too brief. If the practice and theory of 'schizo-analysis' have been swallowed up by the shifting sands of the 'truth of madness', the question of the madness of truth remains intact, and, along with it, that of the schize, that is, of the ὕβρις that constitutes both noesis and the psychic apparatus, something that we can afford *less than ever* to deny. See Marcel Gauchet, 'À la recherche d'une autre histoire de la folie', in Gladys Swain, *Dialogue avec l'insensé: Essais d'histoire de la psychiatrie* (Paris: Gallimard, 1994).

33 This interpretation of the origin of nihilism is obviously not that of Nietzsche. And this is why we must re-evaluate and transvaluate Nietzsche himself, who, having constantly circled around the question of the *pharmakon*, in the end ignored it, just as he did the question of entropy.

34 Stéphane Mallarmé, 'Action Restricted' (from *As for the Book*), trans. Mary Ann Caws, *Selected Poetry and Prose* (New York: New Directions, 1982), p. 77.

35 Bill Evans, *Conversations with Myself* (Verve, 1963).

36 See Bernard Stiegler, *Acting Out*, trans. David Barison, Daniel Ross and Patrick Crogan (Stanford: Stanford University Press, 2009), pp. 12–15.

37 Pierre Aubenque, *Le Problème de l'être chez Aristote* (Paris: Presses Universitaires de France, 1962).

38 *Translator's note*: A text in which the author discusses the concepts of stereotypes and traumatypes is Stiegler, *Symbolic Misery, Volume 2*, ch. 4.

39 Jean Lauxerois, 'À titre amical', afterword in Aristotle, *L'Amicalité* (Paris: À propos, 2002).

40 Peter Sloterdijk, 'Rules for the Human Park: A Response to Heidegger's "Letter on Humanism"', *Not Saved: Essays after Heidegger*, trans. Ian Alexander Moore and Christopher Turner (Cambridge: Polity, 2017), p. 193: 'Books, the poet Jean Paul once remarked, are long letters to friends. With this statement, he appropriately characterized the essence and function of humanism in a quintessential, graceful manner: it is telecommunication that builds friendship in the medium of writing. What from Cicero's time onward has been called *humanitas* belongs, in the narrowest and broadest senses, to the consequences of literacy.' These 'consequences of literacy' are, in the language we are proposing here, the writing of the circuits of transindividuation of the *pharmakon* that is the letter, by the letter and to the letter. *Translator's note*: Where Sloterdijk's German is translated into English as 'consequences of literacy', the French translation is, more specifically, '*conséquence de l'alphabétisation*'.

41 I have explained why this attempt seems to me to be both necessary and insufficient in Stiegler, *Automatic Society, Volume 1*, pp. 127ff.

42 See 'Téléphonie mobile', article in the French Wikipédia (accessed 10 January 2016): 'According to the International Telecommunication Union, four billion mobile telephone subscriptions were current at the end of 2009, a figure equal to 60 per cent of the world's population. An increasing number of people had more than one subscription simultaneously. In January 2011, the number of mobile subscriptions rose to five billion, according to a UN report. The use of mobile phones has exploded in the poorest countries, where the fixed telephone network is often embryonic. In 2008, three out of four subscriptions (three billion) were in developing countries, compared to one in four in 2000.' *Translator's note*: See also the 2017 report by the International Telecommunication Union (which was formed in 1865, before eventually becoming the UN agency responsible for information and communication technologies), *Measuring the Information Society Report, Volume 1* (Geneva: International Telecommunication Union, 2017), available at: https://www.itu.int/en/ITU-D/Statistics/Documents/publications/misr2017/MISR2017_Volume1.pdf, p. 5:

'The total number of mobile-cellular subscriptions has increased from 2.20 billion in 2005 to 5.29 billion in 2010, 7.18 billion in 2015 and an estimated 7.74 billion in 2017. The number of subscriptions per 100 population has grown from 33.9 in 2005 to 76.6 in 2010, 98.2 in 2015 and an estimated 103.5 in 2017.' In other words, there are now more active mobile telephony subscriptions than there are people on earth.

43 See E.R. Dodds, *The Greeks and the Irrational* (Berkeley, Los Angeles and London: University of California Press, 1951), ch. 2.

44 On this point, see the 5 February 2011 lecture on pharmakon.fr.

45 See p. 71.

46 Martin Heidegger, *Being and Time*, trans. Joan Stambaugh, revised by Dennis J. Schmidt (Albany: State University of New York Press, 2010), §6.

47 Michel Foucault, 'Self Writing', trans. Robert Hurley, in Paul Rabinow (ed.), *The Essential Works of Michel Foucault, 1954–1984, Volume 1: Ethics, Subjectivity and Truth* (London: Penguin, 1997).

48 Stéphane Mallarmé, 'The Mystery in Letters', in Anthony Hartley (ed. and trans.), *Mallarmé* (London: Penguin, 1965), p. 203, translation modified.

49 Mallarmé, 'When My Old Books are Closed on Paphos' Name…', *Selected Poetry and Prose*, p. 61.

50 In Mallarmé's *Entre quatre murs*, written between 1859 and 1860, we find the poems 'Rêve antique' [Old Dream], 'La chanson du fol' [Song of the Mad], 'La colère d'Allah!' [The Wrath of Allah], but we do not find 'Ma bibliothèque', which he had indicated in his plan for the volume.

51 *Surprehension* or surprise is *what exceeds comprehension*, just as *reason* is *what exceeds the understanding*. Today, at IRI, in order to specify a graphic language intended to bring change to the formats recommended by the World Wide Web Consortium, I work with Vincent Puig, Anne Alombert and Simon Lincelles, as we attempt to rethink the question of categories starting from the question of annotation.

If it is true that: (a) the theory of the categories lies at the base of Aristotle's onto-theology; and if (b) as Auroux has shown, the latter stems from grammatization inasmuch as it sets in motion what will become Greek grammar (this is the subject of the debate that Derrida will open up with Émile Benveniste; see Jacques Derrida, *Margins of Philosophy*, trans. Alan Bass [Chicago: University of Chicago Press, 1982], p. 179); and if (c), with digital disruption, we are living through a true *revolution of those categorization technologies* that lie at the centre of every algorithmic machine; then (d) questions concerning annotation, as that which makes it possible to analytically disengage the 'parts of speech' as well as the categories of the predication of being, are *primordial* in the strict sense – and it is with these questions that we must begin (again).

52 René Thom, *Mathematical Models of Morphogenesis*, trans. W.M. Brookes and D. Rand (Chichester: Ellis Horwood, 1983).

53 Alain Bashung, 'Vertige de l'amour' (1981), music video available at: https://www.youtube.com/watch?v=CwQe9XrsW88.

54 Edmund Husserl, *On the Phenomenology of the Consciousness of Internal Time*, trans. John Barnett Brough (Dordrecht: Kluwer, 1991).

55 Jacques Derrida, *Edmund Husserl's Origin of Geometry: An Introduction*, trans. John P. Leavey, Jr. (Lincoln and London: University of Nebraska Press, 1978).

56 Jacques Derrida, *Speech and Phenomena*, trans. David B. Allison (Evanston: Northwestern University Press, 1973).

57 Marcel Proust, *In Search of Lost Time VI: Time Regained*, trans. Lydia Davis (London: Vintage, 2000), pp. 273–4.

58 Italo Calvino, *If On a Winter's Night a Traveller*, trans. William Weaver (London: Vintage, 1998).

59 Jean-Michel Frodon, *La Projection nationale. Cinéma et nation* (Paris: Seuil, 1998), and see my commentary in Stiegler, *Technics and Time, 3*, ch. 3.

13 Death Drive, Moral Philosophy and Denial

1 Without '*verguenza*'. See Bernard Stiegler, *Constituer l'Europe 1. Dans un monde sans vergogne* (Paris: Galilée, 2005).

2 See Bernard Stiegler, *The Neganthropocene*, trans. Daniel Ross (London: Open Humanities Press, 2018), ch. 4.

3 In Durkheim's sense.

4 I recall again here that 'organological' describes what organically and organologically *binds* the artificial organs, and that the integration involved here is that of the psychic, technical and social systems within which one instance does not short-circuit the others *insofar as they together form the neganthropic function that is reason*.

5 See Gilles Dostaler and Bernard Maris, *Capitalisme et pulsion de mort* (Paris: Albin Michel, 2009), p. 14: 'After destroying nature, we will become our own victims, as the culmination of the unconscious hate we bear. Have we not heard an Islamist leader justify the attacks and the coming victory of jihad by using this absurd phrase: "You will never love life as much as we love death"?' *Translator's note*: Hence see, for example, Raffaello Pantucci, *'We Love Death as You Love Life': Britain's Suburban Terrorists* (London: Hurst, 2015).

6 Georges Bataille, *Prehistoric Painting: Lascaux, or The Birth of Art*, trans. Austryn Wainhouse (Geneva: Skira, 1955).

7 See Marc Azéma, *La Préhistoire du cinéma. Origines paleolithiques de la narration graphique et du cinématographe* (Paris: Errance, 2011).

8 Martin Heidegger, 'The Turn', *Bremen and Freiburg Lectures: Insight Into That Which Is and Basic Principles of Thinking*, trans. Andrew J. Mitchell (Bloomington and Indianapolis: Indiana University Press, 2012), p. 65.

9 Edmund Husserl, *Ideas Pertaining to a Pure Phenomenology and to a Phenomenological Philosophy. Second Book: Studies in the Phenomenology of Constitution*, trans. Richard Rojcewicz and André Schuwer (Dordrecht: Kluwer, 1989), p. 250, translation modified.

10 This is Maurice Blanchot's question.

11 Establishing axioms as values.

12 Plato, *Meno*.

13 Such a support, which is the *hypokeimenon*, but which is never *prōton*, because it is not an origin but, precisely, a *default* of origin, such a *hypokeimenon* is always constituted through an adjustment between *hypomnēmata* that are themselves *pharmaka*, and which therefore require therapies that themselves form knowledge. It is from such knowledge that the *hypokeimenon* stems, as that which, from generation to generation, maintains a différance that is also an individuation struggling against an intergenerational indifférance to the transgenerational, and which, as noetic différance, takes care of life by means other than life.

14 It was in Martin Heidegger, *Kant and the Problem of Metaphysics*, fifth edition, trans. Richard Taft (Bloomington and Indianapolis: Indiana University Press, 1997), that Heidegger showed that the *Critique of Pure Reason*'s synthesis of recognition is the synthesis of the future.

15 That this organic solidarity is also, and always already, organological is what Durkheim did not think, but it is what his nephew, Marcel Mauss, introduced into the mind of André Leroi-Gourhan.

16 *Oikonomia*, here, refers to the work of Giorgio Agamben, which – through an approach that should be articulated with that of Pierre Legendre – roots 'economy' in its current sense with the Trinitarian becoming of monotheism, that is, in the division of the one into three, and hence as a question of spirit. It is not possible for me, here, to conduct an analysis of the work in which this thesis is developed, which is Giorgio Agamben, *The Kingdom and the Glory: For a Theological Genealogy and Economy of Government*, trans. Lorenzo Chiesa with Matteo Mandarini (Stanford: Stanford University Press, 2011).

17 I will return to this point in Bernard Stiegler, *Mystagogies*, forthcoming, Inshallah.

A partial version already exists in the journal *Boundary 2*. See Bernard Stiegler, 'The Proletarianization of Sensibility', trans. Arne De Boever, *Boundary 2* 44 (2017), pp. 5–18; Stiegler, 'Kant, Art and Time', trans. Stephen Barker with Arne De Boever, *Boundary 2* 44 (2017), pp. 19–34; and Stiegler, 'The Quarrel of the Amateurs', trans. Robert Hughes, *Boundary 2* 44 (2017), pp. 35–52.

18 As is made clear in Sigmund Freud, *Moses and Monotheism: Three Essays*, in Volume 23 of James Strachey (ed. and trans.), *The Standard Edition of the Complete Psychological Works of Sigmund Freud* (London: Hogarth, 1953–74).

19 That is, on the designation of a *pharmakos*. See Bernard Stiegler, *Pharmacologie du Front national* (Paris: Flammarion, 2013), §§3, 6, 7, 14, 43, 47, 49, 64, 66, 68 and 75.

20 Immanuel Kant, *Critique of Pure Reason*, trans. Paul Guyer and Allen W. Wood (Cambridge: Cambridge University Press, 1998).

21 'Still tragic' means that this dialogue, which is among the earliest of the Platonic dialogues and hence considered to be 'Socratic', remains close to the fundamentally tragic sensibility of Socrates himself.

22 See the article by Laurent Jaffro, 'Foucault et le problème de l'éducation morale', *Le Télémaque* 29 (2006), pp. 111–24, previously mentioned.

23 See §111.

24 This was explored by the so-called moralists towards the end of the Ancien Régime.

25 Fethi Benslama, interviewed in 'Pour les désespérés, l'islamisme radical est un produit excitant', *Le Monde* (12 November 2015), available at: http://www.lemonde.fr/societe/article/2015/11/12/pour-les-desesperes-l-islamisme-radical-est-un-produit-excitant_4808430_3224.html.

26 Ibid.

27 Ibid.

28 See Bernard Stiegler, *Uncontrollable Societies of Disaffected Individuals: Disbelief and Discredit, Volume 2*, trans. Daniel Ross (Cambridge: Polity, 2013).

29 Mark Hunyadi, *La Tyrannie des modes de vie: sur le paradoxe moral de notre temps* (Lormont: Le Bord de l'Eau, 2015).

30 See Bernard Stiegler, *Taking Care of Youth and the Generations*, trans. Stephen Barker (Stanford: Stanford University Press, 2010), ch. 1, esp. p. 8. *Translator's note*: And see Bernard Stiegler, *States of Shock: Stupidity and Knowledge in the Twenty-First Century*, trans. Daniel Ross (Cambridge: Polity, 2015), p. 26 and §27, esp. pp. 68–9 and including the translator's note appended to note 29 on p. 239.

31 See Bernard Stiegler, *Automatic Society, Volume 1: The Future of Work*, trans. Daniel Ross (Cambridge: Polity, 2016).

32 *Translator's note*: Jérôme Cahuzac.

33 Frédéric Lordon, *Imperium. Structures et affects des corps politiques* (Paris: La Fabrique, 2015), p. 175.

34 Sigmund Freud, *The Future of an Illusion*, in Volume 21 of Strachey, *The Standard Edition of the Complete Psychological Works of Sigmund Freud*, p. 15.

35 See Sigmund Freud, 'Instincts and their Vicissitudes', in Volume 14 of Strachey, *The Standard Edition of the Complete Psychological Works of Sigmund Freud*. *Translator's note*: See esp. pp. 122–3. It is necessary to recall that Strachey translates *Trieb* as 'instinct' rather than 'drive', which is unfortunate because, as Strachey himself points out, this word 'is in any case not used here in the sense which seems at the moment to be the most current among biologists' ('Editor's note', p. 111).

36 We will see that this is what Dostaler and Maris, like Marcuse, do not understand.

37 See Donald Winnicott, *Playing and Reality* (Abingdon: Routledge, 2005), and my commentary in the 'Introduction' to Bernard Stiegler, *What Makes Life Worth Living: On Pharmacology*, trans. Daniel Ross (Cambridge: Polity, 2013).

38 Lordon, *Imperium*, p. 176.

39 Ibid., p. 118.

40 Hunyadi, *La Tyrannie des modes de vie*, pp. 43–4.

41 See Stiegler, *Automatic Society, Volume 1*, ch. 4.
42 During the 1980s, the 'conservative revolution', which makes government 'the problem and not the solution' [as Ronald Reagan said – *trans.*], takes disinhibition to extremes by liquidating public power, which, since the beginning of the industrial revolution, has had the goal of preventing the social systems from being destroyed by the uncontrolled growth of the industrial technological system. The conservative revolution will then be the ideology that accompanies the systematic replacement of public power through the marketing of behavioural prescriptions. This will result in the liquidation of all forms of knowledge, including scientific knowledge, as I have begun to argue in *Automatic Society, Volume 1* by discussing Chris Anderson's 'The End of Theory'.
43 Hunyadı, *La Tyrannie des modes de vie*, p. 44.
44 Ibid.
45 See Stiegler, *Automatic Society, Volume 1*, p. 111.
46 In Stiegler, *La Société automatique 2. L'avenir du savoir*, I will try to show why the rules that are produced and consented to communally at the heart of what Elinor Ostrom, Charlotte Hess and Benjamin Coriat describe as the commons are, equally, all the forms of knowledge that take care of their objects, and why an economy of contribution is negentropic precisely in the sense that it constitutes the macro-economic model that rests on the valuation of knowledge thus conceived.
47 Hunyadi, *La Tyrannie des modes de vie*, p. 32.

14 Nonconformism, 'Uncoolness' and *Libido Sciendi* at the University

1 Isidore Isou, *Traité d'économie nucléaire 1. Le soulèvement de la jeunesse* (Paris: Maurice Lemaître, 1957).
2 Adam Smith is obviously, along with Bernard Mandeville and his *Fable of the Bees* to which Dany-Robert Dufour so frequently refers, a turning point in the history of conventions and of what he calls 'propriety' [*convenance*, in the French translation]. See, for example, Adam Smith, *The Theory of Moral Sentiments* (Cambridge: Cambridge University Press, 2002), part 1, section 1, ch. 3. As for the behaviour of bees and the lessons to be drawn from them with respect to neganthropic mores to come, see Yann Moulier Boutang, *L'Abeille et l'Économiste* (Paris: Carnets Nord, 2010), and my commentary in Bernard Stiegler, *Automatic Society, Volume 1: The Future of Work*, trans. Daniel Ross (Cambridge: Polity, 2016), ch. 8.
3 Herbert Marcuse, *Eros and Civilization: A Philosophical Inquiry into Freud* (London: Abacus, 1972). *Translator's note*: The first edition was published in 1955.
4 Herbert Marcuse, *One-Dimensional Man: Studies in the Ideology of Advanced Industrial Society* (Boston: Beacon, 1964).
5 Guy Debord, *Society of the Spectacle*, no translator credited (Detroit: Black & Red, 1983).
6 Luc Boltanski and Ève Chiapello, *The New Spirit of Capitalism*, trans. Gregory Elliott (London and New York: Verso, 2005).
7 *Translator's note*: See René Descartes, *Meditations on First Philosophy, with Selections from the Objections and Replies*, trans. John Cottingham (Cambridge: Cambridge University Press, 1986), p. 12, translation modified: 'I had to seriously undertake, once in my life, to rid myself of all the opinions I had received into my set of beliefs [*créance*] up until that moment, and to begin afresh from the foundations.' And see Bernard Stiegler, *States of Shock: Stupidity and Knowledge in the Twenty-First Century*, trans. Daniel Ross (Cambridge: Polity, 2015), §48.
8 Friedrich Engels, *Dialectic of Nature*, trans. Clemens Dutt, in Karl Marx and Friedrich Engels, *Collected Works, Volume 25* (London: Lawrence & Wishart, 1987), p. 319.
9 Karl Polanyi, *The Great Transformation: The Political and Economic Origins of Our Time* (Boston: Beacon, 2001).

10 Jean-François Billeter, *Contre François Jullien* (Paris: Allia, 2006).
11 See Immanuel Kant, 'An Answer to the Question: "What is Enlightenment?"', *Political Writings*, trans. H.B. Nisbet (Cambridge: Cambridge University Press, 1991), and my commentary in Bernard Stiegler, *Taking Care of Youth and the Generations*, trans. Daniel Ross (Stanford: Stanford University Press, 2010), pp. 20–2.
12 Stiegler, *Automatic Society, Volume 1*, pp. 87–8.
13 Karl Marx and Friedrich Engels, *The Communist Manifesto*, trans. Samuel Moore and Friedrich Engels (London: Penguin, 1967), p. 82.
14 Ibid., translation modified.
15 Ibid., pp. 82–3.
16 Ibid., p. 83.
17 Ibid.
18 Ibid., p. 84.
19 This, at least, is what I have tried to show in *States of Shock*, by drawing attention to the fact that the working-class proletariat will not overthrow capital. We can no longer describe this revolutionary process of overthrowing the 'bourgeois' revolution in terms of a dialectic where the power of the negative will achieve a revolutionary synthesis: the proletariat *is not* revolutionary. The 'revolutionary' power is *de-proletarianization* inasmuch as it gives rise to a new era of value: practical value, irreducible to use value, not soluble into exchange value and productive of negan-thropic bifurcations.
20 Marx and Engels, *The Communist Manifesto*, p. 88.
21 Ibid., p. 87.
22 Karl Marx, *Grundrisse: Foundations of the Critique of Political Economy (Rough Draft)*, trans. Martin Nicolaus (London: Pelican, 1973).
23 In particular, where *Capital* analyses the *value* and *nature* of *work*.
24 This is what Lyotard argues, notably in Jean-François Lyotard, 'The Tomb of the Intellectual', *Political Writings*, trans. Bill Readings and Kevin Paul Geiman (London: UCL Press, 1993). But he maintains this argument without being able to think it: he observes the fact, nothing more. On this point, see Stiegler, *States of Shock*, ch. 4.
25 Engels, *Dialectic of Nature*, p. 334.
26 Augustine, *Confessions*, trans. R.S. Pine-Coffin (London: Penguin, 1961), Book 10, §35, p. 242.
27 See Bernard Stiegler and Ariel Kyrou, *L'emploi est mort, vive le travail!* (Paris: Mille et une nuits, 2015) and Stiegler, *Automatic Society, Volume 1*.
28 This is what Socrates teaches, and what by the same token he takes to the extreme by drinking the *pharmakon* while sacrificing himself to the god of the *pharmakon* – and it is on such examples that the West will be built, like all civilizations. As for Christian exemplarity and Christianity in general, they have been transformed over the course of their disinhibiting 'globalatinization', eventually constituting the vector of capitalism as Weber conceived it – the rule thus establishing its secularization, and the example proving to itself be pharmacological.
29 Mark Hunyadi, *La Tyrannie des modes de vie: sur le paradoxe moral de notre temps* (Lormont: Le Bord de l'Eau, 2015), p. 8.
30 This will be a key theme in Bernard Stiegler, *La Technique et le temps 6*.
31 Gilles Dostaler and Bernard Maris, *Capitalisme et pulsion de mort* (Paris: Albin Michel, 2009), p. 7. *Translator's note*: It has not been possible to locate this precise phrase, 'morbid desire for liquidity', in Keynes's work, however it may be that their reference is to a passage from which they quote directly later in the work, and which expresses this notion: see John Maynard Keynes, 'Economic Possibilities for Our Grandchildren', *Essays in Persuasion* (Houndmills, Basingstoke: Palgrave Macmillan, 2010), p. 329: 'The love of money as a possession – as distinguished from the love of money as a means to the enjoyments and realities of life – will be recognised

for what it is, a somewhat disgusting morbidity, one of those semi-criminal, semi-pathological propensities which one hands over with a shudder to the specialists in mental disease.'

32 Dostaler and Maris, *Capitalisme et pulsion de mort*, p. 33.
33 Moreover, we cannot exactly say that man 'represses pleasure': he defers it, which is something other than a repression. Repression is something else: it consists in hiding a desire that thereby remains unconscious.
34 Bernard Stiegler, *The Lost Spirit of Capitalism: Disbelief and Discredit, Volume 3*, trans. Daniel Ross (Cambridge: Polity, 2014), pp. 48–58.
35 See Sigmund Freud, *Beyond the Pleasure Principle*, in Volume 18 of James Strachey (ed. and trans.), *The Standard Edition of the Complete Psychological Works of Sigmund Freud* (London: Hogarth, 1953–74), pp. 19–20, and especially Sigmund Freud, *The Ego and the Id*, in Volume 19 of Strachey, *The Standard Edition of the Complete Psychological Works of Sigmund Freud*, pp. 16–18.
36 *Translator's note*: See Alain Minc, *La mondialisation heureuse* (Paris: Plon, 1997).
37 Dostaler and Maris, *Capitalisme et pulsion de mort*, p. 8.
38 Ibid.
39 Ibid.
40 I began an analysis of this transformation of the very idea of science, to which the industrial revolution gave rise, in Bernard Stiegler, *Technics and Time, 3: Cinematic Time and the Question of Malaise*, trans. Stephen Barker (Stanford: Stanford University Press, 2011).
41 Dostaler and Maris, *Capitalisme et pulsion de mort*, p. 8.
42 Ibid., p. 9.
43 *Translator's note*: The reference is to Sigmund Freud, *Civilization and Its Discontents*, in Volume 21 of Strachey, *The Standard Edition of the Complete Psychological Works of Sigmund Freud*, pp. 96 and 123.
44 Dostaler and Maris, *Capitalisme et pulsion de mort*, pp. 12–13.
45 *Translator's note*: See Freud, *Civilization and Its Discontents*, p. 90: 'With every tool man is perfecting his own organs, whether motor or sensory, or is removing the limits to their functioning.'
46 *Translator's note*: The reference here is to Sigmund Freud, 'Instincts and their Vicissitudes', in Volume 14 of Strachey, *The Standard Edition of the Complete Psychological Works of Sigmund Freud*, which is the non-literal English translation of the German title, 'Triebe und Triebschicksale', translated into French more literally as 'Pulsions et destins des pulsions'.
47 Dostaler and Maris, *Capitalisme et pulsion de mort*, p. 10.
48 Ibid., p. 13.
49 See ibid., p. 9: 'Capitalism exists only by infinitely accruing surpluses. And waste, the "accursed share" that Bataille describes from time to time, claims its due.'
50 Ibid., p. 17: 'Brown is probably the first to highlight the convergences between the ideas of Freud and Keynes on money, capitalism and death.' See Norman O. Brown, *Life Against Death: The Psychoanalytical Meaning of History*, 2nd edition (Hanover, NH: Wesleyan University Press, 1985).
51 Dostaler and Maris, *Capitalisme et pulsion de mort*, p. 22: 'Keynes [...] considered [the love of money] as the main moral problem of his time.'
52 Ibid., p. 19.
53 Ibid., p. 20.
54 Ibid., p. 34.
55 Ibid., my italics.
56 Freud, *The Ego and the Id*, p. 30, translation modified.
57 Anatole Bailly, *Dictionnaire Grec-Français* (Paris: Hachette, 1903) p. 992. *Translator's note*: This may be compared with Henry George Liddell and Robert Scott, *A Greek-English Lexicon*, 4th edition (Oxford: Oxford University Press, 1855), p. 658: 'a holding, keeping hold of, preservation, [...] a holding in'.

58 Bernard Stiegler, *For a New Critique of Political Economy*, trans. Daniel Ross (Cambridge: Polity, 2010).
59 Dostaler and Maris, *Capitalisme et pulsion de mort*, p. 36. For the quotation from Marcuse, see Marcuse, *Eros and Civilization*, p. 73.
60 Freud, *Civilization and Its Discontents*, p. 145, quoted in Dostaler and Maris, *Capitalisme et pulsion de mort*, p. 127.
61 Freud, *Civilization and Its Discontents*, p. 90.
62 Bernard Stiegler, *What Makes Life Worth Living: On Pharmacology*, trans. Daniel Ross (Cambridge: Polity, 2013).
63 Paul Valéry, 'The Crisis of the Mind', *The Collected Works of Paul Valéry, Volume 10: History and Politics*, trans. Denise Folliot and Jackson Matthews (New York: Bollingen, 1962).
64 Paul Valéry, 'Freedom of the Mind', in ibid.
65 Edmund Husserl, *The Crisis of European Sciences and Transcendental Phenomenology: An Introduction to Phenomenological Philosophy*, trans. David Carr (Evanston: Northwestern University Press, 1970).
66 Valéry, 'Freedom of the Mind', p. 190, translation modified.
67 Dostaler and Maris, *Capitalisme et pulsion de mort*, p. 128.
68 Freud, *Civilization and Its Discontents*, p. 145. *Translator's note*: Note that this was the final line of the text as it went to the printers in November 1929 (see the 'Editor's Introduction', p. 59), but, by the time the reprint edition was published in 1931, Freud felt compelled to add one further sentence, 'But who can foresee with what success and with what result?' As Strachey remarks, this addition was reflective of 'the menace of Hitler [that] was already beginning to be apparent' (p. 145).
69 Keynes, 'Economic Possibilities for Our Grandchildren', pp. 330–1, my italics.
70 Dostaler and Maris, *Capitalisme et pulsion de mort*, p. 140, my italics.
71 Here, we should dwell at length on Keynes's analyses of the crisis of the 1930s, where he evokes, according to the paraphrase by Dostaler and Maris, 'an extremely rapid process of technical improvement' (ibid., p. 129), just as did Freud in the same year, adding that 'this means in the long run *that mankind is solving its economic problem*' (Keynes, 'Economic Possibilities for Our Grandchildren', p. 325, quoted in Dostaler and Maris, *Capitalisme et pulsion de mort*, p. 129). It would also be necessary to challenge Keynes and Smith with regard to the absorption of 'moral and material energies' by capitalism and its technology, which requires us to 'create the organisation to use them, capable of reducing the economic problem [...] to a position of secondary importance' (Keynes, 'Economic Possibilities for Our Grandchildren', p. xviii, quoted in Dostaler and Maris, *Capitalisme et pulsion de mort*, p. 130).
72 See ch. 15, n. 42.
73 Catherine Malabou, *The New Wounded: From Neurosis to Brain Damage*, trans. Steven Miller (New York: Fordham University Press, 2012).

15 The Wounds of Truth: Panic, Cowardice, Courage

1 This sub-chapter is a reminder and a development of what I have posited in §9 with respect to denial [*denegation*].
2 *Translator's note*: On *dénégation* and *déni*, see the translator's note in ch. 2, n. 31.
3 Claude Lévi-Strauss, *Tristes Tropiques*, trans. John Weightman and Doreen Weightman (Harmondsworth, Middlesex: Penguin, 1976), pp. 542–3.
4 See §34.
5 Engels, for example, is concerned about the future of nature in *Dialectics of Nature*, as is Marx, as we have already seen with Fressoz, but Fressoz and Bonneuil also point to many other cases.
6 Such as the climatologists of the IPCC, the biologists who signed Anthony D. Barnosky et al., 'Approaching a State Shift in Earth's Biosphere', *Nature* 486 (7 June 2012), pp. 52–8, as well as the scholars surrounding Stephen Hawking, and so on.

7 The work of Hidetaka Ishida, presented at the Institut de recherche et d'innovation, suggests that this is something we have cause to fear. *Translator's note*: This denial with respect to Fukushima, and the possibility that such denial could lead to a repetition, is also the theme of the Japanese film, *Kibô no kuni* (*The Land of Hope*, 2012), directed by Sion Sono.

8 *Translator's note*: This is a reference to the 1935 song, 'Tout va très bien, Madame la Marquise', by Paul Misraki, and see Bernard Stiegler, *States of Shock: Stupidity and Knowledge in the Twenty-First Century*, trans. Daniel Ross (Cambridge: Polity, 2015), §10.

9 See pp. 255–6 and Gilles Dostaler and Bernard Maris, *Capitalisme et pulsion de mort* (Paris: Albin Michel, 2009), p. 140.

10 Immanuel Kant, 'On a Supposed Right to Lie because of Philanthropic Concerns', *Grounding for the Metaphysics of Morals, with On a Supposed Right to Lie Because of Philanthropic Concerns*, 3rd edition, trans. James W. Ellington (Indianapolis: Hackett, 1993).

11 This divorce destroys the possibility of transcendental apperception, that is, the possibility of unifying the soul's experience, which would also be a transcendental affinity, that is, a coherence of the world to which this experience would bear witness.

12 Paul Valéry, 'The Crisis of the Mind', *The Collected Works of Paul Valéry, Volume 10: History and Politics*, trans. Denise Folliot and Jackson Matthews (New York: Bollingen, 1962).

13 As summarized in the article in French Wikipédia (no access date), Milgram's experiment 'sought to assess the obedience of an individual before an authority that he or she deemed to be legitimate, and to analyse the process of submission to authority, particularly when it demands actions that pose problems of conscience for the subject'.

14 *Translator's note*: This programme quite faithfully adapted the Milgram experiment to the game show context, as can be seen by reading the description in David Chazan, 'Row over "torture" on French TV', *BBC News* (18 March 2010), available at: http://news.bbc.co.uk/2/hi/europe/8573755.stm.

15 Paul Quilès, 'Pourquoi j'ai porté plainte contre le "Jeu de la mort"', *Rue89* (26 March 2010), available at: https://tempsreel.nouvelobs.com/rue89/rue89-medias/20100326.RUE5735/pourquoi-j-ai-porte-plainte-contre-le-jeu-de-la-mort.html.

16 *Translator's note*: The phrase 'éléphants du PS' is a pejorative term for important figures in the French Socialist Party. Both Quilès and Lienemann were or are important figures in the Parti socialiste, and both were former French government ministers.

17 *Translator's note*: '*Noyer le poisson*' is an idiomatic expression whose meaning is equivalent to the English idiomatic expressions 'to cloud the issue' and 'to muddy the waters', but with the added sense of going over and over something until all resistance is overcome. It apparently derives from fishing: having hooked a fish, the line is pulled, lifting its head repeatedly out of the water (into the air, 'drowning' it) until it is too exhausted to struggle any further.

18 In this instance the Socialist Party, of which the founder and CEO of Skyrock, Pierre Bellanger, is a member, and one cannot help but wonder what this radio station, so highly prized by the politicians of the 'modern left', has to offer them.

19 *Translator's note*: Denis Robert is a French investigative journalist who exposed what is known as the Clearstream affair, that is, the uncovering of the fact that this giant trade settlement institution was heavily involved in financial and organized crime. As a result of his work exposing this affair, Robert found himself the subject of dozens of lawsuits and charges, leading to a decade of wrangling, appeals and legal fees, as well as the prospect of prison. The ultimate outcome of all of these lawsuits and trials was a set of decisions that overwhelmingly vindicated the journalist, who by that time, of course, had nevertheless been forced to pay an enormous price for his investigative activities.

20 *Translator's note*: The French term '*lampiste*', 'lamplighter', referred in the nineteenth century to the lowliest railway worker, and by extension refers to the underling who can be held responsible for the mistakes of his superiors, the fall guy that bosses can pin the blame on while themselves remaining in the shadows.

21 This is what I have debated at length in Bernard Stiegler, *Pharmacologie du Front national* (Paris: Flammarion, 2013).

22 Michel Foucault, *The Courage of Truth (The Government of Self and Others II): Lectures at the Collège de France, 1983–1984*, trans. Graham Burchell (New York: Palgrave Macmillan, 2011), p. 45.

23 Ibid., p. 46.

24 See pp. 103 and 156.

25 Foucault, *The Courage of Truth*, p. 35.

26 '*Sapere aude*', 'dare to know', is Kant's injunction in 'What is Enlightenment?'

27 This is Valéry's injunction in *Le Cimetière marin*.

28 Foucault, *The Courage of Truth*, p. 67.

29 On the question of locality, place and taking place in exosomatization, see §137.

30 See Bernard Stiegler, *States of Shock: Stupidity and Knowledge in the Twenty-First Century*, trans. Daniel Ross (Cambridge: Polity, 2015), ch. 3.

31 See Henri Bergson, *Creative Evolution*, trans. Arthur Mitchell (New York: Modern Library, 1944), and Henri Bergson, *The Two Sources of Morality and Religion*, trans. R. Ashley Audra and Cloudesley Brereton, with W. Horsfall Carter (Notre Dame: University of Notre Dame Press, 1977).

32 Michel Crozier, Samuel P. Huntington and Joji Watanuki, *The Crisis of Democracy: Report on the Governability of Democracies to the Trilateral Commission* (New York: New York University Press, 1975).

33 Jacques Rancière, *Hatred of Democracy*, trans. Steve Corcoran (London and New York: Verso, 2006), p. 7, translation modified.

34 See Jacques Rancière, 'Non, le peuple n'est pas une masse brutale et ignorante', *Libération* (3 January 2011), available at: http://www.liberation.fr/france/2011/01/03/non-le-peuple-n-est-pas-une-masse-brutale-et-ignorante_704326: 'The notion of populism constructs a people characterized by the formidable amalgam of a capacity – the brute power of great numbers – and an incapacity – the ignorance attributed to that same great number. For this, a third trait, racism, is essential. It is a matter of showing democrats, always suspected of being naïve [*angélisme*], what, deep down, people really are: a pack, inhabited by a primary drive to reject, which at the same time targets rulers whom it declares traitors, without understanding the complexity of political mechanisms, and foreigners whom it fears out of atavistic attachment to a set of living conditions threatened by demographic, economic and social evolution. The notion of populism puts back on stage an image of the people elaborated in the late nineteenth century by thinkers such as Hippolyte Taine and Gustave Le Bon, frightened by the Paris Commune and the rise of the workers' movement: that of ignorant crowds impressed by the resounding words of "leaders" and led to extreme violence by the uncontrolled circulation of rumours and contagious fears.'

35 What is at stake, here, is the crowd, not the people. It is one thing to point out that Le Bon or Taine may misinterpret the fact that crowds can behave in a herd-like way. It is another thing altogether to deny this fact: it is in itself an intellectual regression. In saying that a crowd is regressive, did this amount to a crime against the people's sovereignty? No: the people are unfortunately and *evidently* regressive, *as is* Rancière, as is the *whole world*. Regression is a tendency that is constitutive for any 'progression'. As for placing Freud on the side of the 'reactionaries', this is something that Michel Onfray will then be delighted to repeat in a bestseller widely promoted by the highly reactionary magazine *Le Point*. *Translator's note*: The reference here is to Michel Onfray, *Le Crépuscule d'une idole. L'Affabulation freudienne* (Paris: Grasset, 2010).

36 See Gilbert Simondon, *L'Individuation psychique et collective* (Paris: Aubier, 2007),

pp. 21–2, and my commentary in the preface to that book, translated as Bernard Stiegler, 'The Uncanniness of Thought and the Metaphysics of Penelope', trans. Arne De Boever, with Greg Flanders and Alicia Harrison, *Parrhesia* 23 (2015), p. 63, available at: http://www.parrhesiajournal.org/parrhesia23/parrhesia23_stiegler.pdf.

37 This is what Rancière refuses to admit, by relying on Jacotot. I have already explained elsewhere, however, why such a position remains highly superficial.

38 Which is also to say, those who do not exercise power effectively. In this respect, Frédéric Lordon's analyses fail to take into account this system for capturing the *imperium* of the multitude that is exercised by economic power, and not political power, via the analogical and digital *pharmaka* of our time. Everything he says is very forceful, but what it describes is above all that which would stem from the *imperium* of the epoch of books. What is lacking is an account of what leads to the absence of epoch as the epoch of a 'new form of barbarism'.

39 Jacques Rancière, *The Politics of Aesthetics: The Distribution of the Sensible*, trans. Gabriel Rockhill (London and New York: Bloomsbury Academic, 2004). *Symbolic Misery* is a kind of commentary on this work.

40 *Translator's note*: See Cynthia Fleury, *Les Pathologies de la démocratie* (Paris: Fayard, 2005) and Cynthia Fleury, *La Fin du courage: la reconquête d'une vertu démocratique* (Paris: Fayard, 2010).

41 See my contribution to the debate opened by *Libération* on the notion of populism, in Bernard Stiegler, 'Les médias analogiques ont engendré un nouveau populisme', *Libération* (5 January 2011), available at: http://www.liberation.fr/france/2011/01/05/les-medias-analogiques-ont-engendre-un-nouveau-populisme_704860.

42 The 'extropians' have their manifesto (see 'The Extropist Manifesto', available at: http://extropism.tumblr.com/post/393563122/the-extropist-manifesto) and their leader, Max More. See Antonio A. Casilli, 'Le debat sur le nouveau corps dans le cyberculture: le cas des extropiens', in Olivier Sirost (ed.), *Le Corps extreme dans les sociétés occidentales* (Paris: L'Harmattan, 2005), p. 297:

> In 1987, Max O'Connor, a 22-year-old English bodybuilder studying philosophy at Oxford and passionate about far-out scientific theories about colonization in space and cryogenics, decided to leave England in order to pursue his doctoral research at the less traditional University of Southern California. Together with a group of friends, he became interested in the issues involved in new technologies and practices to do with extending human lifespan. In 1988, he changed his name to Max More and began publishing a new journal, *Extropy*, whose initial subtitle was 'Vaccine for Future Shock'.

In an internal note for Ars Industrialis, Anne Alombert writes:

> Extropianism is a transhumanist tendency – if not its foundation. It presents itself as a framework of thought, a set of values and perspectives that aim to inspire and generate innovative and optimistic thinking about contemporary social, organizational and technological transformations, to which traditional value systems are no longer well-adapted. It is a matter of proactively codifying, in (evolutionary) principles, ideas of the promotion and affirmation of life: these principles are not presented as eternal truths but as tools through which sense can be made of the ongoing transformation of the human condition. [...] Extropy is here assimilated to order and growth, and it is thought in opposition to entropy: this is not the idea of a play of entropic/negentropic tendencies whose composition secures existence, and it contains no mention of singularity, bifurcation or the diversification and differentiation that all seem necessary in order to conceive the production of 'negentropy'.
>   Extropy is associated with intelligence conceived as creativity, order, critical thinking, ingenuity and unlimited energy, opposed to irrational faith and

dogmatic tradition: valuing information and imagination (the most precious materials in the universe).

The journal strongly proclaims its confidence in the individual, against any authority or law that constrains free initiative. The extropians define themselves as libertarians, in the anarcho-capitalist tradition of Nozick and von Hayek.

43 See p. 230.

44 Sigmund Freud, 'Negation', in Volume 19 of James Strachey (ed. and trans.), *The Standard Edition of the Complete Psychological Works of Sigmund Freud* (London: Hogarth, 1953–74), pp. 235–6.

45 *Germen* and *soma* are the terms by which August Weismann, a contemporary of Freud, established the primary concepts of genetics.

46 See Karl Marx and Friedrich Engels, *The German Ideology*, no translator credited (Moscow: Progress Press, 1976), and my commentary in Stiegler, *Pharmacologie du Front national*.

47 I have explained this point in Bernard Stiegler, *Automatic Society, Volume 1: The Future of Work*, trans. Daniel Ross (Cambridge: Polity, 2016), pp. 60 and 69–70, and in Bernard Stiegler, *The Neganthropocene*, trans. Daniel Ross (London: Open Humanities Press, 2018), ch. 10.

48 Jakob von Uexküll, *A Foray into the World of Animals and Humans, with, A Theory of Meaning*, trans. Joseph D. O'Neil (Minneapolis and London: University of Minnesota Press, 2010).

49 Sigmund Freud, *Civilization and Its Discontents*, in Volume 21 of Strachey, *The Standard Edition of the Complete Psychological Works of Sigmund Freud*, p. 90.

50 See this chapter, n. 42.

51 See Sigmund Freud, 'The Acquisition and Control of Fire', in Volume 22 of Strachey, *The Standard Edition of the Complete Psychological Works of Sigmund Freud*.

52 *Translator's note*: It is worth mentioning in passing that the date of printing of the French edition of this book is April 2016 (and so these lines were obviously written some considerable time prior to that), that Donald Trump did not become the presumptive nominee of the Republican Party until May, and that, both before and after winning the nomination, Trump was considered by the vast majority of pundits to have little or no serious chance of achieving the presidency, right up until the day in November when he did just that. We may well consider that this predictive failure is itself symptomatic evidence of one of the kinds of contemporary denial to which the author here refers.

53 *Translator's note*: The German word used by Kant here is *Gängelwagen* (and, in the French translation, *voiture d'enfant*, as discussed in the next note). The translation of this term as 'leading-strings' by H.B. Nisbet certainly conveys Kant's meaning, but has the dual disadvantage of referring to a seventeenth- and eighteenth-century technique of 'aiding' children who are learning to walk that is no longer practised today (although, on the other hand, the use of a 'lead' or a 'harness' on young children seems to have become more common), and losing the connection that Stiegler is making between a more 'vehicular' baby walker (one on wheels, say) and the automobile driven by adults. For that reason, the translation has been changed to 'child's buggy', even though the device that is intended is most often referred to as a baby walker.

54 *Gängelwagen*, a word that Cyril Morana, the French translator of Kant's text, also renders by '*youpala*', the baby walker into which infants who are learning to walk are placed.

55 Immanuel Kant, 'An Answer to the Question: "What is Enlightenment?"', *Political Writings*, trans. H.B. Nisbet (Cambridge: Cambridge University Press, 1991), p. 54, translation modified.

56 See Alain Badiou, 'L'hypothèse communiste – interview d'Alain Badiou par Pierre Gaultier', *Le Grand Soir* (6 August 2009), available at: https://www.legrandsoir.

info/L-hypothese-communiste-interview-d-Alain-Badiou-par-Pierre.html: 'We must not allow ourselves to be distracted from this demand by millenarian diversions, the main one being, today, ecology. It would be so useful for our opponents in crisis if we all had to reconcile to save the planet! A threatened planet united against violent political division, what a godsend! Capitalism will itself become, in order to save itself, more ecologically friendly. It will all be sustainable development banks, water purity holding companies and pension funds for the whales. I'm not afraid to say it: ecology is the new opium of the people. And, as ever, this opium has its official philosopher, in this case Sloterdijk. To be affirmationist is also to refuse to be intimidated by the manoeuvres carried out around "nature". We must clearly state that humanity is an animal species that tries to exceed its animality, a natural whole that tries to de-naturalize itself.'

57 René Char, *À une sérénité crispée* (Paris: Gallimard, 1951), p. 21. *Translator's note*: These lines from Char were also quoted by Georges Bataille in his 1951 review of Char's volume. See Georges Bataille, 'René Char and the Force of Poetry', *The Absence of Myth: Writings on Surrealism*, trans. Michael Richardson (London and New York: Verso, 1994), p. 132.

58 I have stressed and argued this point in Bernard Stiegler, *Taking Care of Youth and the Generations*, trans. Stephen Barker (Stanford: Stanford University Press, 2010).

59 Jacques Derrida, *Of Spirit: Heidegger and the Question*, trans. Geoffrey Bennington and Rachel Bowlby (Chicago and London: University of Chicago Press, 1989).

60 Alain Rey (ed.), *Dictionnaire historique de la langue française*, vol. 2 (Paris: Le Robert, 2012), p. 2418.

61 In the Anthropocene, a geological era and a theologico-calendrical era seem to merge and to be confounded into a kind of atheological era whose calendarity and cardinality have become totally computational. On the notions of calendarity and cardinality, see Bernard Stiegler, *Technics and Time, 3: Cinematic Time and the Question of Malaise*, trans. Stephen Barker (Stanford: Stanford University Press, 2011). The notions of kingdom, age, era and epoch require a transdisciplinary dialogue between geologists, archaeologists, historians, geographers (there is no era without an area), philosophers, anthropologists and theologians. The beginnings of such a project have been developed in Krzysztof Pomian, *L'Ordre du temps* (Paris: Gallimard, 1984).

62 *Translator's note*: On 'calculative thinking' [*rechnendes Denken*], and on the way Heidegger opposes it to 'meditative thinking' [*besinnliches Denken*], see, for example, Martin Heidegger, *Discourse on Thinking*, trans. John M. Anderson and E. Hans Freund (New York: Harper & Row, 1969), pp. 46, 50 and 52–3.

63 It is as the *Abbau* of metaphysics that Heidegger describes his enterprise. The word *Abbau* means reduction, suppression, decomposition, elimination, disassimilation, and Gérard Granel translated it as deconstruction – a term taken up by Derrida to describe his own thought. *Translator's note*: In English translations of Heidegger's work, *Abbau* and *abbauen* are most frequently translated as 'dismantling' or 'deconstruction'. In Rodolphe Gasché, *The Tain of the Mirror: Derrida and the Philosophy of Reflection* (Cambridge, MA and London: Harvard University Press, 1986), p. 113, Gasché notes Heidegger's use of the phrase *kritischer Abbau* in Martin Heidegger, *The Basic Problems of Phenomenology*, trans. Albert Hofstadter, revised edition (Bloomington and Indianapolis: Indiana University Press, 1982), pp. 22–3. While Gasché is undoubtedly right that the phrase arises, in Heidegger's work, in debate with the Husserlian notion of the reduction, it in fact appears at least four years earlier, in lectures not published in German until 1988 (hence after Gasché's book). This debate with phenomenology, and with the question of its historicity and the necessity of the deconstruction of the tradition, is outlined in a rather full and clear way in the 1923 lectures published as Martin Heidegger, *Ontology – The Hermeneutics of Facticity*, trans. John van Buren (Bloomington and Indianapolis: Indiana University Press, 1999), §15. For the use of the term in later Heidegger,

see the 1955 lecture at Cérisy-la-Salle published as Martin Heidegger, *What is Philosophy?*, trans. Jean T. Wilde and William Kluback (Albany: NCUP, 1958), pp. 71–3, and see also, from the same year, Martin Heidegger, 'On the Question of Being', trans. William McNeill, *Pathmarks* (Cambridge: Cambridge University Press, 1998), p. 315.

64 Martin Heidegger, *What is Called Thinking?*, trans. J. Glenn Gray (New York: Harper & Row, 1968), pp. 7–8.

65 See Bernard Stiegler, *Technics and Time, 1: The Fault of Epimetheus*, trans. Richard Beardsworth and George Collins (Stanford: Stanford University Press, 1998), pp. 272–5.

66 Heidegger, *Discourse on Thinking*, p. 56.

67 Ibid.

68 Martin Heidegger, 'The Turn', *Bremen and Freiburg Lectures: Insight Into That Which Is and Basic Principles of Thinking*, trans. Andrew J. Mitchell (Bloomington and Indianapolis: Indiana University Press, 2012), pp. 69–70, translation modified. And see also Martin Heidegger, 'Time and Being', *On Time and Being*, trans. Joan Stambaugh (New York: Harper & Row, 1972), p. 2, where Heidegger states: 'We want to say something about the attempt to think Being without regard to its being grounded in terms of beings. The attempt to think Being without beings becomes necessary because otherwise, it seems to me, there is no longer any possibility of explicitly bringing into view the Being of what *is* today all over the earth, let alone of adequately determining the relation of man to what has been called "Being" up to now.'

69 See Alfred North Whitehead, *The Function of Reason* (Princeton: Princeton University Press, 1929).

Conclusion: Let's Make a Dream

1 Madeleine de Souvré, Marquise de Sablé, Maxim 8, quoted in Margot Kruse, 'Sagesse et folie dans l'oeuvre des moralistes', *Cahiers de l'Association internationale des études françaises* 30 (1978), p. 130, available at: http://www.persee.fr/doc/caief_0571-5865_1978_num_30_1_1166.

2 Balthasar Gracian, *The Art of Worldly Wisdom*, trans. Joseph Jacobs (London and New York: Macmillan, 1892), originally published in 1647.

3 Kruse, 'Sagesse et folie dans l'oeuvre des moralistes', p. 129.

4 It is completed when, in the eighteenth century, Voltaire defines reason as the *opposite* of madness.

5 François de La Rochefoucault, *Collected Maxims and Other Reflections*, trans. E.H. Blackmore, A.M. Blackmore and Francine Giguère (Oxford and New York: Oxford University Press, 2007), p. 159, quoted in Kruse, 'Sagesse et folie dans l'oeuvre des moralistes', p. 129.

6 Jonathan Crary, *24/7: Late Capitalism and the Ends of Sleep* (London and New York: Verso, 2013).

7 See p. 288.

8 The First Epistle of Paul the Apostle to the Corinthians 1:18: 'For the preaching of the Crosse is to them that perish, foolishnesse; but unto us which are saved, it is the power of God.' Quoted from *The Holy Bible: Quatercentenary Edition* (Oxford and New York: Oxford University Press, 2010), no page numbers.

9 Kruse, 'Sagesse et folie dans l'oeuvre des moralistes', p. 126. *Translator's note*: For the quotation from Erasmus, see Desiderius Erasmus, *The Praise of Folly*, trans. Hoyt Hopewell Hudson (Princeton and Oxford: Princeton University Press, 2015), p. 51, where the translators render this *deux sortes de démence* as the need 'to distinguish madness from madness', while for the quotation from Seneca, see Seneca, 'On the Tranquillity of the Mind', *Dialogues and Essays*, trans. John Davie (Oxford and New York: Oxford University Press, 2007), ch. 17, p. 139, translation modified.

10 See pp. 136 and 144–5.
11 Michel Foucault, 'Dream, Imagination and Existence', trans. Forrest Williams, in *Dream and Existence*, reprint of a special issue of *Review of Existential Psychology & Psychiatry* 19 (1) (1986), p. 31.
12 Ibid. p. 51.
13 See Binswanger, 'Dream and Existence', trans. Jacob Needleman, in *Dream and Existence*, p. 98.
14 It is a question of going beyond Freud through 'a method of interpretation [...] that reinstates acts of expression in their fullness. The hermeneutic journey should not stop at the verbal sequences which have preoccupied psychoanalysis' (Foucault, 'Dream, Imagination and Existence', p. 42).
15 Ibid., p. 51, translation modified.
16 Ibid., p. 31.
17 Ibid., p. 67. Here, it would clearly be necessary to approach Simondon's *Imagination et invention* – and it is likely that Simondon and Foucault were moving in the same circles at the time the latter was writing this introduction.
18 I have proposed a reading of Simondon's *Imagination et invention*, a major work that is often misunderstood, in 'Dans le cycle des images – cercle ou spirale? Imagination, invention et transindividuation', in Vincent Bontems (ed.), *Gilbert Simondon ou l'invention du futur* (Paris: Klincksieck, 2016).
19 See Bernard Stiegler, *Automatic Society, Volume 1: The Future of Work*, trans. Daniel Ross (Cambridge: Polity, 2016), ch. 8.
20 See in particular Martin Heidegger, *Identity and Difference*, trans. Joan Stambaugh (New York: Harper & Row, 1969), p. 39, translation modified: 'a leap that departs from being as the ground of beings, and thus springs into the abyss, into the *Abgrund*'.
21 This is evident in *Discourse on Thinking* but it was already the case in *Being and Time*, as I have tried to show in Bernard Stiegler, *Technics and Time, 1: The Fault of Epimetheus*, trans. Richard Beardsworth and George Collins (Stanford: Stanford University Press, 1998), pp. 272–6.
22 Martin Heidegger, *Discourse on Thinking*, trans. John M. Anderson and E. Hans Freund (New York: Harper & Row, 1969), pp. 48–9.
23 See Jacques Derrida, 'Shibboleth: For Paul Celan', trans. Joshua Wilner, revised by Thomas Dutoit, *Sovereignties in Question: The Poetics of Paul Celan* (New York: Fordham University Press, 2005).
24 Heidegger opposes speech and writing in his letter to Jean Beaufret, while noting in passing that 'written composition exerts a wholesome pressure toward deliberate linguistic formulation'. See Martin Heidegger, 'Letter on "Humanism"', trans. Frank A. Capuzzi, with John Glenn Gray, revised by William McNeill and David Farrell Krell, *Pathmarks* (Cambridge: Cambridge University Press, 1998), p. 241.
25 Henri Bergson, *The Two Sources of Morality and Religion*, trans. R. Ashley Audra and Cloudesley Brereton, with W. Horsfall Carter (Notre Dame: University of Notre Dame Press, 1977).
26 Gustave Flaubert, *Bouvard and Pecuchet*, trans. Mark Polizzotti (Champaign: Dalkey Archive Press, 2005).
27 Gladys Swain, *Dialogue avec l'insensé: Essais d'histoire de la psychiatrie* (Paris: Gallimard, 1994), the introduction to which is Marcel Gauchet, 'À la recherche d'une autre histoire de la folie'.
28 Gauchet, 'À la recherche d'une autre histoire de la folie', p. lvii.
29 It is through this discourse that the transhumanist ultralibertarians, who have gathered in universities at the initiative of Raymond Kurzweil and with the support of Google and NASA, want, in particular, to conceal the major issue entailed by this new era, which is the submission of the future of the species to a logic of calculation that is in reality based on the disavowal of ὕβρις, and of its psycho-pathological and socio-pathological consequences.

30 Pope Francis, *Laudato Si'* (dated 24 May 2015; published 18 June 2015), available at: http://w2.vatican.va/content/francesco/en/encyclicals/documents/papa-francesco_20150524_enciclica-laudato-si.html.
31 Nick Srnicek and Alex Williams, '#Accelerate: Manifesto for an Accelerationist Politics', in Robin Mackay and Armen Avanessian (eds), *#Accelerate: The Accelerationist Reader* (Windsor Quarry: Urbanomic, 2014). The manifesto contained in this reader is also available online at: https://syntheticedifice.files.wordpress.com/2013/06/accelerate.pdf.
32 As for the status of technics, technology and nature in *Laudato Si'*, this is something to which it will be necessary to return.
33 Srnicek and Williams, '#Accelerate: Manifesto for an Accelerationist Politics', p. 350.
34 Ibid., p. 351.
35 Ibid., pp. 349–50.
36 Ibid., p. 352.
37 Ibid., p. 353.
38 Srnicek and Williams write, 'Indeed, as even Lenin wrote in the 1918 text *"'Left Wing' Childishness"'*, after which they reproduce the following quotation from Lenin's text:

> Socialism is inconceivable without large-scale capitalist engineering based on the latest discoveries of modern science. It is inconceivable without planned state organisation which keeps tens of millions of people to the strictest observance of a unified standard in production and distribution. We Marxists have always spoken of this, and it is not worth while wasting two seconds talking to people who do not understand even this (anarchists and a good half of the Left Socialist-Revolutionaries).

This quotation is from V.I. Lenin, '"Left-Wing" Childishness and the Petty-Bourgeois Mentality', trans. Clemens Dutt, in Volume 27 of V.I. Lenin, *Collected Works* (Moscow: Progress Press, 1960–70), p. 333, available online at: https://www.marxists.org/archive/lenin/works/1918/may/09.htm.
39 These theses are developed in detail in Bernard Stiegler, *States of Shock: Stupidity and Knowledge in the Twenty-First Century*, trans. Daniel Ross (Cambridge: Polity, 2015), Bernard Stiegler, *Pharmacologie du Front national* (Paris: Flammarion, 2013), and Stiegler, *Automatic Society, Volume 1*.
40 Srnicek and Williams, '#Accelerate: Manifesto for an Accelerationist Politics', p. 357.
41 See Michel Bauwens, *Sauver le monde* (Paris: Les Liens qui libèrent, 2015). We discussed these questions with Michel Bauwens, Benjamin Coriat, Franck Cormerais, Mark Hunyadi, Frédéric Sultan and Carlo Vercellone on 30 January 2016 at the Théâtre Gérard Philipe de Saint-Denis. Video available online at: http://www.arsindustrialis.org/communs-contribution-nouvelle-puissance-publique.
42 See Les Entretiens du nouveau monde industriel 2015: https://enmi-conf.org/wp/enmi15.
43 It is such a negentropic web in the service of the formation and sharing of such knowledge that we are attempting to begin to develop in a large-scale way on the territory of the Plaine Commune region, with the goal of constituting a contributory regional economy, founded on a contributory income, turning this territory into a place of learning at the forefront of digital urbanity, and developing large-scale contributory approaches to research, together with the Maison des sciences de l'homme de Saint-Denis.
44 Elinor Ostrom, *Governing the Commons: The Evolution of Institutions for Collective Action* (Cambridge: Cambridge University Press, 1990).
45 Benjamin Coriat (ed.), *Le Retour des communs* (Paris: Les Liens qui libèrent, 2015).
46 See the journal *Études digitales*, and my specifications on this point in Stiegler, *Automatic Society, Volume 1*.

47 This point has been partially developed in Stiegler, *Automatic Society, Volume 1*, and will be further specified in Bernard Stiegler, *La Société automatique 2. L'avenir du savoir*, forthcoming.
48 Srnicek and Williams, '#Accelerate: Manifesto for an Accelerationist Politics', p. 361.
49 Ibid., pp. 361–2.
50 Ibid., p. 362.
51 The question of the status of sublimation in the constitution of desire as economy is here the entire issue. I will return to this via Sophie de Mijolla-Mellor, *Le Choix de la sublimation* (Paris: PUF, 2009).
52 See Bernard Stiegler, *The Lost Spirit of Capitalism: Disbelief and Discredit, Volume 3*, trans. Daniel Ross (Cambridge: Polity, 2014).
53 Bernard Stiegler, 'How I Became a Philosopher', *Acting Out*, trans. David Barison, Daniel Ross and Patrick Crogan (Stanford: Stanford University Press, 2009).
54 Lévi-Strauss declared, in an interview with Laurent Lemire for the programme Campus and broadcast on France 2 on 28 October 2004: 'The human race lives under a regime of a kind that poisons itself from within, and *I think about the present and about the world in which my experience is coming to an end: it is not a world that I love.*' See Bernard Stiegler, *Constituer l'Europe 1. Dans un monde sans vergogne* (Paris: Galilée, 2005), pp. 35–6.
55 See Edmund Husserl, 'Foundational Investigations of the Phenomenological Origin of the Spatiality of Nature: The Originary Ark, the Earth, Does Not Move', trans. Fred Kersten, revised by Leonard Lawlor, in Maurice Merleau-Ponty, *Husserl at the Limits of Phenomenology, Including Texts by Edmund Husserl*, edited by Leonard Lawlor and Bettina Bergo (Evanston: Northwestern University Press, 2002).
56 Denoetization is obviously, and more than anything else, an object of denial. Thomas Berns bore witness to this in the intervention he presented during the 2015 Entretiens du nouveau monde industrial, held at the Centre Pompidou on 14–15 December 2015 and available at: https://enmi-conf.org/wp/enmi15/session-3. The present work is wholly a response to this lecture, and this is why I have dedicated it to Thomas Berns.
57 It is not Heidegger who here uses the word 'desire', but Jean Beaufret, and he does so in order to translate *Möglichkeit*. This translation is obviously very interesting and highly problematic, especially since the question of Heidegger's relationship to desire is a vast one – and Binswanger is here clearly a touchstone on the path beyond Heidegger, as well as beyond Freud. As for the way in which Heidegger ignores psychoanalysis, this is clearly an extraordinary symptom of the nature of his own denial.
58 See Jacobus da Voragine, *The Golden Legend: Readings on the Saints*, trans. William Granger Ryan (Princeton and Oxford: Princeton University Press, 2012), pp. 126–30.
59 See Gustave Flaubert, 'The Legend of Saint Julian the Hospitaller', *Three Tales*, trans. A.J. Krailsheimer (Oxford: Oxford University Press, 1991).
60 Hidetaka Ishida, 'Culture de soi au Japon: le jardin de pierres', unpublished lecture (Paris, 1999).
61 Jean-Marie André, *L'Otium dans la vie morale et intellectuelle romaine: des origins à l'époque augustéenne* (Paris: puf, 1966).
62 Hegel is the one who rethinks πόλεμος starting from its exteriorization – and as a contradiction in a process that is a '*geschichtlich*' dialectic. It is for this reason that Hegel is a thinker of history. But since he does not conceive that the process of interiorization that makes synthesis possible is itself organological and pharmacological, he turns exosomatization into a 'stage' that absolute knowledge absorbs and interiorizes in totality. In this way, Hegel 'preconceives' psychotic capitalism.
63 Genesis 3, quoted from *The Holy Bible: Quatercentenary Edition* (Oxford and New York: Oxford University Press, 2010), no page numbers: 'And the eyes of them both were opened, & they knew that they *were* naked, and they sewed figge leaues together, and made themselues aprons. [...] And the woman said, The Serpent

beguiled me, and I did eate.' The serpent, which splits into the caduceus of Hermes and of Asclepius, in Greek mythology symbolizes the *pharmakon*. In Genesis, it is temptation that brings with it shame. Before the serpent tempted Eve, 'they were both naked, the man & his wife, and were not ashamed' (Genesis 2:25).

64 On *promētheia* and *epimētheia*, see Stiegler, *Technics and Time, 1*, part II, ch. 1, 'Prometheus's Liver'.

65 *Translator's note*: The role of memory tape in computing machines is first described by Turing in 1936, but, for the idealized account of limitless memory tape, see, for example, Alan Turing, 'Intelligent Machinery' (1948), in B. Jack Copeland (ed.), *The Essential Turing: Seminal Writings in Computing, Logic, Philosophy, Artificial Intelligence, and Artificial Life: Plus The Secrets of Enigma* (Oxford: Oxford University Press, 2004), p. 413.

66 Alan Turing, 'On Computable Numbers, with an Application to the Entscheidungsproblem', in ibid.

67 Alan Turing, 'Computing Machinery and Intelligence', in ibid.

68 See Jean Lassègue, *Turing* (Paris: Les Belles Lettres, 1998), and Francis Bailly and Giuseppe Longo, *Mathematics and the Natural Sciences: The Physical Singularity of Life*, no translator credited (London: Imperial College Press, 2011), where the question passes through life as a potential for anti-entropy. See also David Bates, 'Penser l'automaticité au seuil du numérique', in Bernard Stiegler (ed.), *Digital Studies. Organologie des savoirs et technologies de la connaissance* (Paris: FYP, 2014), and David Bates, 'Automaticity, Plasticity, and the Deviant Origins of Artificial Intelligence', in David Bates and Nima Bassiri (eds), *Plasticity and Pathology: On the Formation of the Neural Subject* (New York: Fordham University Press, 2015).

69 Chris Anderson, 'The End of Theory: The Data Deluge Makes the Scientific Method Obsolete', *Wired* (23 June 2008), available at: http://archive.wired.com/science/discoveries/magazine/16-07/pb_theory.

70 Martin Heidegger, 'Der Spiegel Interview with Martin Heidegger', in Günther Neske and Emil Kettering (eds), *Martin Heidegger and National Socialism: Questions and Answers*, trans. Lisa Harries (New York: Paragon House, 1990), p. 57.

71 Concerning infidelity, which is here firstly that of the milieu – in Canguilhem's sense – see pp. 96 and 192.

72 This is what Graham Harman seems to want to do in *The Quadruple Object* (Winchester: Zero Books, 2011), but I am unable to grasp where the rereading he proposes of Heidegger's later works leads.

73 I have evoked this 'death of the sun' in Stiegler, *Technics and Time, 1*, pp. 88–91.

74 Heidegger, *Discourse on Thinking*, p. 53. *Translator's note*: The German term that John M. Anderson and E. Hans Freund translate as 'autochthony' is *Bodenständigkeit*, rendered in French as '*terre natale*', 'native land'. Other candidates for an English translation of *Bodenständigkeit* are 'groundedness' (which retains the connection to *Boden*, ground) and 'rootedness' (despite the lack of connection to *Wurzel*, root, that can indeed be found in other Heideggerian terms). See also Robert Metcalf, 'Rethinking "Bodenständigkeit" in the Technological Age', *Research in Phenomenology* 42 (2012), pp. 49–66.

75 See p. 280.

76 René Char, *À une sérénité crispée* (Paris: Gallimard, 1951), p. 21.

77 Günther Anders, *Die Antiquiertheit des Menschen 1: Über die Seele im Zeitalter der zweiten industriellen Revolution* (Munich: C.H. Beck, 2010), p. 234.

78 Ibid., p. 236.

79 Ibid.

80 Ibid., p. 259.

81 Ibid., p. 239.

82 Günther Anders, 'Vorwort zur 5. Auflage, 1979' [Preface to the 5th Edition, 1979], in ibid., p. viii.

83 Anders, *Die Antiquiertheit des Menschen*, pp. 240–1.

84 Eugène Minkowski, *Vers une cosmologie: Fragments philosophiques* (Paris: Payot, 1999), p. 97.
85 Ibid.
86 Ibid., p. 99.
87 Michel de Montaigne, 'On Educating Children', *The Complete Essays*, trans. M.A. Screech (London: Penguin, 2003), p. 176, translation modified.
88 Italo Calvino, *Invisible Cities*, trans. William Weaver (San Diego, New York and London: Harvest, 1974), p. 165.
89 *Translator's note*: Note that the French translation of the passage from Calvino just quoted ends with '*lui faire place*', translated into English as 'give them space'.
90 See §92.
91 Maurice Godelier, *The Metamorphoses of Kinship*, trans. Nora Scott (London and New York: Verso, 2011).
92 The advertising for this software states: 'Dragon NaturallySpeaking 13 Premium is a voice recognition solution that allows you to use your voice to save time and be more efficient on your PC. Whether at the office, the university or at home, Dragon Premium converts all your speech into text, and executes your voice commands much faster than if you were using the keyboard or mouse. You can create and edit documents, send emails, search the web, or even chat on social networks, without ever touching your keyboard! Dragon Premium lets you work wirelessly, transcribe notes dictated while on the go, and use voice-based shortcuts to reach unparalleled levels of productivity. So, forget the keyboard. And speak up!'
93 Philippe-Alain Michaud, *Aby Warburg and the Image in Motion*, trans. Sophie Hawkes (New York: Zone Books, 2004).
94 Aby Warburg, 'A Lecture on Serpent Ritual', *Journal of the Warburg Institute* 2 (1939), pp. 277–92.
95 This infinite speed that occurs in becoming and wholly against it, right up against it, as the function of reason, is what can be produced only through a politics of technology that puts algorithms functioning at twice the speed of Olympian fire into the service of the Epimethean Titans that we are before and after having been Prometheans – and this is what Anders does not see in his analysis of Promethean shame. Any noesis is always *epimētheia*, and putting the speed of algorithms into the service of this *epimētheia* means developing a new political economy, where the time saved by the productivity gains arising from full and generalized automation are put into the service of the production of neganthropy, which produces a practical value beyond use value and exchange value, neganthropogenesis becoming the value of values.
96 Marcel Proust, *Remembrance of Things Past, Volume Two: The Guermantes Way, Cities of the Plain*, trans. C.K. Scott Moncrieff and Terence Kilmartin (London: Penguin, 1983), p. 743.
97 Pierre Jacquemain, 'L'ex-conseiller de Myriam El Khomri explique pourquoi il claque la porte', *Le Monde* (1 January 2016), available at: http://www.lemonde.fr/idees/article/2016/03/01/droit-du-travail-lex-conseiller-de-myriam-el-Khomri-explique-pourquoi-il-claque-la-porte_4874575_3232.html.

A Conversation about Christianity

1 Blaise Pascal, *Pensées*, A.J. Krailsheimer (Harmondsworth, Middlesex: Penguin, 1966), pp. 64–5, translation modified. *Translator's note*: Nancy also discusses this quotation from Pascal in Jean-Luc Nancy, *The Truth of Democracy*, trans. Pascale-Anne Briault and Michael Naas (New York: Fordham University Press, 2010), esp. p. 11.
2 *Translator's note*: Parc Astérix is a theme park outside of Paris that opened in 1989 and is based on the comic books of Goscinny and Uderzo. It has rides and attractions supposedly related to the culture of the Gauls, the Romans and so on.
3 Jean-Luc Nancy, 'The Judeo-Christian (on Faith)', *Dis-Enclosure: The Deconstruction*

of Christianity, trans. Bettina Bergo, Gabriel Malenfant and Michael B. Smith (New York: Fordham University Press, 2008).

4 Philippe Lacoue-Labarthe, a philosopher and a friend of Jean-Luc Nancy, who died in 2007.

5 *Translator's note*: This has now been published, and translated as Jean-Luc Nancy, *Adoration: The Deconstruction of Christianity II*, trans. John McKeane (New York: Fordham University Press, 2013).

6 *Translator's note*: See Jean-Jacques Rousseau, *The Social Contract*, trans. Maurice Cranston (London: Penguin, 1968), Book IV, ch. 8.

7 Bernard Stiegler, *The Decadence of Industrial Democracies: Disbelief and Discredit, Volume 1*, trans. Daniel Ross (Cambridge: Polity, 2011); Stiegler, *Uncontrollable Societies of Disaffected Individuals: Disbelief and Discredit, Volume 2*, trans. Daniel Ross (Cambridge: Polity, 2013); and Stiegler, *The Lost Spirit of Capitalism: Disbelief and Discredit, Volume 3*, trans. Daniel Ross (Cambridge: Polity, 2014).

8 *Translator's note*: See Stiegler, *The Decadence of Industrial Democracies*, ch. 2, §14.

9 Jean-Christophe Bailly, *The Animal Side*, trans. Catherine Porter (New York: Fordham University Press, 2011).

10 Plato, *Theaetetus*, 176b.

11 *Translator's note*: The everyday sense of this phrase, '*créances douteuses*', is 'bad debts', as the author indicates in the next sentence, but the reference is also undoubtedly to the more obsolete but etymologically literal sense of *créance* as belief, as used, for instance, by Descartes. See René Descartes, *Meditations on First Philosophy, with Selections from the Objections and Replies*, trans. John Cottingham (Cambridge: Cambridge University Press, 1986), p. 12, translation modified: 'I had to seriously undertake, once in my life, to rid myself of all the opinions I had received into my set of beliefs [*créance*] up until that moment, and to begin afresh from the foundations.' And see Bernard Stiegler, *States of Shock: Stupidity and Knowledge in the Twenty-First Century*, trans. Daniel Ross (Cambridge: Polity, 2015), §48.

12 *Translator's note*: See Sigmund Freud, *Civilization and Its Discontents*, in Volume 21 of James Strachey (ed. and trans.), *The Standard Edition of the Complete Psychological Works of Sigmund Freud* (London: Hogarth, 1953–74), ch. 5, esp. p. 112.

13 Aristotle, *Metaphysics*, 982b 30–1.

14 Plato, *Protagoras*, 341e–344c.

15 Martin Heidegger, *Plato's Sophist*, trans. Richard Rojcewicz and André Schuwer (Bloomington and Indianapolis: Indiana University Press, 1997), p. 92.

16 *Translator's note*: See also Stiegler, *The Decadence of Industrial Democracies*, p. 134, and Bernard Stiegler, 'The Theater of Individuation: Phase-Shift and Resolution in Simondon and Heidegger', trans. Kristina Lebedeva, in Arne De Boever et al. (eds), *Gilbert Simondon: Being and Technology* (Edinburgh: Edinburgh University Press, 2012), p. 192.

Index